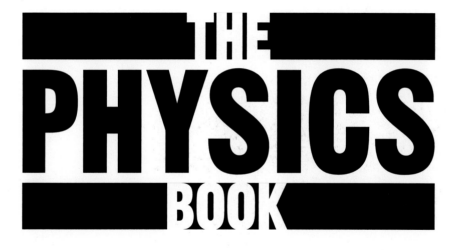

THE PHYSICS BOOK

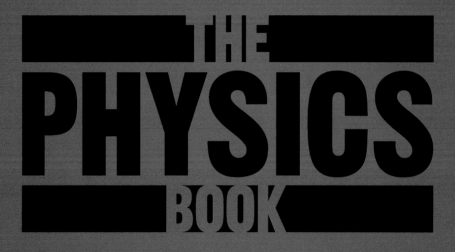

THE PHYSICS BOOK

DK LONDON

SENIOR ART EDITOR
Gillian Andrews

SENIOR EDITORS
Camilla Hallinan, Laura Sandford

EDITORS
John Andrews, Jessica Cawthra, Joy Evatt,
Claire Gell, Richard Gilbert, Tim Harris,
Janet Mohun, Victoria Pyke, Dorothy Stannard,
Rachel Warren Chadd

US EDITOR
Megan Douglass

ILLUSTRATIONS
James Graham

**JACKET DESIGN
DEVELOPMENT MANAGER**
Sophia MTT

PRODUCER, PRE-PRODUCTION
Gillian Reid

PRODUCER
Nancy-Jane Maun

SENIOR MANAGING ART EDITOR
Lee Griffiths

MANAGING EDITOR
Gareth Jones

ASSOCIATE PUBLISHING DIRECTOR
Liz Wheeler

ART DIRECTOR
Karen Self

DESIGN DIRECTOR
Philip Ormerod

PUBLISHING DIRECTOR
Jonathan Metcalf

DK DELHI

PROJECT ART EDITOR
Pooja Pipil

ART EDITORS
Meenal Goel, Debjyoti Mukherjee

ASSISTANT ART EDITOR
Nobina Chakravorty

SENIOR EDITOR
Suefa Lee

ASSISTANT EDITOR
Aashirwad Jain

SENIOR JACKET DESIGNER
Suhita Dharamjit

SENIOR DTP DESIGNER
Neeraj Bhatia

DTP DESIGNER
Anita Yadav

PROJECT PICTURE RESEARCHER
Deepak Negi

PICTURE RESEARCH MANAGER
Taiyaba Khatoon

PRE-PRODUCTION MANAGER
Balwant Singh

PRODUCTION MANAGER
Pankaj Sharma

MANAGING ART EDITOR
Sudakshina Basu

SENIOR MANAGING EDITOR
Rohan Sinha

original styling by
STUDIO 8

First American Edition, 2020
Published in the United States by DK Publishing
1745 Broadway, 20th Floor, New York, NY 10019

Copyright © 2020 Dorling Kindersley Limited
DK, a Division of Penguin Random House LLC
23 24 25 26 10 9 8 7 6 5 4 3 2 1
027–335981–Jul/2023

A catalog record for this book
is available from the Library of Congress.
ISBN: 978-0-7440-8268-5

DK books are available at special discounts when
purchased in bulk for sales promotions, premiums,
fund-raising, or educational use. For details, contact:
DK Publishing Special Markets, 1745 Broadway,
20th Floor, New York, NY 10019
SpecialSales@dk.com

Printed in China

For the curious
www.dk.com

MIX
Paper | Supporting
responsible forestry
FSC™ C018179

This book was made with Forest Stewardship
Council™ certified paper – one small step in
DK's commitment to a sustainable future.
**For more information go to
www.dk.com/our-green-pledge**

CONTRIBUTORS

DR. BEN STILL, CONSULTANT EDITOR

A prize-winning science communicator, particle physicist, and author, Ben teaches high school physics and is also a visiting research fellow at Queen Mary University of London. After a master's degree in rocket science, a PhD in particle physics, and years of research, he stepped into the world of outreach and education in 2014. He is the author of a growing collection of popular science books and travels the world teaching particle physics using LEGO®.

JOHN FARNDON

John Farndon has been short-listed five times for the Royal Society's Young People's Science Book Prize, among other awards. A widely published author of popular books on science and nature, he has written around 1,000 books on a range of subjects, including internationally acclaimed titles such as *The Oceans Atlas*, *Do You Think You're Clever?*, and *Do Not Open*, and has contributed to major books such as *Science* and *Science Year By Year*.

TIM HARRIS

A widely published author on science and nature for both children and adults, Tim Harris has written more than 100 mostly educational reference books and contributed to many others. These include *An Illustrated History of Engineering*, *Physics Matters!*, *Great Scientists*, *Exploring the Solar System*, and *Routes of Science*.

HILARY LAMB

Hilary Lamb studied physics at the University of Bristol and science communication at Imperial College London. She is a staff journalist at *Engineering & Technology Magazine*, covering science and technology, and has written for previous DK titles, including *How Technology Works* and *Explanatorium of Science*.

JONATHAN O'CALLAGHAN

With a background in astrophysics, Jonathan O'Callaghan has been a space and science journalist for almost a decade. His work has appeared in numerous publications including *New Scientist*, *Wired*, *Scientific American*, and *Forbes*. He has also appeared as a space expert on several radio and television shows, and is currently working on a series of educational science books for children.

MUKUL PATEL

Mukul Patel studied natural sciences at King's College Cambridge and mathematics at Imperial College London. He is the author of *We've Got Your Number*, a children's math book, and over the last 25 years has contributed to numerous other books across scientific and technological fields for a general audience. He is currently investigating ethical issues in AI.

ROBERT SNEDDEN

Robert Snedden has been involved in publishing for 40 years, researching and writing science and technology books for young people on topics ranging from medical ethics to space exploration, engineering, computers, and the internet. He has also contributed to histories of mathematics, engineering, biology, and evolution, and written books for an adult audience on breakthroughs in mathematics and medicine and the works of Albert Einstein.

GILES SPARROW

A popular science author specializing in physics and astronomy, Giles Sparrow studied astronomy at University College London and science communication at Imperial College London. He is the author of books including *Physics in Minutes*, *Physics Squared*, *The Genius Test* and *What Shape Is Space?*, as well as DK's *Spaceflight*, and has contributed to bestselling DK titles including *Universe* and *Science*.

JIM AL-KHALILI, FOREWORD

An academic, author, and broadcaster, Jim Al-Khalili FRS holds a dual professorship in theoretical physics and the public engagement in science at the University of Surrey. He has written 12 books on popular science, translated into over 20 languages. A regular presenter on British TV, he is also the host of the Radio 4 program *The Life Scientific*. He is a recipient of the Royal Society Michael Faraday Medal, the Institute of Physics Kelvin Medal, and the Stephen Hawking Medal for science communication.

6

CONTENTS

ELECTRICITY AND MAGNETISM
TWO FORCES UNITE

SOUND AND LIGHT
THE PROPERTIES OF WAVES

THE QUANTUM WORLD
OUR UNCERTAIN UNIVERSE

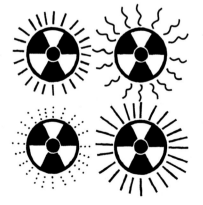

FOREWORD

I fell in love with physics as a boy when I discovered that this was the subject that best provided answers to many of the questions I had about the world around me—questions like how magnets worked, whether space went on forever, why rainbows form, and how we know what the inside of an atom or the inside of a star looks like. I also realized that by studying physics I could get a better grip on some of the more profound questions swirling around in my head, such as: What is the nature of time? What is it like to fall into a black hole? How did the universe begin and how might it end?

Now, decades later, I have answers to some of my questions, but I continue to search for answers to new ones. Physics, you see, is a living subject. Although there are many things we now know with confidence about the laws of nature, and we have used this knowledge to develop technologies that have transformed our world, there is still much more we do not yet know. That is what makes physics, for me, the most exciting area of knowledge of all. In fact, I sometimes wonder why everyone isn't as in love with physics as I am.

But to bring the subject alive—to convey that sense of wonder—requires much more than collecting together a mountain of dry facts. Explaining how our world works is about telling stories; it is about acknowledging how we have come to know what we know about the universe, and it is about sharing in the joy of discovery made by the many great scientists who first unlocked nature's secrets. How we have come to our current understanding of physics can be as important and as joyful as the knowledge itself.

This is why I have always had a fascination with the history of physics. I often think it a shame that we are not taught at school about how concepts and ideas in science first developed. We are expected to simply accept them unquestioningly. But physics, and indeed the whole of science, isn't like that. We ask questions about how the world works and we develop theories and hypotheses. At the same time,

we make observations and conduct experiments, revising and improving on what we know. Often, we take wrong turns or discover after many years that a particular description or theory is wrong, or only an approximation of reality. Sometimes, new discoveries are made that shock us and force us to revise our view entirely.

One beautiful example of this that has happened in my lifetime was the discovery, in 1998, that the universe is expanding at an accelerating pace, leading to the idea of so-called dark energy. Until recently, this was regarded as a complete mystery. What was this invisible field that acted to stretch space against the pull of gravity? Gradually, we are learning that this is most likely something called the vacuum energy. You might wonder how changing the name of something (from "dark energy" to "vacuum energy") can constitute an advance in our understanding. But the concept of vacuum energy is not new. Einstein had suggested it a hundred years ago, then changed his mind when he thought he'd made a mistake, calling it his "biggest blunder." It is stories like this that, for me, make physics so joyous.

This is also why *The Physics Book* is so enjoyable. Each topic is made more accessible and readable with the introduction of key figures, fascinating anecdotes, and the timeline of the development of the ideas. Not only is this a more honest account of the way science progresses, it is also a more effective way of bringing the subject alive.

I hope you enjoy the book as much as I do.

Jim Al-Khalili

We humans have a heightened sense of our surroundings. We evolved this way to outmaneuver stronger and faster predators. To achieve this, we have had to predict the behavior of both the living and the inanimate world. Knowledge gained from our experiences was passed down through generations via an ever-evolving system of language, and our cognitive prowess and ability to use tools took our species to the top of the food chain.

We spread out of Africa from around 60,000 years ago, extending our abilities to survive in inhospitable locations through sheer ingenuity. Our ancestors developed techniques to allow them to grow plentiful food for their families, and settled into communities.

Experimental methods

Early societies drew meaning from unrelated events, saw patterns that did not exist, and spun mythologies. They also developed new tools and methods of working, which required advanced knowledge of the inner workings of the world—be it the seasons or the annual flooding of the Nile—in order to expand resources. In some regions, there were periods of relative peace and abundance. In these civilized societies, some people were free to wonder about our place in the universe. First the Greeks, then the Romans tried to make sense of the world through patterns they observed in nature. Thales of Miletus, Socrates, Plato, Aristotle, and others began to reject supernatural explanations and produce rational answers in the quest to create absolute knowledge—they began to experiment.

At the fall of the Roman Empire, so many of these ideas were lost to the Western world, which fell into a dark age of religious wars, but they continued to flourish in the Arab world and Asia. Scholars there continued to ask questions and conduct experiments. The language of mathematics was invented to document this new-found knowledge. Ibn al-Haytham and Ibn Sahl were just two of the Arab scholars who kept the flame of scientific knowledge alive in the 10th and 11th centuries, yet their discoveries, particularly in the fields of optics and astronomy, were ignored for centuries outside the Islamic world.

A new age of ideas

With global trade and exploration came the exchange of ideas. Merchants and mariners carried books, stories, and technological marvels from east to west. Ideas from this wealth of culture drew Europe out of the dark ages and into a new age of enlightenment known as the Renaissance. A revolution of our world view began as ideas from ancient civilizations became updated or outmoded, replaced by new ideas of our place in the universe. A new generation of experimenters poked and prodded nature to extract her secrets. In Poland and Italy, Copernicus and Galileo challenged ideas that had been considered sacrosanct for two millennia—and they suffered harsh persecution as a result.

Then, in England in the 17th century, Isaac Newton's laws of motion established the basis of

> " Whosoever studies works of science must … examine tests and explanations with the greatest precision.
> **Ibn al-Haytham**

classical physics, which was to reign supreme for more than two centuries. Understanding motion allowed us to build new tools—machines—able to harness energy in many forms to do work. Steam engines and water mills were two of the most important of these—they ushered in the Industrial Revolution (1760–1840).

The evolution of physics

In the 19th century, the results of experiments were tried and tested numerous times by a new international network of scientists. They shared their findings through papers, explaining the patterns they observed in the language of mathematics. Others built models from which they attempted to explain these empirical equations of correlation. Models simplified the complexities of nature into digestible chunks, easily described by simple geometries and relationships. These models made predictions about new behaviors in nature, which were tested by a new wave of pioneering experimentalists—if the predictions were proven true, the models were deemed laws which all of nature seemed to obey. The relationship of heat and energy was explored by French physicist Sadi Carnot and others, founding the new science of thermodynamics. British physicist James Clerk Maxwell produced equations to describe the close relationship of electricity and magnetism—electromagnetism.

By 1900, it seemed that there were laws to cover all the great phenomena of the physical world. Then, in the first decade of the 20th century, a series of discoveries sent shock waves through the scientific community, challenging former "truths" and giving birth to modern physics. A German, Max Planck, uncovered the world of quantum physics. Then fellow countryman Albert Einstein revealed his theory of relativity. Others discovered the structure of the atom and uncovered the role of even smaller, subatomic particles. In so doing, they launched the study of particle physics. New discoveries weren't confined to the microscopic—more advanced telescopes opened up the study of the universe.

Within a few generations, humanity went from living at the center of the universe to residing on a speck of dust on the edge of one galaxy among billions. Not only had we seen inside the heart of matter and released the energy within, we had charted the seas of space with light that had been traveling since soon after the Big Bang.

> "
> One cannot help but be in awe when he contemplates the mysteries of eternity, of life, of the marvellous structure of reality.
> **Albert Einstein**
> "

Physics has evolved over the years as a science, branching out and breaching new horizons as discoveries are made. Arguably, its main areas of concern now lie at the fringes of our physical world, at scales both larger than life and smaller than atoms. Modern physics has found applications in many other fields, including new technology, chemistry, biology, and astronomy. This book presents the biggest ideas in physics, beginning with the everyday and ancient, then moving through classical physics into the tiny atomic world, and ending with the vast expanse of space. ∎

MEASURE
AND MOTI
PHYSICS AND THE
EVERYDAY WORLD

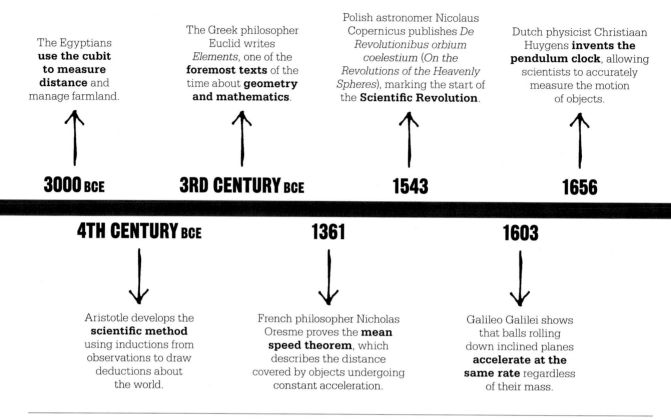

The Egyptians **use the cubit to measure distance** and manage farmland.

The Greek philosopher Euclid writes *Elements*, one of the **foremost texts** of the time about **geometry and mathematics**.

Polish astronomer Nicolaus Copernicus publishes *De Revolutionibus orbium coelestium* (*On the Revolutions of the Heavenly Spheres*), marking the start of the **Scientific Revolution**.

Dutch physicist Christiaan Huygens **invents the pendulum clock**, allowing scientists to accurately measure the motion of objects.

3000 BCE   **3RD CENTURY BCE**   **1543**   **1656**

4TH CENTURY BCE   **1361**   **1603**

Aristotle develops the **scientific method** using inductions from observations to draw deductions about the world.

French philosopher Nicholas Oresme proves the **mean speed theorem**, which describes the distance covered by objects undergoing constant acceleration.

Galileo Galilei shows that balls rolling down inclined planes **accelerate at the same rate** regardless of their mass.

O ur survival instincts have made us creatures of comparison. Our ancient struggle to survive by ensuring that we found enough food for our family or reproduced with the correct mate has been supplanted. These primal instincts have evolved with our society into modern equivalents such as wealth and power. We cannot help but measure ourselves, others, and the world around us by metrics. Some of these measures are interpretive, focusing upon personality traits that we benchmark against our own feelings. Others, such as height, weight, or age, are absolutes.

For many people in the ancient and modern world alike, a measure of success was wealth. To amass fortune, adventurers traded goods across the globe. Merchants would purchase plentiful goods cheaply in one location before transporting and selling them for a higher price in another location where that commodity was scarce. As trade in goods grew to become global, local leaders began taxing trade and imposing standard prices. To enforce this, they needed standard measures of physical things to allow them to make comparisons.

Language of measurement

Realizing that each person's experience is relative, the ancient Egyptians devised systems that could be communicated without bias from one person to another. They developed the first system of metrics, a standard method for measuring the world around them. The Egyptian cubit allowed engineers to plan buildings that were unrivalled for millennia and devise farming systems to feed the burgeoning population. As trade with ancient Egypt became global, the idea of a common language of measurement spread around the world.

The Scientific Revolution (1543–1700) brought about a new need for these metrics. For the scientist, metrics were to be used not for trading goods but as a tool with which nature could be understood. Distrusting their instincts, scientists developed controlled environments in which they tested connections between different behaviors—they experimented. Early experiments focused on the movement of everyday objects, which had a direct effect upon daily life. Scientists discovered patterns

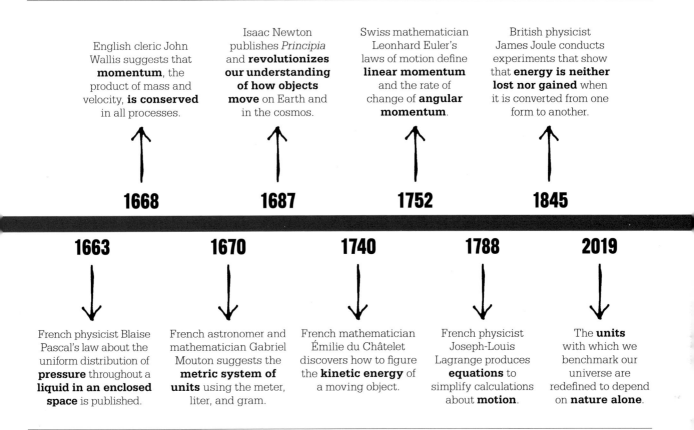

English cleric John Wallis suggests that **momentum**, the product of mass and velocity, **is conserved** in all processes.

Isaac Newton publishes *Principia* and **revolutionizes our understanding of how objects move** on Earth and in the cosmos.

Swiss mathematician Leonhard Euler's laws of motion define **linear momentum** and the rate of change of **angular momentum**.

British physicist James Joule conducts experiments that show that **energy is neither lost nor gained** when it is converted from one form to another.

1668   **1687**   **1752**   **1845**

1663   **1670**   **1740**   **1788**   **2019**

French physicist Blaise Pascal's law about the uniform distribution of **pressure** throughout a **liquid in an enclosed space** is published.

French astronomer and mathematician Gabriel Mouton suggests the **metric system of units** using the meter, liter, and gram.

French mathematician Émilie du Châtelet discovers how to figure the **kinetic energy** of a moving object.

French physicist Joseph-Louis Lagrange produces **equations** to simplify calculations about **motion**.

The **units** with which we benchmark our universe are redefined to depend on **nature alone**.

in linear, circular, and repetitive oscillating motion. These patterns became immortalized in the language of mathematics, a gift from ancient civilizations that had then been developed in the Islamic world for centuries. Mathematics offered an unambiguous way of sharing the outcomes of experiments and allowed scientists to make predictions and test these predictions with new experiments. With a common language and metrics, science marched forward. These pioneers discovered links between distance, time, and speed and set out their own repeatable and tested explanation of nature.

Measuring motion

Scientific theories progressed rapidly and with them the language of mathematics changed. Building on his laws of motion, English physicist Isaac Newton invented calculus, which brought a new ability to describe the change in systems over time, not just calculate single snapshots. To explain the acceleration of falling objects, and eventually the nature of heat, ideas of an unseen entity called energy began to emerge. Our world could no longer be defined by distance, time, and mass alone, and new metrics were needed to benchmark the measurement of energy.

Scientists use metrics to convey the results of experiments. Metrics provide an unambiguous language that enables scientists to interpret the results of an experiment and repeat the experiment to check that their conclusions are correct. Today, scientists use the Système

international (SI) collection of metrics to convey their results. The value of each of these SI metrics and their link to the world around us are defined and decided upon by an international group of scientists known as metrologists.

This first chapter charts these early years of the science we today call physics, the way in which the science operates through experimentation, and how results from these tests are shared across the world. From the falling objects that Italian polymath Galileo Galilei used to study acceleration to the oscillating pendulums that paved the way to accurate timekeeping, this is the story of how scientists began to measure distance, time, energy, and motion, revolutionizing our understanding of what makes the world work. ∎

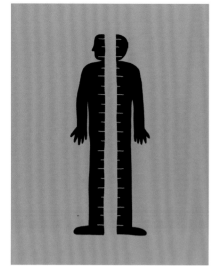

MAN IS THE MEASURE OF ALL THINGS
MEASURING DISTANCE

IN CONTEXT

KEY CIVILIZATION
Ancient Egypt

BEFORE
c. 3500 BCE Administrators use a system of measuring field sizes in ancient Mesopotamia.

c. 3100 BCE Officials in ancient Egypt use knotted cords—pre-stretched ropes tied at regular intervals—to measure land and survey building foundations.

AFTER
1585 In the Netherlands, Simon Stevin proposes a decimal system of numbers.

1799 France adopts the prototype platinum meter.

1875 Signed by 17 nations, the Meter Convention agrees a consistent length for the unit.

1960 The eleventh General Conference on Weights and Measures sets the metric system as the International System of Units ("SI," from the French Système international).

When people began to build structures on an organized scale, they needed a way to measure height and length. The earliest measuring devices are likely to have been primitive wooden sticks scored with notches, with no accepted consistency in unit length. The first widespread unit was the "cubit," which emerged in the 4th and 3rd millennia BCE among the peoples of Egypt, Mesopotamia, and the Indus Valley. The term cubit derives from the Latin for elbow, *cubitum*, and was the distance from the elbow

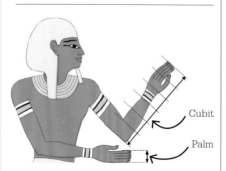

The Egyptian royal cubit was based on the length of the forearm, measured from the elbow to the middle fingertip. Cubits were subdivided into 28 digits (each a finger's breadth in length) and a series of intermediary units, such as palms and hands.

to the tip of the outstretched middle finger. Of course, not everyone has the same length of forearm and middle finger, so this "standard" was only approximate.

Imperial measure
As prodigious architects and builders of monuments on a grand scale, the ancient Egyptians needed a standard unit of distance. Fittingly, the royal cubit of the Old Kingdom of ancient Egypt is the first known standardized cubit measure in the world. In use since at least 2700 BCE, it was 20.6–20.8 in (523–529 mm) long and was divided into 28 equal digits, each based on a finger's breadth.

Archaeological excavations of pyramids have revealed cubit rods of wood, slate, basalt, and bronze, which would have been used as measures by craftsmen and architects. The Great Pyramid at Giza, where a cubit rod was found in the King's Chamber, was built to be 280 cubits in height, with a base of 440 cubits squared. The Egyptians further subdivided cubits into palms (4 digits), hands (5 digits), small spans (12 digits), large spans (14 digits, or half a cubit), and *t'sers* (16 digits or

See also: Free falling 32–35 ▪ Measuring time 38–39 ▪ SI units and physical constants 58–63 ▪ Heat and transfers 80–81

4 palms). The *khet* (100 cubits) was used to measure field boundaries and the *ater* (20,000 cubits) to define larger distances.

Cubits of various length were used across the Middle East. The Assyrians used cubits in c. 700 BCE, while the Hebrew Bible contains plentiful references to cubits—particularly in the Book of Exodus's account of the construction of the Tabernacle, the sacred tent that housed the Ten Commandments. The ancient Greeks developed their own 24-unit cubit, as well as the *stade* (plural *stadia*), a new unit representing 300 cubits. In the 3rd century BCE, the Greek scholar Eratosthenes (c. 276 BCE–c. 194 BCE) estimated the circumference of Earth at 250,000 stadia, a figure he later refined to 252,000 stadia. The Romans also adopted the cubit, along with the inch—an adult

Cubit rods—such as this example from the 18th dynasty in ancient Egypt, c. 14th century BCE—were used widely in the ancient world to achieve consistent measurements.

male's thumb—foot, and mile. The Roman mile was 1,000 paces, or *mille passus*, each of which was five Roman feet. Roman colonial expansion from the 3rd century BCE to the 3rd century CE introduced these units to much of western Asia and Europe, including England, where the mile was redefined as 5,280 feet in 1593 by Queen Elizabeth I.

Going metric

In his 1585 pamphlet *De Thiende* (*The Art of Tenths*), Flemish physicist Simon Stevin proposed a decimal system of measurement, forecasting that, in time, it would be widely accepted. More than two centuries later, work on the metric system was begun by a committee of the French Academy of Sciences, with the meter being defined as one ten-millionth of the distance from Earth's equator to the North Pole. Known as the prototype platinum meter, the measurement was adopted in 1799.

International recognition was not achieved until 1960, when the Système international (SI) set the meter as the base unit for distance. It was agreed that 1 meter (m) is equal to 1,000 millimeters (mm) or 100 centimeters (cm). ▪

Changing definitions

In 1668, English clergyman John Wilkins followed Stevin's proposal of a decimal-based unit of length with a novel definition: he suggested that 1 meter should be set as the distance of a two-second pendulum swing. Dutch physicist Christiaan Huygens (1629–1695) calculated this to be 39.26 in (997 mm).

In 1889, an alloy bar of platinum (90%) and iridium (10%) was cast to represent the definitive 1-meter length, but because it expanded and contracted very slightly at different temperatures, it was accurate only at the melting point of ice. This bar is still kept at the International Bureau of Weights and Measures in Paris, France. When SI definitions were adopted in 1960, the meter was redefined in terms of the wavelength of electromagnetic emissions from a krypton atom. In 1983, yet another definition was adopted: the distance that light travels in a vacuum in 1/299,792,458 of a second.

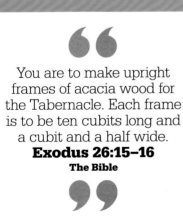

> You are to make upright frames of acacia wood for the Tabernacle. Each frame is to be ten cubits long and a cubit and a half wide.
> **Exodus 26:15–16**
> **The Bible**

> A mile shall contain eight furlongs, every furlong forty poles, and every pole shall contain sixteen foot and a half.
> **Queen Elizabeth I**

A PRUDENT QUESTION IS ONE HALF OF WISDOM

THE SCIENTIFIC METHOD

Careful observation and a questioning attitude to findings are central to the scientific method of investigation, which underpins physics and all the sciences. Since it is easy for prior knowledge and assumptions to distort the interpretation of data, the scientific method follows a set procedure. A hypothesis is drawn up on the basis of findings, and then tested experimentally. If this hypothesis fails, it can be revised and reexamined, but if it is robust, it is shared for peer review—independent evaluation by experts.

People have always sought to understand the world around them, and the need to find food and

See also: Free falling 32–35 ▪ SI units and physical constants 58–63 ▪ Focusing light 170–175 ▪ Models of the universe 272–273 ▪ Dark matter 302–305

The starting point for the scientific method is an **observation**.

Scientists form a **hypothesis** (a theory to **explain** the observation).

An **experiment** is carried out to test the hypothesis.

Data from the experiment is collected.

If the data **supports** the hypothesis, the experiment is **repeated** to make sure the results are correct.

If the data **refutes** the hypothesis, the hypothesis is **revised**.

The hypothesis is eventually accepted as **fact**.

Aristotle

The son of the court physician of the Macedonian royal family, Aristotle was raised by a guardian after his parents died when he was young. At around the age of 17, he joined Plato's Academy in Athens, the foremost center of learning in Greece. Over the next two decades, he studied and wrote about philosophy, astronomy, biology, chemistry, geology, and physics, as well as politics, poetry, and music. He also traveled to Lesvos, where he made ground-breaking observations of the island's botany and zoology.

In c. 343 BCE, Aristotle was invited by Philip II of Macedon to tutor his son, the future Alexander the Great. He established a school at the Lyceum in Athens in 335 BCE, where he wrote many of his most celebrated scientific treatises. Aristotle left Athens in 322 BCE and settled on the island of Euboea, where he died at the age of about 62.

Key works

Metaphysics
On the Heavens
Physics

understand changing weather were matters of life and death long before ideas were written down. In many societies, mythologies developed to explain natural phenomena; elsewhere, it was believed that everything was a gift from the gods and events were preordained.

Early investigations
The world's first civilizations—ancient Mesopotamia, Egypt, Greece, and China—were sufficiently advanced to support "natural philosophers," thinkers who sought to interpret the world and record their findings. One of the first to reject supernatural explanations of natural phenomena was the Greek thinker Thales of Miletus. Later, the philosophers Socrates and Plato introduced debate and argument as a method of advancing understanding, but it was Aristotle—a prolific investigator of physics, biology, and zoology—who began to develop a scientific method of inquiry, applying logical reasoning to observed phenomena. He was an empiricist, someone »

All truths are
easy to understand
once they are
discovered; the point
is to discover them.
Galileo Galilei

who believes that all knowledge is based on experience derived from the senses, and that reason alone is not enough to solve scientific problems—evidence is required.

Traveling widely, Aristotle was the first to make detailed zoological observations, seeking evidence to group living things by behavior and anatomy. He went to sea with fishermen in order to collect and dissect fish and other marine organisms. After discovering that dolphins have lungs, he judged they should be classed with whales, not fish. He separated four-legged animals that give birth to live young (mammals) from those that lay eggs (reptiles and amphibians).

However, in other fields Aristotle was still influenced by traditional ideas that lacked a grounding in good science. He did not challenge the prevailing geocentric idea that the sun and stars rotate around Earth. In the 3rd century BCE, another Greek thinker, Aristarchus of Samos, argued that Earth and the known planets orbit the sun, that stars are very distant equivalents of "our" sun, and that Earth spins on its axis. Though correct, these ideas were dismissed because Aristotle and his student Ptolemy carried

Anatomical drawings from 1543 reflect Vesalius's mastery of dissection and set a new standard for study of the human body, unchanged since the Greek physician Galen (129–216 CE).

greater authority. In fact, the geocentric view of the universe was held to be true—due in part to its enforcement by the Catholic Church, which discouraged ideas that challenged its interpretation of the Bible—until it was superseded in the 17th century by the ideas of Copernicus, Galileo, and Newton.

Testing and observation

Arab scholar Ibn al-Haytham (widely known as "Alhazen") was an early proponent of the scientific method. Working in the 10th and 11th centuries CE, he developed his own method of experimentation to prove or disprove hypotheses. His most important work was in the field of optics, but he also made important contributions to astronomy and mathematics. Al-Haytham experimented with sunlight, light reflected from artificial light sources, and refracted light. For example, he tested—and proved—the hypothesis that every

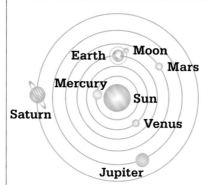

Copernicus's heliocentric model, so-called because it made the sun (*helios* in Greek) the focus of planetary orbits, was endorsed by some scientists but outlawed by the Church.

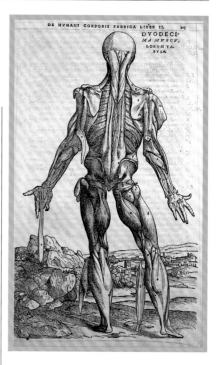

point of a luminous object radiates light along every straight line and in every direction.

Unfortunately, al-Haytham's methods were not adopted beyond the Islamic world, and it would be 500 years before a similar approach emerged independently in Europe, during the Scientific Revolution. But the idea that accepted theories may be challenged, and overthrown if proof of an alternative can be produced, was not the prevailing view in 16th-century Europe. Church authorities rejected many scientific ideas, such as the work of Polish astronomer Nicolaus Copernicus. He made painstaking observations of the night sky with the naked eye, explaining the temporary retrograde ("backward") motion of the planets, which geocentrism had never accounted for. Copernicus realized the phenomenon was due to Earth and the other planets moving around the sun on different orbits. Although Copernicus lacked the tools to prove heliocentrism, his use

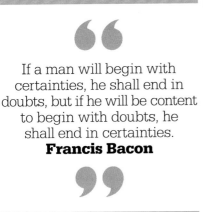

> If a man will begin with certainties, he shall end in doubts, but if he will be content to begin with doubts, he shall end in certainties.
> **Francis Bacon**

of rational argument to challenge accepted thinking set him apart as a true scientist. Around the same time, Flemish anatomist Andreas Vesalius transformed medical thinking with his multi-volume work on the human body in 1543. Just as Copernicus based his theories on detailed observation, Vesalius analyzed what he found when dissecting human body-parts.

Experimental approach

For Italian polymath Galileo Galilei, experimentation was central to the scientific approach. He carefully recorded observations on matters as varied as the movement of the planets, the swing of pendulums, and the speed of falling bodies. He produced theories to explain them, then made more observations to test the theories. He used the new technology of telescopes to study four of the moons orbiting Jupiter, proving Copernicus's heliocentric model—under geocentrism, all objects orbited Earth. In 1633 Galileo was tried by the Church's Roman Inquisition, found guilty of heresy, and placed under house-arrest for the last decade of his life. He continued to publish by smuggling papers to Holland, away from the censorship of the Church.

Later in the 17th century, English philosopher Francis Bacon reinforced the importance of a methodical, skeptical approach to scientific inquiry. Bacon argued that the only means of building true knowledge was to base axioms and laws on observed facts, not relying (even if only partially) on unproven deductions and conjecture. The Baconian method involves making systematic observations to establish verifiable facts; generalizing from a series of facts to create axioms (a process known as "inductivism"), while being careful to avoid generalizing beyond what the facts tell us; then gathering further facts to produce an increasingly complex base of knowledge.

Unproven science

When scientific claims cannot be verified, they are not necessarily wrong. In 1997, scientists at the Gran Sasso laboratory in Italy claimed to have detected evidence of dark matter, which is believed to make up about 27 percent of the universe. The most likely source, they said, were weakly interacting massive particles (WIMPs). These should be detected as tiny flashes of light (scintillations) when a particle strikes the nucleus of a "target" atom. However, despite the best efforts of other research teams to replicate the experiment, no other evidence of dark matter has been found. It is possible that there is an unidentified explanation—or the scintillations could have been produced by helium atoms, which are present in the experiment's photomultiplier tubes. ■

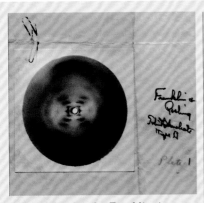

Photo 51, taken by Franklin, is a 1952 X-ray diffraction image of human DNA. The X-shape is due to DNA's double-helix structure.

The scientific method in practice

Deoxyribonucleic acid (DNA) was identified as the carrier of genetic information in the human body in 1944, and its chemical composition was shown to be four different molecules called nucleotides. However, it was unclear how genetic information was stored in DNA. Three scientists—Linus Pauling, Francis Crick, and James Watson—put forward the hypothesis that DNA possessed a helical structure, and realized from work done by other scientists that if that was the case, its X-ray diffraction pattern would be X-shaped. British scientist Rosalind Franklin tested this theory by performing X-ray diffraction on crystallized pure DNA, beginning in 1950. After refining the technique over a period of two years, her analysis revealed an X-shaped pattern (best seen in "Photo 51"), proving that DNA had a helical structure. The Pauling, Crick, Watson hypothesis was proven, forming the starting point for further studies on DNA.

ALL IS NUMBER

THE LANGUAGE OF PHYSICS

IN CONTEXT

KEY FIGURE
Euclid of Alexandria
(c. 325–c. 270 BCE)

BEFORE
3000–300 BCE Ancient
Mesopotamian and Egyptian
civilizations develop number
systems and techniques to
solve mathematical problems.

600–300 BCE Greek scholars,
including Pythagoras and
Thales, formalize mathematics
using logic and proofs.

AFTER
c. 630 CE Indian
mathematician Brahmagupta
uses zero and negative
numbers in arithmetic.

c. 820 CE Persian scholar
al-Khwarizmi sets down
the principles of algebra.

c. 1670 Gottfried Leibniz and
Isaac Newton each develop
calculus, the mathematical
study of continuous change.

P hysics seeks to understand
the universe through
observation, experiment,
and building models and theories.
All of these are intimately entwined
with mathematics. Mathematics is
the language of physics—whether
used in measurement and data
analysis in experimental science,
or to provide rigorous expression
for theories, or to describe the
fundamental "frame of reference" in
which all matter exists and events
take place. The investigation of
space, time, matter, and energy is
only made possible through a prior
understanding of dimension, shape,
symmetry, and change.

Driven by practical needs
The history of mathematics is one
of increasing abstraction. Early
ideas about number and shape
developed over time into the most
general and precise language.
In prehistoric times, before the
advent of writing, herding animals
and trading goods undoubtedly
prompted the earliest attempts
at tallying and counting.
 As complex cultures emerged in
the Middle East and Mesoamerica,
demands for greater precision and

> Number is the ruler of forms
> and ideas, and the cause of
> gods and daemons.
> **Pythagoras**

prediction increased. Power was
tied to knowledge of astronomical
cycles and seasonal patterns,
such as flooding. Agriculture and
architecture required accurate
calendars and land surveys.
The earliest place value number
systems (where a digit's position
in a number indicates its value)
and methods for solving equations
date back more than 3,500 years
to civilizations in Mesopotamia,
Egypt, and (later) Mesoamerica.

Adding logic and analysis
The rise of ancient Greece brought
about a fundamental change in
focus. Number systems and

Euclid

Although his *Elements* were
immensely influential, few details
of Euclid's life are known. He was
born around 325 BCE, in the reign
of Egyptian pharaoh Ptolemy I
and probably died around 270 BCE.
He lived mostly in Alexandria, then
an important center of learning, but
he may also have studied at Plato's
academy in Athens.
 In *Commentary on Euclid*,
written in the 5th century CE, the
Greek philosopher Proclus notes
that Euclid arranged the theorems
of Eudoxus, an earlier Greek
mathematician, and brought
"irrefutable demonstration" to

the loose ideas of other scholars.
Thus, the theorems of the
13 books of Euclid's *Elements*
are not original, but for two
millennia they set the standard
for mathematical exposition.
The earliest surviving editions
of the *Elements* date from the
15th century.

Key works

Elements
Data
Catoptrics
Optics

See also: Measuring distance 18–19 ▪ Measuring time 38–39 ▪ Laws of motion 40–45 ▪ SI units and physical constants 58–63 ▪ Antimatter 246 ▪ The particle zoo and quarks 256–257 ▪ Curving spacetime 280

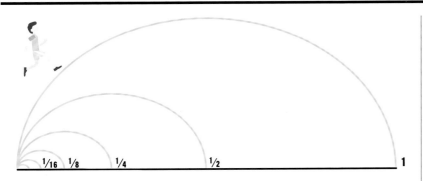

1/16 1/8 1/4 1/2 1

The dichotomy paradox is one of Zeno's paradoxes that show motion to be logically impossible. Before walking a certain distance a person must walk half that distance, before walking half the distance he must walk a quarter of the distance, and so on. Walking any distance will therefore entail an infinite number of stages that take an infinite amount of time to complete.

measurement were no longer simply practical tools; Greek scholars also studied them for their own sake, together with shape and change. Although they inherited much specific mathematical knowledge from earlier cultures, such as elements of Pythagoras's theorem, the Greeks introduced the rigor of logical argument and an approach rooted in philosophy; the ancient Greek word *philosophia* means "love of wisdom."

The ideas of a theorem (a general statement that is true everywhere and for all time) and proof (a formal argument using the laws of logic) are first seen in the geometry of the Greek philosopher Thales of Miletus in the early 6th century BCE. Around the same time, Pythagoras and his followers elevated numbers to be the building blocks of the universe.

For the Pythagoreans, numbers had to be "commensurable"—measurable in terms of ratios or fractions—to preserve the link with nature. This world view was shattered with the discovery of irrational numbers (such as $\sqrt{2}$,

which cannot be exactly expressed as one whole number divided by another) by the Pythagorean philosopher Hippasus; according to legend, he was murdered by scandalized colleagues.

Titans of mathematics

In the 5th century BCE, the Greek philosopher Zeno of Elea devised paradoxes about motion, such as Achilles and the tortoise. This was the idea that, in any race where the pursued has a head start, the pursuer is always catching up—eventually by an infinitesimal amount. Such puzzles, which were logical—if simple to disprove in practice—would worry generations of mathematicians. They were resolved, at least partially, in the 17th century by the development of calculus, a branch of mathematics that deals with continuously changed quantities.

Greek philosophers drew in the sand when teaching geometry, as shown here. Archimedes is said to have been drawing circles in the sand when he was killed by a Roman soldier.

Central to calculus is the idea of calculating infinitesimals (infinitely small quantities), which was anticipated by Archimedes of Syracuse, who lived in the 3rd century BCE. To calculate the approximate volume of a sphere, for instance, he halved it, enclosed the hemisphere in a cylinder, then imagined slicing it horizontally, from the top of the hemisphere, where the radius is infinitesimally small, downward. He knew that the thinner he made his slices, the more accurate the volume would be. Reputed to have shouted "Eureka!" on discovering that the upward buoyant force of an object immersed in water is equal to the weight of the fluid it displaces, Archimedes is notable for applying math to mechanics and other branches of physics in order to solve problems involving levers, screws, pulleys, and pumps.

Archimedes studied in Alexandria, at a school established by Euclid, often known as the "Father of Geometry." It was by »

analyzing geometry itself that Euclid established the template for mathematical argument for the next 2,000 years. His 13-book treatise, *Elements*, introduced the "axiomatic method" for geometry. He defined terms, such as "point," and outlined five axioms (also known as postulates, or self-evident truths), such as "a line segment can be drawn between any two points." From these axioms, he used the laws of logic to deduce theorems.

By today's standards, Euclid's axioms are lacking; there are numerous assumptions that a mathematician would now expect to be stated formally. *Elements* remains, however, a prodigious work, covering not only plane geometry and three-dimensional geometry, but also ratio and proportion, number theory, and the "incommensurables" that Pythagoreans had rejected.

Language and symbols

In ancient Greece and earlier, scholars described and solved algebraic problems (determining unknown quantities given certain known quantities and relationships) in everyday language and by using geometry. The highly-abbreviated,

precise, symbolic language of modern mathematics—which is significantly more effective for analyzing problems and universally understood—is relatively recent. Around 250 CE, however, the Greek mathematician Diophantus of Alexandria introduced the partial use of symbols to solve algebraic problems in his principal work *Arithmetica,* which influenced the development of Arabic algebra after the fall of the Roman Empire.

The study of algebra flourished in the East during the Golden Age of Islam (from the 8th century to the 14th century). Baghdad became the principal seat of learning. Here, at an academic center called the House of Wisdom, mathematicians could study translations of Greek texts on geometry and number theory or Indian works discussing the decimal place-value system. In the early 9th century, Muhammad ibn Musa al-Khwarizmi (from whose name comes the word "algorithm") compiled methods for balancing and solving equations in his book *al Jabr* (the root of the word "algebra"). He popularized the use of Hindu numerals, which evolved into Arabic numerals, but still described his algebraic problems in words.

French mathematician François Viète finally pioneered the use of symbols in equations in his 1591 book, *Introduction to the Analytic Arts*. The language was not yet standard, but mathematicians could now write complicated expressions in a compact form, without resorting to diagrams. In 1637, French philosopher and mathematician René Descartes reunited algebra and geometry by devising the coordinate system.

More abstract numbers

Over millennia, in attempts to solve different problems, mathematicians have extended the number system, expanding the counting numbers 1, 2, 3 ... to include fractions and irrational numbers. The addition of zero and negative numbers indicated increasing abstraction. In ancient number systems, zero had been used as a placeholder—a way to tell 10 from 100, for instance. By around the 7th century CE,

Islamic scholars gather in one of Baghdad's great libraries in this 1237 image by the painter Yahya al-Wasiti. Scholars came to the city from all points of the Islamic Empire, including Persia, Egypt, Arabia, and even Iberia (Spain).

negative numbers were used for representing debts. In 628 CE, the Indian mathematician Brahmagupta was the first to treat negative integers (whole numbers) just like the positive integers for arithmetic. Yet, even 1,000 years later, many European scholars still considered negative numbers unacceptable as formal solutions to equations.

The 16th-century Italian polymath Gerolamo Cardano not only used negative numbers, but, in *Ars Magna*, introduced the idea of complex numbers (combining a real and imaginary number) to solve cubic equations (those with at least one variable to the power of three, such as x^3, but no higher). Complex numbers take the form $a + bi$, where a and b are real numbers and i is the imaginary unit, usually expressed as $i = \sqrt{-1}$. The unit is termed "imaginary" because when squared it is negative, and squaring any real number, whether it is positive or negative, produces a positive number. Although Cardano's contemporary Rafael Bombelli

> A new, a vast, and a powerful language is developed for the future use of analysis, in which to wield its truths so that these may become of more speedy and accurate practical application for the purposes of mankind.
> **Ada Lovelace**
> **British computer scientist**

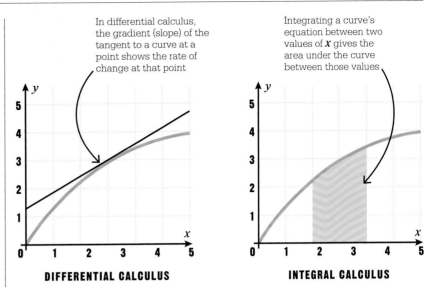

In differential calculus, the gradient (slope) of the tangent to a curve at a point shows the rate of change at that point

Integrating a curve's equation between two values of x gives the area under the curve between those values

DIFFERENTIAL CALCULUS

INTEGRAL CALCULUS

Differential calculus examines the rate of change over time, shown geometrically here as the rate of change of a curve. Integral calculus examines the areas, volumes, or displacement bounded by curves.

set down the first rules for using complex and imaginary numbers, it took a further 200 years before Swiss mathematician Leonhard Euler introduced the symbol i to denote the imaginary unit.

Like negative numbers, complex numbers were met with resistance, right up until the 18th century. Yet they represented a significant advance in mathematics. Not only do they enable the solution of cubic equations but, unlike real numbers, they can be used to solve all higher-order polynomial equations (those involving two or more terms added together and higher powers of a variable x, such as x^4 or x^5). Complex numbers emerge naturally in many branches of physics, such as quantum mechanics and electromagnetism.

Infinitesimal calculus
From the 14th century to the 17th century, together with the increasing use of symbols, many

new methods and techniques emerged. One of the most significant for physics, was the development of "infinitesimal" methods in order to study curves and change. The ancient Greek method of exhaustion—finding the area of a shape by filling it with smaller polygons—was refined in order to compute areas bounded by curves. It finally evolved into a branch of mathematics called integral calculus. In the 17th century, French lawyer Pierre de Fermat's study of tangents to curves inspired the development of differential calculus—the calculation of rates of change.

Around 1670, English physicist Isaac Newton and German philosopher Gottfried Leibniz independently worked out a theory that united integral and differential calculus into infinitesimal calculus. The underlying idea is of approximating a curve (a changing quantity) by »

Euclidean and non-Euclidean geometries

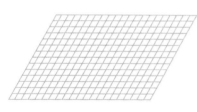

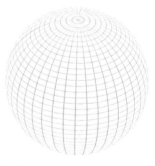

In Euclidean geometry, space is assumed to be "flat." Parallel lines remain at a constant distance from one another and never meet.

In hyperbolic geometry, developed by Bolyai and Lobachevsky, the surface curves like a saddle and lines on the surface curve away from each other.

In elliptic geometry, the surface curves outward like a sphere and parallel lines curve toward each other, eventually intersecting.

considering that it is made up of many straight lines (a series of different, fixed quantities). At the theoretical limit, the curve is identical to an infinite number of infinitesimal approximations.

During the 18th and 19th centuries, applications of calculus in physics exploded. Physicists could now precisely model dynamic (changing) systems, from vibrating strings to the diffusion of heat. The work of 19th-century Scottish physicist James Clerk Maxwell greatly influenced the development of vector calculus, which models change in phenomena that have both quantity and direction. Maxwell also pioneered the use of statistical techniques for the study of large numbers of particles.

Non-Euclidean geometries

The fifth axiom, or postulate, on geometry that Euclid set out in his *Elements*, is also known as the parallel postulate. This was controversial, even in ancient times, as it appears less self-evident than the others, although many theorems depend on it. It states that, given a line and a point that

is not on that line, exactly one line can be drawn through the given point and parallel to the given line. Throughout history, various mathematicians, such as Proclus of Athens in the 5th century or the Arabic mathematician al-Haytham, have attempted in vain to show that the parallel postulate can be derived from the other postulates. In the early 1800s, Hungarian mathematician János Bolyai and Russian mathematician Nicolai Lobachevsky independently developed a version of geometry

Out of nothing I have created a strange new universe. All that I have sent you previously is like a house of cards in comparison with a tower.
János Bolyai
in a letter to his father

(hyperbolic geometry) in which the fifth postulate is false and parallel lines never meet. In their geometry, the surface is not flat as in Euclid's, but curves inward. By contrast, in elliptic geometry and spherical geometry, also described in the 19th century, there are no parallel lines; all lines intersect.

German mathematician Bernhard Riemann and others formalized such non-Euclidean geometries. Einstein used Riemannian theory in his general theory of relativity—the most advanced explanation of gravity— in which mass "bends" spacetime, making it non-Euclidean, although space remains homogeneous (uniform, with the same properties at every point).

Abstract algebra

By the 19th century, algebra had undergone a seismic shift, to become a study of abstract symmetry. French mathematician Évariste Galois was responsible for a key development. In 1830, while investigating certain symmetries exhibited by the roots (solutions) of polynomial equations, he developed a theory of abstract mathematical objects, called groups, to encode different kinds of symmetries. For example, all squares exhibit the same reflectional and rotational symmetries, and so are associated with a particular group. From his research, Galois determined that, unlike for quadratic equations (with a variable to the power of two, such as x^2, but no higher), there is no general formula to solve polynomial equations of degree five (with terms such as x^5) or higher. This was a dramatic result; he had proved that there could be no such formula, no matter what future developments occurred in mathematics.

Subsequently, algebra grew into the abstract study of groups and similar objects, and the symmetries they encoded. In the 20th century, groups and symmetry proved vital for describing natural phenomena at the deepest level. In 1915, German algebraist Emmy Noether connected symmetry in equations with conservation laws, such as the conservation of energy, in physics.

In the 1950s and 1960s, physicists used group theory to develop the Standard Model of particle physics.

Modeling reality

Mathematics is the abstract study of numbers, quantities, and shapes, which physics employs to model reality, express theories, and predict future outcomes—often with astonishing accuracy. For example, the electron g-factor— a measure of its behavior in an electromagnetic field—is computed to be 2.002 319 304 361 6, while the experimentally determined value is 2.002 319 304 362 5 (differing by just one part in a trillion).

Certain mathematical models have endured for centuries, requiring only minor adjustments. For example, German astronomer Johannes Kepler's 1619 model of the solar system, with some refinements by Newton and Einstein, remains valid today. Physicists have applied ideas that mathematicians developed, sometimes much earlier, simply to investigate a pattern; for

Emmy Noether was a highly creative algebraist. She taught at the University of Göttingen in Germany, but as a Jew was forced to leave in 1933. She died in the US in 1935, aged 53.

instance, the application of 19th-century group theory to modern quantum physics. There are also many examples of mathematical structures driving insight into nature. When British physicist Paul Dirac found twice as many expressions as expected in his equations describing the behavior of electrons, consistent with relativity and quantum mechanics, he postulated the existence of an anti-electron; it was duly discovered, years later.

While physicists investigate what "is" in the universe, mathematicians are divided as to whether their study is about nature, or the human mind, or the abstract manipulation of symbols. In a strange historical twist, physicists researching string theory are now suggesting revolutionary advances in pure mathematics to geometers (mathematicians who study geometry). Just exactly how this illuminates the relationship between mathematics, physics, and "reality" is yet to be seen. ∎

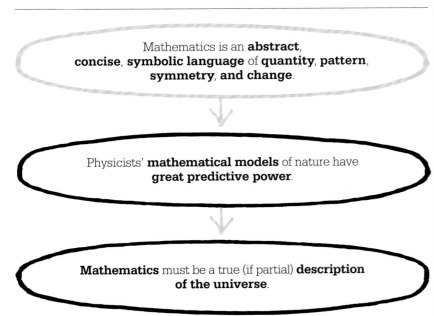

Mathematics is an **abstract, concise, symbolic language** of **quantity, pattern, symmetry, and change**.

↓

Physicists' **mathematical models** of nature have **great predictive power**.

↓

Mathematics must be a true (if partial) **description of the universe**.

BODIES SUFFER NO RESISTANCE BUT FROM THE AIR

FREE FALLING

IN CONTEXT

KEY FIGURE
Galileo Galilei (1564–1642)

BEFORE
c. 350 BCE In *Physics*, Aristotle explains gravity as a force that moves bodies toward their "natural place," down toward the center of Earth.

1576 Giuseppe Moletti writes that objects of different weights free fall at the same rate.

AFTER
1651 Giovanni Riccioli and Francesco Grimaldi measure the time of descent of falling bodies, enabling calculation of their rate of acceleration.

1687 In *Principia*, Isaac Newton expounds gravitational theory in detail.

1971 David Scott shows that a hammer and a feather fall at the same speed on the moon.

When gravity is the only force acting on a moving object, it is said to be in "free fall." A skydiver falling from a plane is not quite in free fall—since air resistance is acting upon him—whereas planets orbiting the sun or another star are. The ancient Greek philosopher Aristotle believed that the downward motion of objects dropped from a height was due to their nature—they were moving toward the center of Earth, their natural place. From Aristotle's time until the Middle Ages, it was accepted as fact that the speed of a free-falling object was proportional to its weight, and inversely proportional to the density

See also: Measuring distance 18–19 ▪ Measuring time 38–39 ▪ Laws of motion 40–45 ▪ Laws of gravity 46–51 ▪ Kinetic energy and potential energy 54

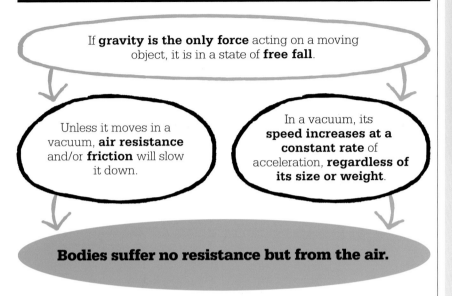

If **gravity is the only force** acting on a moving object, it is in a state of **free fall**.

Unless it moves in a vacuum, **air resistance** and/or **friction** will slow it down.

In a vacuum, its **speed increases at a constant rate** of acceleration, **regardless of its size or weight**.

Bodies suffer no resistance but from the air.

Galileo Galilei

The oldest of six siblings, Galileo was born in Pisa, Italy, in 1564. He enrolled to study medicine at the University of Pisa at the age of 16, but his interests quickly broadened and he was appointed Chair of Mathematics at the University of Padua in 1592. Galileo's contributions to physics, mathematics, astronomy, and engineering single him out as one of the key figures of the Scientific Revolution in 16th- and 17th-century Europe. He created the first thermoscope (an early thermometer), defended the Copernican idea of a heliocentric solar system, and made important discoveries about gravity. Because some of his ideas challenged Church dogma, he was called before the Roman Inquisition in 1633, declared to be a heretic, and sentenced to house arrest until his death in 1642.

Key works

1623 *The Assayer*
1632 *Dialogue Concerning the Two Chief World Systems*
1638 *Discourses and Mathematical Demonstrations Relating to Two New Sciences*

of the medium it was falling through. So, if two objects of different weights are dropped at the same time, the heavier will fall faster and hit the ground before the lighter object. Aristotle also understood that the object's shape and orientation were factors in how quickly it fell, so a piece of unfolded paper would fall more slowly than the same piece of paper rolled into a ball.

Falling spheres

At some time between 1589 and 1592, according to his student and biographer Vincenzo Viviani, Italian polymath Galileo Galilei dropped two spheres of different weight from the Tower of Pisa to test Aristotle's theory. Although it was more likely to have been a thought experiment than a real-life event, Galileo was reportedly excited to discover that the lighter sphere fell to the ground as quickly as the heavier one. This contradicted the Aristotelian view that a heavier

free-falling body will fall more quickly than a lighter one—a view that had recently been challenged by several other scientists.

In 1576, Giuseppe Moletti, Galileo's predecessor in the Chair of Mathematics at the University of Padua, had written that objects of different weights but made of the same material fell to the ground at the same speed. He also believed that bodies of the same volume »

Nature is inexorable and immutable; she never transgresses the laws imposed upon her.
Galileo Galilei

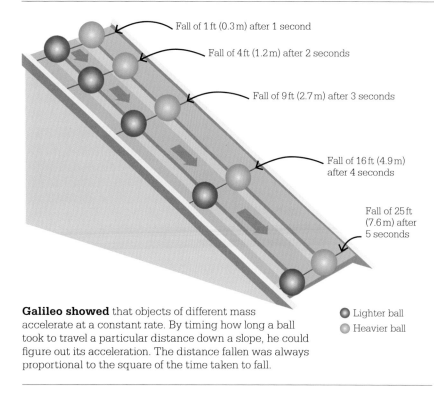

Fall of 1 ft (0.3 m) after 1 second

Fall of 4 ft (1.2 m) after 2 seconds

Fall of 9 ft (2.7 m) after 3 seconds

Fall of 16 ft (4.9 m) after 4 seconds

Fall of 25 ft (7.6 m) after 5 seconds

● Lighter ball
● Heavier ball

Galileo showed that objects of different mass accelerate at a constant rate. By timing how long a ball took to travel a particular distance down a slope, he could figure out its acceleration. The distance fallen was always proportional to the square of the time taken to fall.

but made of different materials fell at the same rate. Ten years later, Dutch scientists Simon Stevin and Jan Cornets de Groot climbed 33 ft (10 m) up a church tower in Delft to release two lead balls, one ten times bigger and heavier than the other. They witnessed them hit the ground at the same time. The age-old idea of heavier objects falling faster than lighter ones was gradually being debunked.

Another of Aristotle's beliefs— that a free-falling object descends at a constant speed—had been challenged earlier still. Around 1361, French mathematician Nicole Oresme had studied the movement of bodies. He discovered that if an object's acceleration is increasing

uniformly, its speed increases in direct proportion to time, and the distance it travels is proportional to the square of the time during which it is accelerating. It was perhaps surprising that Oresme should have challenged the established Aristotelian "truth," which at the time was considered

sacrosanct by the Catholic Church, in which Oresme served as a bishop. It is not known whether Oresme's studies influenced the later work of Galileo.

Balls on ramps

From 1603, Galileo set out to investigate the acceleration of free-falling objects. Unconvinced that they fell at a constant speed, he believed that they accelerated as they fell—but the problem was how to prove it. The technology to accurately record such speeds simply did not exist. Galileo's ingenious solution was to slow down the motion to a measurable speed, by replacing a falling object with a ball rolling down a sloping ramp. He timed the experiment using both a water clock—a device that weighed the water spurting into an urn as the ball traveled— and his own pulse. If he doubled the period of time the ball rolled, he found the distance it traveled was four times as far.

Leaving nothing to chance, Galileo repeated the experiment "a full hundred times" until he had achieved "an accuracy such that the deviation between two observations never exceeded one-

In this fresco by Giuseppe Bezzuoli, Galileo is shown demonstrating his rolling-ball experiment in the presence of the powerful Medici family in Florence.

tenth of a pulse beat." He also changed the incline of the ramp: as it became steeper, the acceleration increased uniformly. Since Galileo's experiments were not carried out in a vacuum, they were imperfect—the moving balls were subject to air resistance and friction from the ramp. Nevertheless, Galileo concluded that in a vacuum, all objects—regardless of weight or shape—would accelerate at a uniform rate: the square of the elapsed time of the fall is proportional to the distance fallen.

Quantifying gravitational acceleration

In spite of Galileo's work, the question of the acceleration of free-falling objects was still contentious in the mid-17th century. From 1640 to 1650, Jesuit priests Giovanni Riccioli and Francesco Grimaldi conducted various investigations in Bologna. Key to their eventual success were Riccioli's time-keeping pendulums—which were as accurate as any available at the time—and a very tall tower. The two priests and their assistants dropped heavy objects from various levels of the 321-ft (98-m) Asinelli Tower, timing their descents.

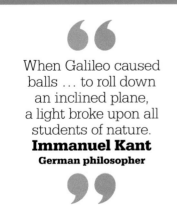

When Galileo caused balls ... to roll down an inclined plane, a light broke upon all students of nature.
Immanuel Kant
German philosopher

The hammer and the feather

In 1971, American astronaut David Scott—commander of the Apollo 15 moon mission—performed a famous free-fall experiment. The fourth NASA expedition to land on the moon, Apollo 15 was capable of a longer stay on the moon than previous expeditions, and its crew was the first to use a Lunar Roving Vehicle.

Apollo 15 also featured a greater focus on science than earlier moon landings. At the end of the mission's final lunar walk, Scott dropped a 3-lb geological hammer and a 1-oz falcon's feather from a height of 5 ft. In the virtual vacuum conditions of the moon's surface, with no air resistance, the ultralight feather fell to the ground at the same speed as the heavy hammer. The experiment was filmed, so this confirmation of Galileo's theory that all objects accelerate at a uniform rate regardless of mass was witnessed by a television audience of millions.

The priests, who described their methodology in detail, repeated the experiments several times.

Riccioli believed that free-falling objects accelerated exponentially, but the results showed him that he was wrong. A series of falling objects were timed by pendulums at the top and bottom of the tower. They fell 15 Roman feet (1 Roman foot = 11.6 in) in 1 second, 60 feet in 2 seconds, 135 feet in 3 seconds, and 240 feet in 4 seconds. The data, published in 1651, proved that the distance of descent was proportional to the square of the length of time the object was falling—confirming Galileo's ramp experiments. And for the first time, due to relatively accurate time-keeping, it was possible to work out the value of acceleration due to gravity: 9.36 (±0.22) m/s². This figure is only about 5 percent less than the range of figures accepted today: around 9.81 m/s².

The value of g (gravity) varies according to a number of factors: it is greater at Earth's poles than at the equator, lower at high altitudes than at sea level, and it varies very slightly according to local geology, for example if there are particularly dense rocks near Earth's surface. If the constant acceleration of an object in free fall near Earth's surface is represented by g, the height at which it is released is z_0 and time is t, then at any stage in its descent, the height of the body above the surface $z = z_0 - \frac{1}{2} gt^2$, where gt is the speed of the body and g its acceleration. A body of mass m at a height z_0 above Earth's surface possesses gravitational potential energy U, which can be calculated by the equation $U = mgz_0$ (mass × acceleration × height above Earth's surface). ∎

In questions of science, the authority of a thousand is not worth the humble reasoning of a single individual.
Galileo Galilei

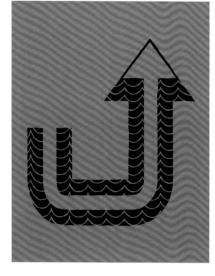

A NEW MACHINE FOR MULTIPLYING FORCES
PRESSURE

IN CONTEXT

KEY FIGURE
Blaise Pascal (1623–1662)

BEFORE
1643 Italian physicist Evangelista Torricelli demonstrates the existence of a vacuum using mercury in a tube; his principle is later used to invent the barometer.

AFTER
1738 In *Hydrodynamica*, Swiss mathematician Daniel Bernoulli argues that energy in a fluid is due to elevation, motion, and pressure.

1796 Joseph Bramah, a British inventor, uses Pascal's law to patent the first hydraulic press.

1851 Scottish–American inventor Richard Dudgeon patents a hydraulic jack.

1906 An oil hydraulic system is installed to raise and lower the guns of the US warship *Virginia*.

While investigating hydraulics (the mechanical properties of liquids), French mathematician and physicist Blaise Pascal made a discovery that would eventually revolutionize many industrial processes. Pascal's law, as it became known, states that if pressure is applied to any part of a liquid in an enclosed space, that pressure is transmitted equally to every part of the fluid, and to the container walls.

The impact of Pascal

Pascal's law means that pressure exerted on a piston at one end of a fluid-filled cylinder produces an equal increase in pressure on another piston at the other end of the cylinder. More significantly, if the cross-section of the second piston is twice that of the first, the force on it will be twice as great. So, a 2.2lb (1 kg) load on the small piston will allow the large piston to lift 4.4lb (2 kg); the larger the ratio of the cross-sections, the more weight the large piston can raise.

Pascal's findings weren't published until 1663, the year after his death, but they would be used by engineers to make the operation of machinery much easier. In 1796, Joseph Bramah applied the principle to construct a hydraulic press that flattened paper, cloth, and steel, doing so more efficiently and powerfully than previous wooden presses. ∎

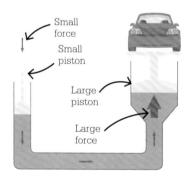

Small force
Small piston
Large piston
Large force

Liquids cannot be compressed and are used to transmit forces in hydraulics systems such as car jacks. A small force applied over a long distance is turned into a larger force over a small distance, which can raise a heavy load.

See also: Laws of motion 40–45 ▪ Stretching and squeezing 72–75 ▪ Fluids 76–79 ▪ The gas laws 82–85

MOTION WILL PERSIST

MOMENTUM

IN CONTEXT

KEY FIGURE
John Wallis (1616–1703)

BEFORE
1518 French natural philosopher Jean Buridan describes "impetus," the measure of which is later understood to be momentum.

1644 In his *Principia Philosophiae* (*Principles of Philosophy*), French scientist René Descartes describes momentum as the "amount of motion."

AFTER
1687 Isaac Newton describes his laws of motion in his three-volume work *Principia*.

1927 German theoretical physicist Werner Heisenberg argues that for a subatomic particle, such as an electron, the more precisely its position is known, the less precisely its momentum can be known, and vice versa.

W hen objects collide, several things happen. They change velocity and direction, and the kinetic energy of motion may be converted to heat or sound.

In 1666, the Royal Society of London challenged scientists to come up with a theory to explain what happens when objects collide. Two years later, three individuals published their theories: from England, John Wallis and Christopher Wren, and from Holland, Christiaan Huygens.

All moving bodies have momentum (the product of their mass and velocity). Stationary bodies have no momentum because their velocity is zero. Wallis, Wren, and Huygens agreed that in an elastic collision (any collision in which no kinetic energy is lost through the creation of heat or noise), momentum is conserved as long as there are no other external forces at work. Truly elastic collisions are rare in nature; the nudging of one billiard ball by another comes close, but there is

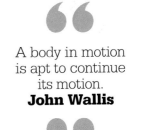

> A body in motion is apt to continue its motion.
> **John Wallis**

still some loss of kinetic energy. In *The Geometrical Treatment of the Mechanics of Motion*, John Wallis went further, correctly arguing that momentum is also conserved in inelastic collisions, where objects become attached after they collide, causing the loss of kinetic energy. One such example is that of a comet striking a planet.

Nowadays, the principles of conservation of momentum have many practical applications, such as determining the speed of vehicles after traffic accidents. ■

See also: Laws of motion 40–45 ▪ Kinetic energy and potential energy 54 ▪ The conservation of energy 55 ▪ Energy and motion 56–57

THE MOST WONDERFUL PRODUCTIONS OF THE MECHANICAL ARTS
MEASURING TIME

IN CONTEXT

KEY FIGURE
Christiaan Huygens
(1629–1695)

BEFORE
c. 1275 The first all-mechanical clock is built.

1505 German clockmaker Peter Henlein uses the force from an uncoiling spring to make the first pocket watch.

1637 Galileo Galilei has the idea for a pendulum clock.

AFTER
c. 1670 The anchor escapement mechanism makes the pendulum clock more accurate.

1761 John Harrison's fourth marine chronometer, H4, passes its sea trials.

1927 The first electronic clock, using quartz crystal, is built.

1955 British physicists Louis Essen and Jack Parry make the first atomic clock.

A pendulum takes the **same time to swing in each direction** because of gravity.

⬇

The **longer** the **pendulum**, the **more slowly** it swings.

➡

The **smaller the swing**, the **more accurately** the pendulum keeps time.

⬇

A pendulum is a **simple timekeeping device**.

⬅

An **escapement mechanism** keeps the pendulum moving.

Two inventions in the mid 1650s heralded the start of the era of precision timekeeping. In 1656, Dutch mathematician, physicist, and inventor Christiaan Huygens built the first pendulum clock. Soon after, the anchor escapement was invented, probably by English scientist Robert Hooke. By the 1670s, the accuracy of timekeeping devices had been revolutionized.

The first entirely mechanical clocks had appeared in Europe in the 13th century, replacing clocks reliant on the movement of the sun, the flow of water, or the burning of a candle. These mechanical clocks relied on a "verge escapement mechanism," which transmitted force from a suspended weight through the timepiece's gear train, a series of toothed wheels. Over the next three centuries, there were incremental advances in the accuracy of these clocks, but they had to be wound regularly and still weren't very accurate.

In 1637, Galileo Galilei had realized the potential for pendulums to provide more accurate clocks. He found that

See also: Free falling 32–35 ▪ Harmonic motion 52–53 ▪ SI units and physical constants 58–63 ▪ Subatomic particles 242–243

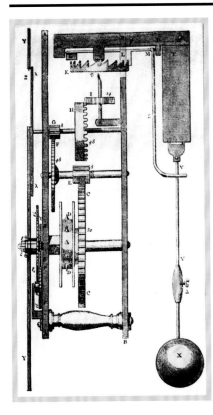

Christiaan Huygens' pendulum clock dramatically improved the accuracy of timekeeping devices. This 17th-century woodcut shows the inner workings of his clock, including toothed gears and pendulum.

this, even the most advanced non-pendulum clocks lost 15 minutes a day; now that margin of error could be reduced to as little as 15 seconds.

Quartz and atomic clocks

Pendulum clocks remained the most accurate form of time measurement until the 1930s, when synchronous electric clocks became available. These counted the oscillations of alternating current coming from electric power supply; a certain number of oscillations translated into movements of the clock's hands.

The first quartz clock was built in 1927, taking advantage of the piezoelectric quality of crystalline quartz. When bent or squeezed, it generates a tiny electric voltage, or conversely, if it is subject to an electric voltage, it vibrates. A battery inside the clock emits the voltage, and the quartz chip vibrates, causing an LCD display to change or a tiny motor to move second, minute, and hour hands.

The first accurate atomic clock, built in 1955, used the cesium-133 isotope. Atomic clocks measure the frequency of regular electromagnetic signals that electrons emit as they change between two different energy levels when bombarded with microwaves. Electrons in an "excited" cesium atom oscillate, or vibrate, 9,192,631,770 times per second, making a clock calibrated on the basis of these oscillations extremely accurate. ∎

a swinging pendulum was almost isochronous, meaning the time it took for the bob at its end to return to its starting point (its period) was roughly the same whatever the length of its swing. A pendulum's swing could produce a more accurate way of keeping time than the existing mechanical clocks. However, he hadn't managed to build one before his death in 1642.

Huygens' first pendulum clock had a swing of 80–100 degrees, which was too great for complete accuracy. The introduction of Hooke's anchor escapement, which maintained the swing of the pendulum by giving it a small push each swing, enabled the use of a longer pendulum with a smaller swing of just 4–6 degrees, which gave much better accuracy. Before

Harrison's marine chronometer

In the early 18th century, even the most accurate pendulum clocks didn't work at sea—a major problem for nautical navigation. With no visible landmarks, calculating a ship's position depended on accurate latitude and longitude readings. While it was easy to gauge latitude (by viewing the position of the sun), longitude could be determined only by knowing the time relative to a fixed point, such as the Greenwich Meridian. Without clocks that worked at sea, this was impossible. Ships were lost and many men died, so, in 1714, the British government offered a prize to encourage the invention of a marine clock.

British inventor John Harrison solved the problem in 1761. His marine chronometer used a fast-beating balance wheel and a temperature-compensated spiral spring to achieve remarkably accurate timekeeping on transatlantic journeys. The device saved lives and revolutionized exploration and trade.

John Harrison's prototype chronometer, H1, underwent sea trials from Britain to Portugal in 1736, losing just a few seconds on the entire voyage.

ALL ACTION HAS A REACTION

LAWS OF MOTION

IN CONTEXT

KEY FIGURES
Gottfried Leibniz (1646–1716),
Isaac Newton (1642–1727)

BEFORE
c. 330 BCE In *Physics*, Aristotle expounds his theory that it takes force to produce motion.

1638 Galileo's *Dialogues Concerning Two New Sciences* is published. It is later described by Albert Einstein as anticipating the work of Leibniz and Newton.

1644 René Descartes publishes *Principles in Philosophy*, which includes laws of motion.

AFTER
1827–1833 William Rowan Hamilton establishes that objects tend to move along the path that requires the least energy.

1907–1915 Einstein proposes his theory of general relativity.

P rior to the late 16th century, there was little understanding of why moving bodies accelerated or decelerated—most people believed that some indeterminate, innate quality made objects fall to the ground or float up to the sky. But this changed at the dawn of the Scientific Revolution, when scientists began to understand that several forces are responsible for changing a moving object's velocity (a combined measure of its speed and direction), including friction, air resistance, and gravity.

Early views

For many centuries, the generally accepted views of motion were those of the ancient Greek philosopher Aristotle, who classified everything in the world according to its elemental composition: earth, water, air, fire, and quintessence, a fifth element that made up the "heavens." For Aristotle, a rock falls to the ground because it has a similar composition to the ground ("earth"). Rain falls to the ground because water's natural place is at Earth's surface. Smoke rises because it is largely made of air. However, the circular movement of celestial objects was not considered to be governed by the elements—rather, they were thought to be guided by the hand of a deity.

Aristotle believed that bodies move only if they are pushed, and once the pushing force is removed, they come to a stop. Some questioned why an arrow unleashed from a bow continues to fly through the air long after direct contact with the bow has ceased, but Aristotle's views went largely unchallenged for more than two millennia.

In 1543, Polish astronomer Nicolaus Copernicus published his theory that Earth was not the center of the universe, but that it and the other planets orbited the sun in a "heliocentric" system. Between 1609 and 1619, German astronomer Johannes Kepler developed his laws of planetary motion, which describe the shape and speed of the orbits of planets. Then, in the 1630s, Galileo challenged Aristotle's views on falling objects, explained that a loosed arrow continues to fly

Gottfried Leibniz

Born in Leipzig (now Germany) in 1646, Leibniz was a great philosopher, mathematician, and physicist. After studying philosophy at the University of Leipzig, he met Christiaan Huygens in Paris and determined to teach himself math and physics. He became a political adviser, historian, and librarian to the royal House of Brunswick in Hanover in 1676, a role that gave him the opportunity to work on a broad range of projects, including the development of infinitesimal calculus. However, he was also accused of having seen Newton's unpublished ideas and passing them off as his own. Although it was later generally accepted that Leibniz had arrived at his ideas independently, he never managed to shake off the scandal during his lifetime. He died in Hanover in 1716.

Key works

1684 "Nova methodus pro maximis et minimis" ("New method for maximums and minimums")
1687 *Essay on Dynamics*

See also: Free falling 32–35 ▪ Laws of gravity 46–51 ▪ Kinetic energy and potential energy 54 ▪ Energy and motion 56–57 ▪ The heavens 270–271 ▪ Models of the universe 272–273 ▪ From classical to special relativity 274

> There is neither more nor less power in an effect than there is in its cause.
> **Gottfried Leibniz**

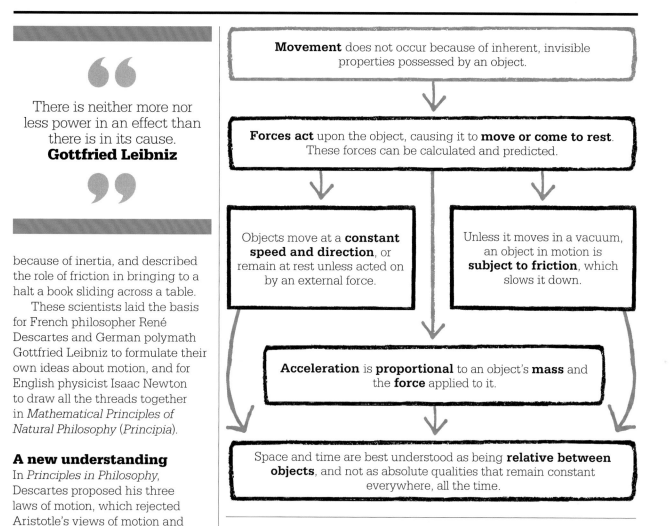

Movement does not occur because of inherent, invisible properties possessed by an object.

Forces act upon the object, causing it to **move or come to rest**. These forces can be calculated and predicted.

Objects move at a **constant speed and direction**, or remain at rest unless acted on by an external force.

Unless it moves in a vacuum, an object in motion is **subject to friction**, which slows it down.

Acceleration is **proportional** to an object's **mass** and the **force** applied to it.

Space and time are best understood as being **relative between objects**, and not as absolute qualities that remain constant everywhere, all the time.

because of inertia, and described the role of friction in bringing to a halt a book sliding across a table.

These scientists laid the basis for French philosopher René Descartes and German polymath Gottfried Leibniz to formulate their own ideas about motion, and for English physicist Isaac Newton to draw all the threads together in *Mathematical Principles of Natural Philosophy* (*Principia*).

A new understanding

In *Principles in Philosophy*, Descartes proposed his three laws of motion, which rejected Aristotle's views of motion and a divinely guided universe, and explained motion in terms of forces, momentum, and collisions. In his 1687 *Essay on Dynamics*, Leibniz produced a critique of Descartes' laws of motion. Realizing that many of Descartes' criticisms of Aristotle were justified, Leibniz went on to develop his own theories on "dynamics," his term for motion and impact, during the 1690s.

Leibniz's work remained unfinished, and he was possibly put off after reading Newton's

thorough laws of motion in *Principia*, which—like *Dynamics*—was also published in 1687. Newton respected Descartes' rejection of Aristotelian ideas, but argued that the Cartesians (followers of Descartes) did not make enough use of the mathematical techniques of Galileo, nor the experimental methods of chemist Robert Boyle. However, Descartes' first two laws of motion won the support of both Newton and Leibniz, and became the basis for Newton's first law of motion.

Newton's three laws of motion (see pp.44–45) clearly explained the forces acting on all bodies, revolutionizing the understanding of the mechanics of the physical world and laying the foundations for classical mechanics (the study of the motion of bodies). Not all of Newton's views were accepted during his lifetime—one of those who raised criticisms was Leibniz himself—but after his death they were largely unchallenged until the early 20th century, just as Aristotle's beliefs about motion »

The bicycle is in motion due to the force supplied by the pedalling of the rider, until the external force of the rock acts upon it, causing it to stop.

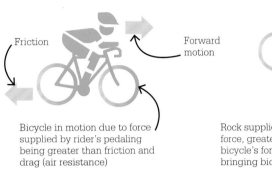

Friction

Forward motion

Bicycle in motion due to force supplied by rider's pedaling being greater than friction and drag (air resistance)

Rider flies over handlebars, since he or she has not been acted on by the external force (the rock)

Rock supplies external force, greater in quantity than bicycle's forward motion, bringing bicycle to a stop

had dominated scientific thinking for the best part of 2,000 years. However, some of Leibniz's views on motion and criticisms of Newton were far ahead of their time, and were given credence by Albert Einstein's general theory of relativity two centuries later.

Law of inertia

Newton's first law of motion, which is sometimes called the law of inertia, explains that an object at rest stays at rest, and an object in motion remains in motion with the same velocity unless acted upon by an external force. For instance, if the front wheel of a bicycle being ridden at speed hits a large rock, the bike is acted upon by an external force, causing it to stop. Unfortunately for the cyclist, he or she will not have been acted upon by the same force and will continue in motion—over the handlebars.

For the first time, Newton's law enabled accurate predictions of motion to be made. Force is defined as a push or pull exerted on one object by another and is measured in Newtons (denoted N, where 1 N is the force required to give a 1 kg mass an acceleration of 1 m/s²). If the strength of all the forces on an

object are known, it is possible to calculate the net external force— the combined total of the external forces—expressed as $\sum F$ ($\sum$ stands for "sum of"). For example, if a ball has a force of 23 N pushing it left, and a force of 12 N pushing it right, $\sum F = 11$ N in a leftward direction. It is not quite as simple as this, since the downward force of gravity will also be acting on the ball, so horizontal and vertical net forces also need to be taken into account.

There are other factors at play. Newton's first law states that a moving object that is not acted upon by outside forces should continue to move in a straight line at a constant velocity. But when a ball is rolled across the floor, for

example, why does it eventually stop? In fact, as the ball rolls it experiences an outside force: friction, which causes it to decelerate. According to Newton's second law, an object will accelerate in the direction of the net force. Since the force of friction is opposite to the direction of travel, this acceleration causes the object to slow and eventually stop. In interstellar space, a spacecraft will continue to move at the same velocity because of an absence of friction and air resistance—unless it is accelerated by the gravitational field of a planet or star, for example.

Change is proportional

Newton's second law is one of the most important in physics, and describes how much an object accelerates when a given net force is applied to it. It states that the rate of change of a body's momentum—the product of its mass and velocity—is proportional to the force applied, and takes place in the direction of the applied force.

This can be expressed as $\sum F = ma$, where F is the net force, a is the acceleration of the object in the direction of the net force, and m is its mass. If the force increases, so does acceleration. Also, the rate of change of momentum is inversely proportional to the mass of the

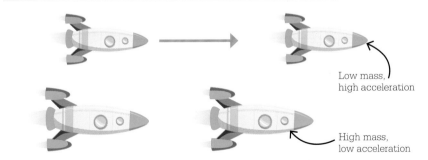

Low mass, high acceleration

High mass, low acceleration

Two rockets with different masses but identical engines will accelerate at different rates. The smaller rocket will accelerate more quickly due to its lower mass.

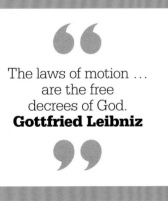

> "The laws of motion … are the free decrees of God.
> **Gottfried Leibniz**

object, so if the object's mass increases, its acceleration decreases. This can be expressed as $a = \sum F/m$. For example, as a rocket's fuel propellant is burned during flight, its mass decreases and—assuming the thrust of its engines remains the same—it will accelerate at an ever-faster rate.

Equal action and reaction
Newton's third law states that for every action there is an equal and opposite reaction. Sitting down, a person exerts a downward force on the chair, and the chair exerts an equal upward force on the person's body. One force is called the action, the other the reaction. A rifle recoils after it is fired due to the opposing forces of such an action–reaction. When the rifle's trigger is pulled, a gunpowder explosion creates hot gases that expand outward, allowing the rifle to push forward on the bullet. But the bullet also pushes backward on the rifle. The force acting on the rifle is the same as the force that acts on the bullet, but because acceleration depends on force and mass (in accordance with Newton's second law), the bullet accelerates much faster than the rifle due to its far smaller mass.

Notions of time, distance, and acceleration are fundamental to an understanding of motion. Newton argued that space and time are entities in their own right, existing independently of matter. In 1715–1716, Leibniz argued in favor of a relationist alternative: in other words, that space and time are systems of relations between objects. While Newton believed that absolute time exists independently of any observer and progresses at a constant pace throughout the universe, Leibniz reasoned that time makes no sense except when understood as the relative movement of bodies. Newton argued that absolute space "remains always similar and immovable," but his German critic argued that it only makes sense as the relative location of objects.

From Leibniz to Einstein
A conundrum raised by Irish bishop and philosopher George Berkeley around 1710 illustrated problems with Newton's concepts of absolute time, space, and velocity. It concerned a spinning sphere: Berkeley questioned whether, if it was rotating in an otherwise empty universe, it could be said to have motion at all. Although Leibniz's criticisms of Newton were generally

> "Motion is really nothing more than change of place. So motion as we experience it is nothing but a relation.
> **Gottfried Leibniz**

dismissed at the time, Einstein's general theory of relativity (1907–1915) made more sense of them two centuries later. While Newton's laws of motion are generally true for macroscopic objects (objects that are visible to the naked eye) under everyday conditions, they break down at very high speeds, at very small scales, and in very strong gravitational fields. ∎

Two Voyager spacecraft were launched in 1977. With no friction or air resistance in space, the craft are still moving through space today, due to Newton's first law of motion.

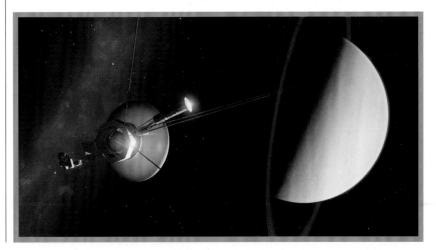

THE FRAME OF THE SYSTEM OF THE WORLD

LAWS OF GRAVITY

IN CONTEXT

KEY FIGURE
Isaac Newton (1642–1727)

BEFORE
1543 Nicolaus Copernicus challenges orthodox thought with a heliocentric model of the solar system.

1609 Johannes Kepler publishes his first two laws of planetary motion in *Astronomia Nova* (*A New Astronomy*), arguing that the planets move freely in elliptical orbits.

AFTER
1859 French astronomer Urbain Le Verrier argues that Mercury's precessionary orbit (the slight variance in its axial rotation) is incompatible with Newtonian mechanics.

1905 In his paper "On the Electrodynamics of Moving Bodies," Einstein introduces his theory of special relativity.

1915 Einstein's theory of general relativity states that gravity affects time, light, and matter.

What hinders the fixed stars from falling upon one another?
Isaac Newton

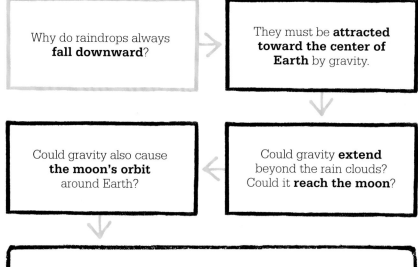

Why do raindrops always **fall downward**?

They must be **attracted toward the center of Earth** by gravity.

Could gravity **extend** beyond the rain clouds? Could it **reach the moon**?

Could gravity also cause **the moon's orbit** around Earth?

If that's the case, perhaps **gravity is universal**.

Published in 1687, Newton's law of universal gravitation remained—alongside his laws of motion—the unchallenged bedrock of "classical mechanics" for more than two centuries. It states that every particle attracts every other particle with a force that is directly proportional to the product of their masses, and inversely proportional to the square of the distance between their centers.

Before the scientific age in which Newton's ideas were formulated, the Western understanding of the natural world had been dominated by the writings of Aristotle. The ancient Greek philosopher had no concept of gravity, believing instead that heavy objects fell to Earth because that was their "natural place," and celestial bodies moved around Earth in circles because they were perfect. Aristotle's geocentric view remained largely unchallenged until the Renaissance, when Polish–Italian astronomer Nicolaus Copernicus argued for a heliocentric model of the solar system, with Earth and the planets orbiting the sun. According to him, "We revolve around the sun like any other planet." His ideas, published in 1543, were based on detailed observations of Mercury, Venus, Mars, Jupiter, and Saturn made with the naked eye.

Astronomical evidence

In 1609, Johannes Kepler published *Astronomia Nova* (*A New Astronomy*) which, as well as providing more support for heliocentrism, described the elliptical (rather than circular) orbits of the planets. Kepler also discovered that the orbital speed of each planet depends on its distance from the sun.

Around the same time, Galileo Galilei was able to support Kepler's view with detailed observations made with the aid of telescopes. When he focused a telescope on Jupiter and saw moons orbiting the giant planet, Galileo uncovered further proof that Aristotle had

See also: Free falling 32–35 ▪ Laws of motion 40–45 ▪ The heavens 270–71 ▪ Models of the universe 272–73
▪ Special relativity 276–79 ▪ The equivalence principle 281 ▪ Gravitational waves 312–15

The New Almagest, a 1651 work by Riccioli, illustrates the tussle between rival models of planetary motion: Tycho Brahe's Earth-centric theory is shown outweighing heliocentrism.

been wrong: if all things orbited Earth, Jupiter's moons could not exist. Galileo also observed the phases of Venus, demonstrating that it orbits the sun.

Galileo also challenged the idea that heavy objects fall to the ground more quickly than light objects. His contention was supported by Italian Jesuit priests Giovanni Battista Riccioli and Francesco Maria Grimaldi, who in the 1640s dropped objects from a Bologna tower and timed their descent to the street below. Their calculations provided reasonably accurate values for the rate of acceleration due to gravity, now known to be 32.15 ft/s^2 (9.8 m/s^2). Their experiment was recreated in 1971 by US astronaut David Scott, who dropped a hammer

and a feather on NASA's Apollo 15 mission. With no air resistance at the moon's surface, the two objects hit the ground at the same time.

Newton's apple

While the story of an apple falling on Isaac Newton's head is apocryphal, watching the fruit drop to the ground did spark his curiosity. By the time Newton started thinking seriously about gravity in the 1660s, much important groundwork had already been achieved. In his seminal work *Principia*, Newton credited the work of Italian physicist Giovanni Borelli (1608–1679) and French astronomer Ismael Bullialdus (1605–1694), who both described the sun's gravity exerting an attractive force. Bullialdus believed incorrectly that the sun's gravity attracted a planet at its aphelion (the point on its orbital curve when it is furthest from the sun) but repulsed it at the perihelion (when it is closest).

The work of Johannes Kepler was probably the greatest influence on Newton's ideas. The German astronomer's third law of orbital

If the Earth should cease to attract its waters to itself all the waters of the sea would be raised and would flow to the body of the moon.
Johannes Kepler

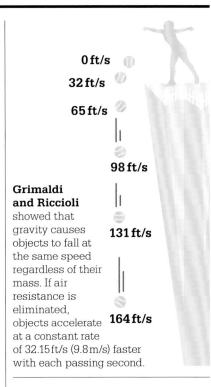

0 ft/s

32 ft/s

65 ft/s

98 ft/s

Grimaldi and Riccioli showed that gravity causes objects to fall at the same speed regardless of their mass. If air resistance is eliminated, objects accelerate at a constant rate of 32.15 ft/s (9.8 m/s) faster with each passing second.

131 ft/s

164 ft/s

motion states that there is an exact mathematical relationship between a planet's distance from the sun and the length of time it takes to complete a full orbit.

In 1670, English natural philosopher Robert Hooke argued that gravitation applies to all celestial bodies and that its power decreases with distance, and—in the absence of any other attractive forces—moves in straight lines. By 1679, he had concluded that the inverse square law applied, so gravity weakens in proportion to the square of the distance from a body. In other words, if the distance between the sun and another body is doubled, the force between them reduces to only one quarter of the original force. However, whether this rule could be applied close to the surface of a large planetary body such as Earth was unknown. »

Universality of free fall

The principle of universality of free fall was empirically discovered by Galileo and others, then mathematically proven by Newton. It states that all materials, heavy or light, fall at the same rate in a uniform gravitational field. Consider two falling bodies of different weight. Since Newton's theory of gravity says that the greater the mass of an object, the greater the gravitational force, the heavy object should fall faster. However, his second law of motion tells us that a larger mass does not accelerate as quickly as a smaller one if the applied force is the same, so it will fall more slowly. In fact, the two cancel each other out, so the light and heavy objects will fall with the same acceleration as long as no other forces—such as air resistance—are present.

The Asinelli tower in Bologna, Italy, was the chosen site for the free-fall experiments of Riccioli and Grimaldi, in which Galileo's theory was put to the test.

Newton argued that gravity is a universal, attractive force that applies to all matter, however large or small. Its strength varies according to how much mass the objects have, and how far they are from each other.

The greater an object's mass, the stronger its gravitational pull

The further apart two objects are, and the less mass they have, the weaker the gravitational pull

Universal gravitation

Newton published his own laws of motion and gravitation in *Principia* in 1687, asserting "Every particle attracts every other particle … with a force directly proportional to the product of their masses." He explained how all matter exerts an attractive force—gravity—that pulls other matter toward its center. It is a universal force, the strength of which depends on the mass of the object. For instance, the sun has a greater gravitational strength than Earth, which in turn has more gravitational strength than the moon, which in turn has more gravitational strength than a ball dropped upon it. Gravitational strength can be expressed with the equation $F = Gm_1m_2/r^2$, where F is the force, m_1 and m_2 are the masses of the two bodies, r is the distance between their centers, and G is the gravitational constant.

Newton continued to refine his views long after the publication of *Principia*. His cannonball thought-experiment speculated on the trajectory of a ball fired from a cannon on top of a very high mountain in an environment in which there was no air resistance. If gravity was also absent, he argued that the cannonball would follow a straight line away from Earth in the direction of fire. Assuming gravity is present, if the speed of the cannonball was relatively slow it would fall back to Earth, but if it was fired at a much faster velocity it would continue to circle Earth in a circular orbit; this would be its orbital velocity. If its velocity was faster still, the ball would continue to travel around Earth in an elliptical orbit. If it reached a speed that was faster than 7 miles/s (11.2 km/s), it would leave Earth's gravitational field and travel onward into outer space.

More than three centuries later, modern-day physics has put Newton's theories into practice. The cannonball phenomenon can be seen when a satellite or spacecraft is launched into orbit. Instead of the gunpowder that

66

Nonsense will fall of its own weight, by a sort of intellectual law of gravitation. And a new truth will go into orbit.
Cecilia Payne-Gaposchkin
British–American astronomer

99

blasted Newton's imaginary projectile, powerful rocket engines lift the satellite from Earth's surface and give it forward speed. When it reaches its orbital velocity, the spacecraft's propulsion ceases and it falls all the way around Earth, never hitting the surface. The angle of the satellite's path is determined by its initial angle and speed. The success of space exploration has relied heavily on Newton's laws of gravitation.

Understanding mass

An object's inertial mass is its inertial resistance to acceleration from any force, gravitational or not. It is defined by Newton's second law of motion as $F = ma$, where F is the force applied, m is its inertial mass, and a is acceleration. If a known force is applied to the object, by measuring its acceleration the inertial mass is found to be F/a. In contrast, according to Newton's law of universal gravitation, gravitational mass is the physical property of an object that causes it to interact with other objects through the gravitational force. Newton was troubled by the question: is the inertial mass of an object the same

as its gravitational mass? Repeated experiments have shown that the two properties are the same, a fact that fascinated Albert Einstein, who used it as a basis for his theory of general relativity.

Reinterpreting gravitation

Newton's ideas on universal gravitation and motion were unchallenged until 1905, when Einstein's theory of special relativity was published. Whereas Newton's theory depended on the assumption that mass, time, and distance remain constant, Einstein's theory treats them as fluid entities that are defined by the observer's frame of reference. A person standing on Earth as it spins on its axis orbits the sun—and moves through the universe in a different frame of reference to an astronaut flying through space in a spacecraft. Einstein's theory of general relativity also states that gravity is not a force, but rather the effect of the distortion of spacetime by massive objects.

Newton's laws are adequate for most day-to-day applications, but they cannot explain the differences in motion, mass, distance, and time that result when bodies are

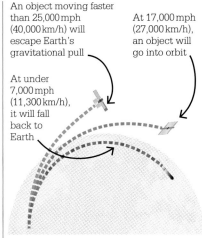

An object moving faster than 25,000 mph (40,000 km/h) will escape Earth's gravitational pull

At 17,000 mph (27,000 km/h), an object will go into orbit

At under 7,000 mph (11,300 km/h), it will fall back to Earth

Newton correctly predicted that objects would orbit Earth if they were launched at the correct speed. If a satellite moves quickly enough, the curvature of its fall is less than that of Earth, so it stays in orbit, never returning to the ground.

observed from two very different frames of reference. In this instance, scientists must rely on Einstein's theories of relativity. Classical mechanics and Einstein's theories of relativity agree as long as an object's speed is low or the gravitational field it experiences is small. ∎

Isaac Newton

Born in the English village of Woolsthorpe on Christmas Day 1642, Newton went to school in Grantham and studied at Cambridge University. In *Principia*, he formulated the laws of universal gravitation and motion, which formed the basis of classical mechanics until the early 20th century when they were partially superseded by Einstein's theories of relativity. Newton also made important contributions to mathematics and optics. Sometimes a controversial character, he had long-running disputes with Gottfried Leibniz over who had discovered calculus, and Robert Hooke regarding the inverse square law. As well as being a keen scientist, Newton was very interested in alchemy and biblical chronology. He died in London in 1727.

Key works

1684 *On the Motion of Bodies in an Orbit*
1687 *Philosophiae Naturalis Principia Mathematica (Mathematical Principles of Natural Philosophy)*

OSCILLATION IS EVERYWHERE
HARMONIC MOTION

Periodic motion—motion repeated in equal time intervals—is found in many natural and artificial phenomena. Studies of pendulums in the 16th and 17th centuries, for instance, helped lay the foundations for Isaac Newton's laws of motion. But groundbreaking though these laws were, physicists still faced major barriers in applying them to real-world problems involving systems (interacting groups of items) that were more complex than Newton's idealized, freely-moving bodies.

Musical oscillations

One particular area of interest was the vibration of musical strings—another form of periodic motion. In Newton's day, the principle that strings vibrate at different frequencies to produce different sounds was well established, but the exact form of the vibrations was unclear. In 1732, Swiss physicist and mathematician Daniel Bernoulli found a means of applying Newton's second law of motion to each segment of a vibrating string. He showed that the force on the string grew as it moved further from the center line (its stationary starting point), and always acted in the opposite direction to the displacement from the center. While tending to restore the string toward the center, it overshot to the other side, creating a repeating cycle.

This type of movement, with a specific relationship between displacement and restoring force, is today known as simple harmonic motion. As well as vibrating strings, it encompasses phenomena such as a swinging pendulum and a weight bouncing on the end of a spring. Bernoulli also discovered that harmonic oscillations plotted

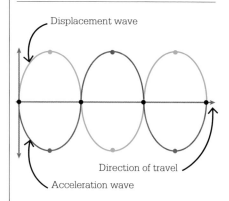

Displacement wave

Direction of travel

Acceleration wave

For any system in simple harmonic motion, the displacement and acceleration can be described by sine wave oscillations that are mirror images of each other.

See also: Measuring time 38–39 ▪ Laws of motion 40–45 ▪ Kinetic energy and potential energy 54 ▪ Music 164–167

> Nothing takes place in the world whose meaning is not that of some maximum or minimum.
> **Leonhard Euler**

on a graph form a sine wave—a mathematical function that is easily manipulated to find solutions to physical problems. Harmonic motion can also be applied in some more surprising places. For instance, both circular motion (for example, a satellite orbiting Earth) and the rotation of objects (Earth spinning on its axis) could be treated as oscillations back and forth in two or more directions.

Using Newton's law

Swiss mathematician and physicist Leonhard Euler was intrigued by the forces that cause ships to pitch (bob up and down lengthwise from bow to stern) and roll (tilt side to side). Around 1736, he realized that the motion of a ship could be split into a translation element (a movement between two places) and a rotational element.

Searching for an equation to describe the rotation part of the movement, Euler built on Daniel

As a young medic in the Russian navy, Leonhard Euler became fascinated by the way in which waves affect the motion of ships.

Bernoulli's earlier work and eventually hit upon a form that mirrored the structure of Newton's second law. In 1752, Euler was the first person to express that famous law in the now familiar equation $F = ma$ (the force acting on a body is equal to its mass multiplied by its acceleration). In parallel, his equation for rotation states that: $L = I\, d\omega/dt$, where L is the torque (the rotational force acting on the object), I is the object's "moment of inertia" (broadly speaking, its resistance to turning), and $d\omega/dt$ is the rate of change of its angular velocity ω (in other words, its "angular acceleration").

Simple harmonic motion has proved to have countless applications, including in fields that were not dreamt of in Euler's time, ranging from harnessing the oscillations of electrical and magnetic fields in electric circuits to mapping the vibrations of electrons between energy levels in atoms. ▪

Leonhard Euler

Born into a religious family in Basel, Switzerland, in 1707, Leonhard Euler was the most important mathematician of his generation, taking an interest in pure math and its many applications—including ship design, mechanics, astronomy, and music theory.

Entering Basel's university at the age of 13, Euler studied under Johann Bernoulli. He spent 14 years teaching and researching at the Imperial Russian Academy in St. Petersburg, before Frederick the Great invited him to Berlin. Despite losing his sight in one eye in 1738 and the other in 1766, Euler continued to work at a prodigious rate, establishing several entirely new areas of mathematical investigation. On his return to St. Petersburg, he continued to work until he died of a brain hemorrhage in 1783.

Key works

1736 *Mechanics*
1744 *A method for finding curved lines enjoying properties of maximum or minimum*
1749 *Naval science*
1765 *Theory on the motion of solid or rigid bodies*

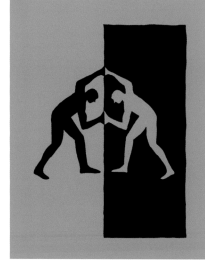

THERE IS NO DESTRUCTION OF FORCE

KINETIC ENERGY AND POTENTIAL ENERGY

IN CONTEXT

KEY FIGURE
Émilie du Châtelet
(1706–1749)

BEFORE
1668 John Wallis proposes a law of conservation of momentum—the first in its modern form.

AFTER
1798 American-born British physicist Benjamin Thompson, Count Rumford, makes measurements that suggest heat is another form of kinetic energy, contributing to the total energy of a system.

1807 British polymath Thomas Young first uses the term "energy" for the *vis viva* investigated by du Châtelet.

1833 Irish mathematician William Rowan Hamilton shows how the evolution of a mechanical system can be thought of in terms of the changing balance between potential and kinetic energies.

Isaac Newton's laws of motion incorporated the fundamental idea that the sum of momentum across all objects involved is the same before as after a collision. He had little to say, however, about the concept of energy as understood today. In the 1680s, Gottfried Leibniz noted that another property of moving bodies, which he called *vis viva* ("living force"), also seemed to be conserved.

Leibniz's idea was widely rejected by Newton's followers, who felt that energy and momentum should be indistinguishable, but it was revived in the 1740s. French philosopher Marquise Émilie du Châtelet, who was working on a translation of Newton's *Principia*, proved *vis viva*'s significance. She repeated an experiment—first carried out by Dutch philosopher Willem 's Gravesande—in which she dropped metal balls of differing weights into clay from various heights and measured the depth of the resulting craters. This showed that a ball traveling twice as fast made a crater four times as deep.

Du Châtelet concluded that each ball's *vis viva* (broadly the same concept as the modern kinetic energy attributed to moving particles) was proportional to its mass but also to the square of its velocity (mv^2). She hypothesized that since *vis viva* was clearly conserved (or transferred wholesale) in such collisions, it must exist in a different form when the weight was suspended before its fall. This form is now known as potential energy and is attributed to an object's position within a force field. ∎

Physics is an immense building that surpasses the powers of a single man.
Émilie du Châtelet

See also: Momentum 37 ▪ Laws of motion 40–45 ▪ Energy and motion 56–57 ▪ Force fields and Maxwell's equations 142–147

ENERGY CAN BE NEITHER CREATED NOR DESTROYED
THE CONSERVATION OF ENERGY

IN CONTEXT

KEY FIGURE
James Joule (1818–1889)

BEFORE
1798 Benjamin Thompson, Count Rumford, uses a cannon barrel immersed in water and bored with a blunted tool to show that heat is created from mechanical motion.

AFTER
1847 In his paper "On the Conservation of Force," German physicist Hermann von Helmholtz explains the convertibility of all forms of energy.

1850 Scottish civil engineer William Rankine is the first to use the phrase "the law of the conservation of energy" to describe the principle.

1905 In his theory of relativity, Albert Einstein introduces his principle of mass–energy equivalence—the idea that every object, even at rest, has energy equivalent to its mass.

The law of conservation of energy states that the total energy of an isolated system remains constant over time. Energy cannot be created or destroyed, but can be transformed from one form to another.

Although German chemist and physicist Julius von Mayer first advanced the idea in 1841, credit is often given to British physicist James Joule. In 1845, Joule published the results of a key experiment. He designed an apparatus with a falling weight to spin a paddle wheel in an insulated cylinder of water, using gravity to do the mechanical work. By measuring the increase in water temperature, he calculated the precise amount of heat an exact amount of mechanical work would create. He also showed that no energy was lost in the conversion.

Joule's discovery that heat had been created mechanically was not widely accepted until 1847, when Hermann von Helmholtz proposed a relationship between mechanics, heat, light, electricity, and magnetism—each a form of energy. Joule's contribution was honored when the standard unit of energy was named after him in 1882. ■

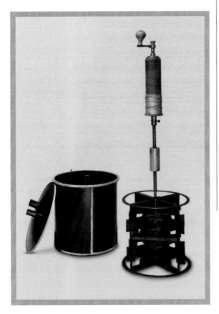

For his experiment, Joule used this vessel, filled with water, and the brass paddle wheel, turned by falling weights. The water's rising temperature showed that mechanical work created heat.

See also: Energy and motion 56–57 ▪ Heat and transfers 80–81 ▪ Internal energy and the first law of thermodynamics 86–89 ▪ Mass and energy 284–285

A NEW TREATISE ON MECHANICS
ENERGY AND MOTION

IN CONTEXT

KEY FIGURE
Joseph-Louis Lagrange
(1736–1813)

BEFORE
1743 French physicist and mathematician Jean Le Rond d'Alembert points out that the inertia of an accelerating body is proportional and opposite to the force causing the acceleration.

1744 Pierre-Louis Maupertuis, a French mathematician, shows that a "principle of least length" for the motion of light can be used to find its equations of motion.

AFTER
1861 James Clerk Maxwell applies the work of Lagrange and William Rowan Hamilton to calculate the effects of electromagnetic force fields.

1925 Erwin Schrödinger derives his wave equation from Hamilton's principle.

Throughout the 18th century, physics advanced considerably from the laws of motion set out by Isaac Newton in 1687. Much of this development was driven by mathematical innovations that made the central principles of Newton's laws easier to apply to a wider range of problems.

A key question was how best to tackle the challenge of systems with constraints, in which bodies are forced to move in a restricted way. An example is the movement of the weight at the end of a fixed pendulum, which can't break free from its swinging rod. Adding any form of constraint complicates Newtonian calculations considerably—at each point in an object's movement, all the forces acting on it must be taken into account and their net effect found.

Langrangian equations
In 1788, French mathematician and astronomer Joseph-Louis Lagrange proposed a radical new approach he called "analytical mechanics." He put forward two mathematical techniques that allowed the laws of motion to be more easily used in a wider variety of situations. The "Lagrangian

equations of the first kind" were simply a structure of equation that allowed constraints to be considered as separate elements in determining the movement of an object or objects.

Even more significant were the equations "of the second kind," which abandoned the "Cartesian coordinates" implicit in Newton's laws. René Descartes' pinning-down of location in three dimensions (commonly denoted x, y, and z) is intuitively easy to interpret, but makes all but the simplest problems in Newtonian

> 66
>
> Newton was the greatest genius that ever existed, and the most fortunate, for we cannot find more than once a system of the world to establish.
> **Joseph-Louis Lagrange**
>
> 99

See also: Laws of motion 40–45 ▪ The conservation of energy 55 ▪ Force fields and Maxwell's equations 142–147 ▪ Reflection and refraction 168–169

Newton's laws of motion describe motion in **Cartesian coordinates** (3-D *x*-, *y*-, and *z*-coordinates).

It is very **difficult to calculate complex problems of motion** using Cartesian coordinates.

Joseph-Louis Lagrange **devised equations** that allowed problems of motion to be solved using the **most appropriate coordinate system**.

This revealed that **objects usually move** along the path that requires **the least energy**.

Joseph-Louis Lagrange

Born in Turin, Italy, in 1736, Lagrange studied law before becoming interested in mathematics at the age of 17. From then on, he taught himself and developed his knowledge rapidly, lecturing on mathematics and ballistics at Turin's military academy. He subsequently became a founding member of the Turin Academy of Sciences, and he published work that attracted the attention of others, including Leonhard Euler.

In 1766, he moved to Berlin, where he succeeded Euler as head of mathematics at the Academy of Science. There, he produced his most important work on analytical mechanics and addressed astronomical problems such as the gravitational relationship between three bodies. He moved to Paris in 1786, where he spent the remainder of his career until his death in 1813.

Key works

1758–73 *Miscellanea Taurinensia*: papers published by the Turin Academy of Sciences
1788–89 *Analytical Mechanics*

---

physics very hard to calculate. The method developed by Lagrange allowed calculations to be made with whatever coordinate system was most appropriate for the problem under investigation. The generalization set out by Lagrange for equations of the second kind was not just a mathematical tool, but also pointed the way to a deeper understanding of the nature of dynamic systems.

Solving systems
Between 1827 and 1833, Irish mathematician William Rowan Hamilton expanded on Lagrange's work and took mechanics to a new level. Drawing on the "principle of least time" in optics, first proposed by French mathematician Pierre de Fermat in the 17th century, Hamilton developed a method to calculate the equations of motion for any system based on a principle of least (or stationary) action. This is the idea that objects, just like light rays, will tend to move along

the path that requires the least energy. Using this principle, he proved that any mechanical system could be described by solving it with a mathematical method similar to identifying the turning points on a graph.

Finally, in 1833, Hamilton set out a powerful new approach to mechanics through equations that described the evolution of a mechanical system over time, in terms of generalized coordinates and the total energy of the system (denoted H and now known as the "Hamiltonian"). Hamilton's equations allowed the system's balance of kinetic and potential energies to be calculated for a particular time, and therefore predicted not just the trajectories of objects, but their exact locations. Alongside his general principle of "least time," they would prove to have applications in several other areas of physics, including gravitation, electromagnetism, and even quantum physics. ▪

WE MUST LOOK TO THE HEAVENS FOR THE MEASURE OF THE EARTH

SI UNITS AND PHYSICAL CONSTANTS

> Each molecule, throughout
> the universe, bears impressed
> on it the stamp of a metric
> system as distinctly as
> does the meter of the
> Archives at Paris.
> **James Clerk Maxwell**

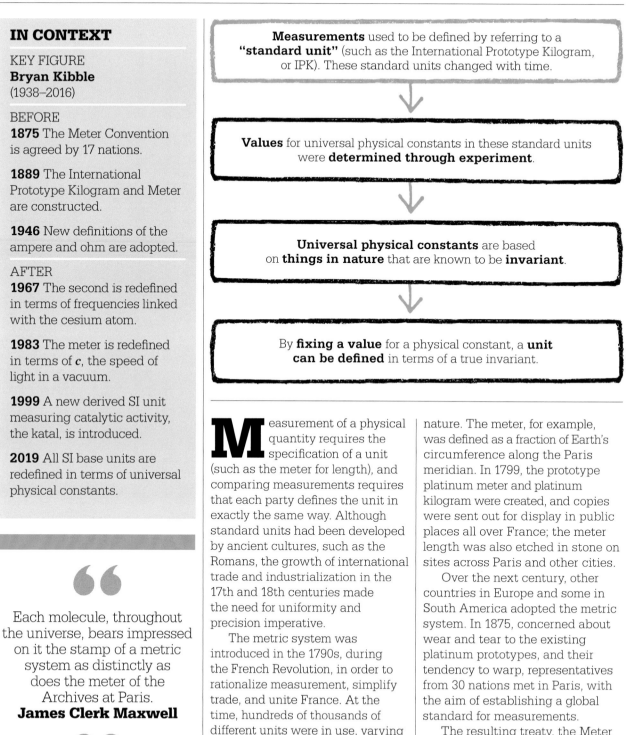

Measurements used to be defined by referring to a **"standard unit"** (such as the International Prototype Kilogram, or IPK). These standard units changed with time.

Values for universal physical constants in these standard units were **determined through experiment**.

Universal physical constants are based on **things in nature** that are known to be **invariant**.

By **fixing a value** for a physical constant, a **unit can be defined** in terms of a true invariant.

Measurement of a physical quantity requires the specification of a unit (such as the meter for length), and comparing measurements requires that each party defines the unit in exactly the same way. Although standard units had been developed by ancient cultures, such as the Romans, the growth of international trade and industrialization in the 17th and 18th centuries made the need for uniformity and precision imperative.

The metric system was introduced in the 1790s, during the French Revolution, in order to rationalize measurement, simplify trade, and unite France. At the time, hundreds of thousands of different units were in use, varying from village to village. The idea was to replace them with universal, permanent standards for length, area, mass, and volume based on

nature. The meter, for example, was defined as a fraction of Earth's circumference along the Paris meridian. In 1799, the prototype platinum meter and platinum kilogram were created, and copies were sent out for display in public places all over France; the meter length was also etched in stone on sites across Paris and other cities.

Over the next century, other countries in Europe and some in South America adopted the metric system. In 1875, concerned about wear and tear to the existing platinum prototypes, and their tendency to warp, representatives from 30 nations met in Paris, with the aim of establishing a global standard for measurements.

The resulting treaty, the Meter Convention (Convention du Mètre), stipulated new prototypes for the meter and kilogram, made out of a platinum–iridium alloy. These

See also: Measuring distance 18–19 ▪ Measuring time 38–39 ▪ The development of statistical mechanics 104–111 ▪ Electric charge 124–127 ▪ The speed of light 275

were kept in Paris, and copies were produced for the national standards institutes of the 17 signatory nations. The Convention outlined procedures for periodically calibrating the national standards against the new prototypes, and also set up the Bureau International des Poids et Mesures (International Bureau of Weights and Measures), or BIPM, to oversee them.

The SI (Système international, or International System) version of the metric system, initiated in 1948, was approved by signatory nations in Paris in 1960. Since then, it has been used for almost all scientific and technological and many everyday measurements. There are still exceptions, such as road distances in the UK and US, but even British imperial and US customary units such as the yard and the pound have been defined in terms of metric standards.

Units of 10

With traditional systems of units that use ratios of 2, 3, and their multiples—for example, 12 inches

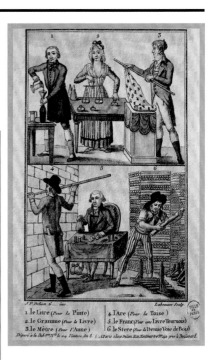

The metric system came into use in France around 1795. This engraving by L.F. Labrousse shows people using the new decimal units to measure things, and includes a list of metric units, each followed by the unit being replaced.

make a foot—some everyday sums are easy, but more complicated arithmetic can be unwieldy. The metric system specifies only decimal ratios (counting in units of 10), making arithmetic much easier; it is clear that 1/10th of 1/100th of a meter is 1/1000th of a meter.

The metric system also specifies names of prefixes and abbreviations for many multiples, such as kilo- (k) for multiplication by 1,000, centi- (c) for one hundredth, and micro- (μ) for one millionth. The prefixes allowed by the SI range from yocto- (y), meaning 10^{-24}, to yotta- (Y), meaning 10^{24}.

CGS, MKS, and SI

In 1832, German mathematician Carl Gauss proposed a system of measurement based on three fundamental units of length, mass,

and time. Gauss's idea was that all physical quantities could be measured in these units, or combinations of them. Each fundamental quantity would have one unit, unlike some traditional systems that used several different units for a quantity (for example, »

The IPK is just 4 cm tall and is kept under three glass bell jars at the BIPM (Bureau International des Poids et Mesures) in Paris, France.

The IPK

For 130 years, the kilogram was defined by a platinum–iridium cylinder, the IPK (International Prototype Kilogram) or "Le Grand K." Copies of the cylinder were held by national metrology institutes around the world, including the NPL (National Physical Laboratory, UK) and NIST (National Institute of Standards and Technology, US), and compared with the IPK about once every 40 years.

Although the platinum–iridium alloy is extremely stable, over time, discrepancies of up to 50 μg

emerged between cylinders. Since other base units depended on the definition of the kilogram, this drift affected measurements of many quantities. As scientists and industry demanded greater precision for experiments and technology, the instability of the IPK became a serious problem.

In 1960, when the meter was redefined in terms of a particular wavelength of light emitted by a krypton atom, the kilogram became the only base unit whose standard depended on a physical object. With the redefinition of SI in 2019, this is no longer the case.

SI base units

Today, SI base units are defined in terms of physical constants whose numerical values are fixed, and—apart from the second and the mole—of definitions of other base units.

Time	Second (s)	The second (s) is defined by fixing Δv_{cs}, the unperturbed ground-state hyperfine transition frequency of the cesium-133 atom, to be 9 192 631 770 Hz (i.e. 9 192 631 770 s^{-1}).
Length	Meter (m)	The meter (m) is defined by fixing c, the speed of light in a vacuum, to be 299 792 458 m s^{-1}, where the second is defined in terms of Δv_{cs}.
Mass	Kilogram (kg)	The kilogram (kg) is defined by fixing h, the Planck constant, to be 6.626 070 15 × 10^{-34} J s, (i.e. 6.626 070 15 × 10^{-34} kg m^2 s^{-1}, where the meter and the second are defined in terms of c and Δv_{cs}).
Electric current	Ampere (A)	The ampere (A) is defined by fixing e, the elementary charge, to be 1.602 176 634 × 10^{-19} C (i.e. 1.602 176 634 × 10^{-19} A s, where the second is defined in terms of Δv_{cs}).
Thermo-dynamic temperature	Kelvin (K)	The kelvin (K) is defined by fixing k, the Boltzmann constant, to be 1.380 649 × 10^{-23} J K^{-1} (i.e. 1.380 649 × 10^{-23} kg m^2 s^{-2} K^{-1}, where the kilogram, meter, and second are defined in terms of h, c and Δv_{cs}).
Amount of substance	Mole (mol)	The mole (mol) is defined by fixing N^A, the Avogadro constant, to be exactly 6.022 140 76 × 10^{23} mol^{-1} (i.e. one mole of a substance contains 6.02214076 × 10^{23} particles such as atoms, molecules, or electrons).
Luminous intensity	Candela (cd)	The candela (cd) is defined by fixing K$_{cd}$, the luminous efficacy of radiation of frequency 540 × 10^{12} Hz, to be 683 lm W^{-1} (i.e. 683 cd sr kg^{-1} m^{-2} s^3, where sr is the solid angle in steradians, and kilogram, meter, and second are defined in terms of h, c, and Δv_{cs}).

inch, yard, and furlong for length). In 1873, British physicists proposed the centimeter, gram, and second (CGS) as the fundamental units. This CGS system worked well for many years, but gradually gave way to the MKS (meter, kilogram, and second) system. Both were superseded by the SI, which included standardized units in newer areas of study, such as electricity and magnetism.

Base and derived SI units
The SI specifies seven "base units" (and abbreviations) to measure seven fundamental quantities, including the meter (m) for length, kilogram (kg) for mass, and second (s) for time. These fundamental quantities are considered to be independent of each other, although the definitions of their units are not—for instance, length and time are independent, but the definition of the meter is dependent on the definition of the second.

Other quantities are measured in "derived units," which are combinations of base units, according to the relationship between the quantities. For example, speed, which is distance per unit time, is measured in meters per second (m s^{-1}). As well as these derived units, there are currently 22 "derived units with special names"—including force, which is measured in newtons (N), where 1 N = 1 kg m s^{-2}.

Increasing accuracy
As theory and technology have advanced, SI base units have been redefined. Modern metrology—the science of measurement—depends on instruments of great precision. British metrologist Bryan Kibble's development of the moving-coil watt balance in 1975 greatly increased the accuracy with which the ampere could be defined. The watt balance compares the power developed by a moving mass to the current and voltage in an electromagnetic coil.

Kibble went on to collaborate with Ian Robinson at the UK's National Physical Laboratory (NPL) in 1978, creating a practical instrument—the Mark I—that enabled the ampere to be measured with unprecedented accuracy. The Mark II balance followed in 1990. Constructed in a vacuum chamber, this instrument made it possible to measure Planck's constant accurately enough to allow for the redefinition of the kilogram. Later models of Kibble's balance have contributed significantly to the most recent version of the SI.

Historically, definitions were made in terms of physical artifacts (such as the IPK) or measured properties (such as the frequency of radiation emitted by a particular type of atom), and included one or more universal physical constants. These constants (such as c, the speed of light in a vacuum, or Δv_{cs}, a frequency associated with an electron moving between particular energy levels—the "hyperfine transition"—in a cesium atom)

It is natural for man to relate the units of distance by which he travels to the dimensions of the globe that he inhabits.

Pierre-Simon Laplace
French mathematician and philosopher

are natural invariants. In other words, universal physical constants are the same over all time and space, and so are more stable than any experimental determination of them, or any material artifact.

SI units redefined

The 2019 redefinition of SI units in terms of fundamental physical constants was a philosophical shift. Prior to 2019, definitions of units were explicit. For example, since 1967, the second had been defined as 9 192 631 770 cycles of the radiation emitted by the cesium hyperfine transition. This number was obtained experimentally, by comparing Δv_{cs} with the most rigorous definition of the second then existing, which was based on Earth's orbit around the sun. Today, the definition is subtly different.

The constant—here, the value of Δv_{cs}—is first explicitly defined (as 9 192 631 770). This expresses our confidence that Δv_{cs} never changes. It does not really matter what numerical value is assigned to it because the size of the unit it is measured in is arbitrary. However, there is an existing, convenient unit—the second—that can be refined, so it is assigned a value that makes the newly defined second as close as possible to the second by the old definition. In other words, instead of having a fixed definition of the second, and measuring Δv_{cs} relative to it, metrologists fix a convenient number for Δv_{cs} and define the second relative to that.

Under the old definition of the kilogram, the IPK was considered a constant. Under the new definition, the value of Planck's constant (6.626 070 15 × 10^{-34} joule-seconds)

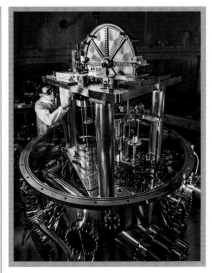

The Kibble balance at NIST produces incredibly accurate measurements, and has contributed to the recent redefinition of all base measurement units in terms of physical constants.

is fixed, and the kilogram has been redefined to fit with this numerical value.

The new SI now has a stronger foundation for its redefinition of units. For practical purposes, most are unchanged, but their stability and precision at very small or very large scales is markedly improved. ∎

Bryan Kibble

Born in 1938, British physicist and metrologist Bryan Kibble showed an early aptitude for science and won a scholarship to study at the University of Oxford, where he was awarded a DPhil in atomic spectroscopy in 1964. After a short post-doctoral period in Canada, he returned to the UK in 1967 and worked as a researcher at the National Physical Laboratory (NPL) until 1998.

Kibble made several significant contributions to metrology over his career, the greatest of which was his development of the moving-coil watt balance, which enabled measurements (initially of the ampere) to be made with great accuracy without reference to a physical artifact. After his death in 2016, the watt balance was renamed the Kibble balance in his honor.

Key works

1984 *Coaxial AC Bridges* (with G.H. Raynor)
2011 *Coaxial Electrical Circuits for Interference-Free Measurements* (with Shakil Awan and Jürgen Schurr)

ENERGY AND MATT

MATERIALS
AND HEAT

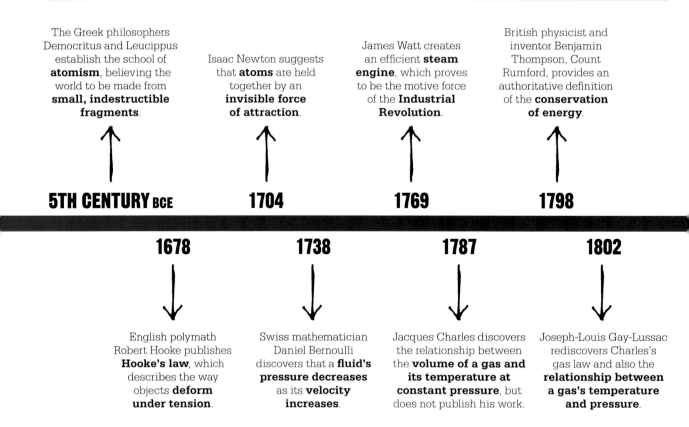

The Greek philosophers Democritus and Leucippus establish the school of **atomism**, believing the world to be made from **small, indestructible fragments**.

Isaac Newton suggests that **atoms** are held together by an **invisible force of attraction**.

James Watt creates an efficient **steam engine**, which proves to be the motive force of the **Industrial Revolution**.

British physicist and inventor Benjamin Thompson, Count Rumford, provides an authoritative definition of the **conservation of energy**.

5TH CENTURY BCE   **1704**   **1769**   **1798**

1678   **1738**   **1787**   **1802**

English polymath Robert Hooke publishes **Hooke's law**, which describes the way objects **deform under tension**.

Swiss mathematician Daniel Bernoulli discovers that a **fluid's pressure decreases** as its **velocity increases**.

Jacques Charles discovers the relationship between the **volume of a gas and its temperature at constant pressure**, but does not publish his work.

Joseph-Louis Gay-Lussac rediscovers Charles's gas law and also the **relationship between a gas's temperature and pressure**.

Some things in our universe are tangible, things we can touch and hold in our hands. Others seem ethereal and unreal until we observe the effect they have on those objects we hold. Our universe is constructed from tangible matter but it is governed by the exchange of intangible energy.

Matter is the name given to anything in nature with shape, form, and mass. The natural philosophers of ancient Greece were the first to propose that matter was made from many small building blocks called atoms. Atoms collect together to form materials, made of one or more different atoms combined in various ways. Such differing microscopic structures give these materials very different properties, some stretchy and elastic, others hard and brittle.

Long before the Greeks, early humans had used the materials around them in order to achieve the desired task. Every now and then, a new material was discovered, mostly by accident but sometimes through trial-and-error experiments. Adding coke (carbon) to iron produced steel, a stronger but more brittle metal that made better blades than iron alone.

The age of experimentation
In Europe during the 17th century, experimentation gave way to laws and theories, and these ideas led toward new materials and methods. During the European Industrial Revolution (1760–1840), engineers selected materials to build machines that could withstand large forces and temperatures. These machines were powered by water in its gaseous form of steam. Heat was the key to creating steam from water. In the 1760s, Scottish engineers Joseph Black and James Watt made the important discovery that heat is a quantity, while temperature is a measurement. Understanding how heat is transferred and how fluids move became crucial to success in the industrial world, with engineers and physicists vying to build the biggest and best machines.

Experiments with the physical properties of gases began with the creation of the vacuum pump by Otto von Guericke in Germany in 1650. Over the next century, chemists Robert Boyle in England and Jacques Charles and Joseph-Louis Gay-Lussac in France discovered three laws that related

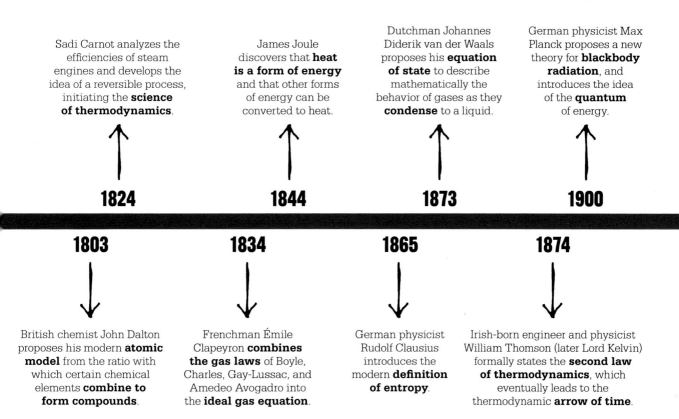

Sadi Carnot analyzes the efficiencies of steam engines and develops the idea of a reversible process, initiating the **science of thermodynamics**.

James Joule discovers that **heat is a form of energy** and that other forms of energy can be converted to heat.

Dutchman Johannes Diderik van der Waals proposes his **equation of state** to describe mathematically the behavior of gases as they **condense** to a liquid.

German physicist Max Planck proposes a new theory for **blackbody radiation**, and introduces the idea of the **quantum** of energy.

1824   **1844**   **1873**   **1900**

1803   **1834**   **1865**   **1874**

British chemist John Dalton proposes his modern **atomic model** from the ratio with which certain chemical elements **combine to form compounds**.

Frenchman Émile Clapeyron **combines the gas laws** of Boyle, Charles, Gay-Lussac, and Amedeo Avogadro into the **ideal gas equation**.

German physicist Rudolf Clausius introduces the modern **definition of entropy**.

Irish-born engineer and physicist William Thomson (later Lord Kelvin) formally states the **second law of thermodynamics**, which eventually leads to the thermodynamic **arrow of time**.

the temperature, volume, and pressure of a gas. In 1834, these laws were combined into a single equation to simply show the relationship between gas pressure, volume, and temperature.

Experiments conducted by British physicist James Joule showed that heat and mechanical work are interchangeable forms of the same thing, which we today call energy. Industrialists desired mechanical work in exchange for heat. Vast quantities of fossil fuels, mainly coal, were burned to boil water and create steam. Heat increased the internal energy of the steam before it expanded and performed mechanical work, pushing pistons and turning turbines. The relationship between heat, energy, and work was set out in the first law of thermodynamics.

Physicists designed new heat engines to squeeze as much work as they could from every last bit of heat. Frenchman Sadi Carnot discovered the most efficient way to achieve this theoretically, placing an upper limit on the amount of work obtainable for each unit of heat exchanged between two reservoirs at different temperatures. He confirmed that heat only moved spontaneously from hot to cold. Machines were imagined that did the opposite, but these refrigerators were only constructed years later.

Entropy and kinetic theory
The single direction of heat transfer from hot to cold suggested an underlying law of nature, and the idea of entropy emerged. Entropy describes the amount of disorder there is among the underlying

particles that make up a system. Heat flowing only from hot to cold was a specialized example of the second law of thermodynamics, which states that the entropy and disorder of an isolated system can only ever increase.

The variables of temperature, volume, pressure, and entropy seem to only be averages of microscopic processes involving innumerable particles. The transition from microscopic huge numbers to a singular macroscopic number was achieved through kinetic theory. Physicists were then able to model complex systems in a simplified way and link the kinetic energy of particles in a gas to its temperature. Understanding matter in all its states has helped physicists solve some of the deepest mysteries of the universe. ∎

THE FIRST PRINCIPLES OF THE UNIVERSE

MODELS OF MATTER

IN CONTEXT

KEY FIGURE
Democritus (c. 460–370 BCE)

BEFORE
c. 500 BCE In ancient Greece, Heraclitus declares that everything is in a state of flux.

AFTER
c. 300 BCE Epicurus adds the concept of atomic "swerve" to atomism, allowing for some behavior to be unpredictable.

1658 French clergyman Pierre Gassendi's *Syntagma philosophicum* (*Philosophical Treatise*), which attempts to marry atomism with Christianity, is published posthumously.

1661 Anglo-Irish physicist Robert Boyle defines elements in *The Sceptical Chymist*.

1803 John Dalton puts forward his atomic theory, based on empirical evidence.

A mong the several mysteries that scholars have contemplated over millennia is the question of what everything is made of. Ancient philosophers—from Greece to Japan—tended to think of all matter as being made from a limited set of simple substances ("elements"), usually earth, air or wind, fire, and water, which combined in different proportions and arrangements to create all material things.

Different cultures imagined these systems of elements in different ways, with some linking them to deities (as in Babylonian mythology) or tying them into

See also: Changes of state and making bonds 100–103 ▪ Atomic theory 236–237 ▪ The nucleus 240–241 ▪ Subatomic particles 242–243

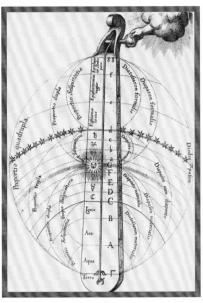

The classical system of elements focused on earth, water, air, and fire. This illustration, from a manuscript dated c. 1617, shows these elements within a divine universe.

grander philosophical frameworks (such as the Chinese philosophy of Wu Xing).

On the Indian subcontinent, for instance, as early as the 8th century BCE, the Vedic sage Aruni had described "particles too small to be seen [which] mass together into the substances and objects of experience." A number of other Indian philosophers had independently developed their own atomic theories.

A materialist approach

In the 5th century BCE, the Greek philosopher Democritus and his teacher Leucippus also took a more materialist approach to these systems of elements. Democritus, who valued rational reasoning over observation alone, based his theory

of atomism on the idea that it must be impossible to continue dividing matter eternally. He argued that all matter must therefore be made up of tiny particles that are too small to see. He named these particles "atoms" from the word *atomos*, meaning uncuttable.

According to Democritus, atoms are infinite and eternal. The properties of an object depend not just on the size and shape of its atoms but how these atoms are assembled. Objects could, he stated, change over time through shifts in their atomic arrangement. For example, he proposed bitter foods were made of jagged atoms that tore the tongue as they were chewed; sweet foods, on the other hand, consisted of smooth atoms that flowed smoothly over the tongue.

While modern atomic theory looks very different from the theory put forward by Leucippus and Democritus almost 2,500 years ago, their idea that the properties of substances are affected by how atoms are arranged remains relevant today. »

By convention sweet and by convention bitter, by convention hot, by convention cold, by convention color; but in reality atoms and void.
Democritus

Democritus

Democritus was born into a wealthy family in Abdera, in the historic region of Thrace in southeast Europe around 460 BCE. He traveled at length through parts of western Asia and Egypt as a young man before arriving in Greece to familiarize himself with natural philosophy.

Democritus acknowledged his teacher Leucippus as his greatest influence, and classicists have sometimes struggled to distinguish between their contributions to philosophy—particularly as none of the original works has survived to this day.

Best known for formulating "atomism," Democritus is also recognized as an early pioneer in aesthetics, geometry, and epistemology. He believed that rational reasoning was a necessary tool to seek truth because observations made through human senses would always be subjective.

Democritus was a modest man and is said to have taken a humorous approach to scholarship, giving him his nickname: "The Laughing Philosopher." He died around 370 BCE.

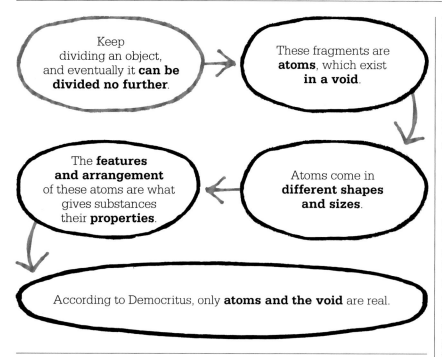

Keep dividing an object, and eventually it **can be divided no further**.

These fragments are **atoms**, which exist **in a void**.

Atoms come in **different shapes and sizes**.

The **features and arrangement** of these atoms are what gives substances their **properties**.

According to Democritus, only **atoms and the void** are real.

Around 300 BCE, the Greek philosopher Epicurus refined Democritus's ideas, by proposing the notion of atomic "swerve." This idea, that atoms can deviate from their expected actions, introduces unpredictability at the atomic scale and allowed for the preservation of "free will," a core belief held by Epicurus. Atomic swerve could be seen as an ancient iteration of the uncertainty at the core of quantum mechanics: since all objects have wave-like properties, it is impossible to accurately measure their position and momentum at the same time.

The rejection of atomism

Some of the most influential Greek philosophers rejected atomism and instead backed the theory of four or five fundamental elements. In 4th-century Athens, Plato proposed that everything was composed of five geometric solids (the Platonic solids), which gave types of matter their characteristics. For instance,

fire was made up of tiny tetrahedra, which—with their sharp points and edges—made it more mobile than earth, which was made of stable and squat cubes. Plato's student Aristotle—who loathed Democritus and reputedly wanted to burn his works—proposed that there were five elements (adding the celestial element of "ether") and no basic unit of matter. Although Western

Europe effectively abandoned the concept of atomism for several centuries, Islamic philosophers such as Al-Ghazali (1058–1111) developed their own distinct forms of atomism. Indian Buddhist philosophers such as Dhamakirti in the 7th century described atoms as point-like bursts of energy.

Atomism revival

The birth of the Renaissance in 14th-century Italy revived classical arts, science, and politics across Europe. It also saw the rebirth of the theory of atomism, as had been described by Leucippus and Democritus. Atomism was controversial, however, because of its link with Epicureanism, which many people believed violated strict Christian teachings. In the 17th century, French clergyman Pierre Gassendi dedicated himself to reconciling Christianity with Epicureanism, including atomism. He put forward a version of Epicurean atomism in which atoms have some of the physical features of the objects they make up, such as solidity and weight. Most importantly, Gassendi's theory stated that God had created a finite number of

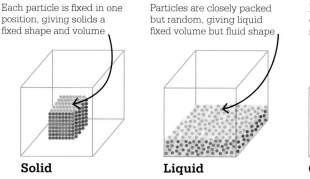

Each particle is fixed in one position, giving solids a fixed shape and volume

Particles are closely packed but random, giving liquid fixed volume but fluid shape

Particles move freely, giving gas no fixed shape or volume

Solid　　**Liquid**　　**Gas**

Dalton's atomic theory proposed that solids, liquids, and gases consist of particles (atoms or molecules). The motion of the particles and distances between them vary.

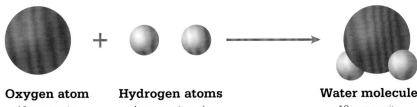

Oxygen atom
16 mass units

Hydrogen atoms
1 mass unit each

Water molecule
18 mass units

John Dalton proposed that atoms combine to produce molecules in simple ratios of mass. For example, two atoms of hydrogen (each with a mass of 1) combine with one of oxygen (with a mass of 16) to create a water molecule with a mass of 18.

atoms at the beginning of the universe, arguing that everything could be made up of these atoms and still be ruled by God. This idea helped steer atomism back into the mainstream among European scholars, aided by the endorsement of Isaac Newton and Robert Boyle.

In 1661, Boyle published *The Sceptical Chymist*, which rejected Aristotle's theory of five elements and instead defined elements as "perfectly unmingled bodies." According to Boyle, many different elements such as mercury and sulfur were made of many particles of different shapes and sizes.

How elements combine

In 1803, British physicist John Dalton created a basic model of how atoms combine to form these elements. It was the first model built from a scientific basis. From his experiments, Dalton noticed that the same pairs of elements, such as hydrogen and oxygen, could be combined in different ways to form various compounds. They always did so with whole number mass ratios (see diagram above). He concluded that each element was composed of its own atom with unique mass and other properties. According to Dalton's atomic theory, atoms cannot be divided, created, or destroyed, but

they can be bonded or broken apart from other atoms to form new substances.

Dalton's theory was confirmed in 1905, when Albert Einstein used mathematics to explain the phenomenon of Brownian motion—the jiggling of tiny pollen grains in water—using atomic theory. According to Einstein, the pollen is constantly bombarded by the random motion of many atoms. Disputes over this model were settled in 1911 when French physicist Jean Perrin verified that atoms were responsible for Brownian motion. The concept that atoms are bonded to or split from other atoms to form different substances is simple but remains useful for understanding everyday phenomena, such as how atoms of iron and oxygen combine to form rust.

States of matter

Plato taught that the consistency of substances depended on the geometric shapes from which they were made, but Dalton's atomic theory more accurately explains the states of matter. As illustrated opposite, atoms in solids are closely packed, giving them a stable shape and size; atoms in liquids are weakly connected, giving them indefinite shapes but mostly stable

size; and atoms in gases are mobile and distant from each other, resulting in a substance with no fixed shape or volume.

The atom is divisible

Atoms are the smallest ordinary object to have the properties of an element. However, they are no longer considered indivisible. In the two centuries since Dalton built modern atomic theory, it has been adapted to explain new discoveries. For instance, plasma—the fourth basic state of matter after solids, liquids, and gases—can only be fully explained if atoms can be divided further. Plasma is created when electrons are delocalized (stripped away) from their atoms.

In the late 19th and early 20th century, scientists found atoms are made up of several subatomic particles—electrons, protons, and neutrons; and those neutrons and protons are composed of even smaller, subatomic particles. This more complex model has allowed physicists to understand phenomena that Democritus and Dalton could never have imagined, such as radioactive beta decay and matter–antimatter annihilation. ∎

[Epicurus] supposes not only all mixt bodies, but all others to be produced by the various and casual occursions of atoms, moving themselves to and fro ... in the ... infinite vacuum.
Robert Boyle

AS THE EXTENSION, SO THE FORCE

STRETCHING AND SQUEEZING

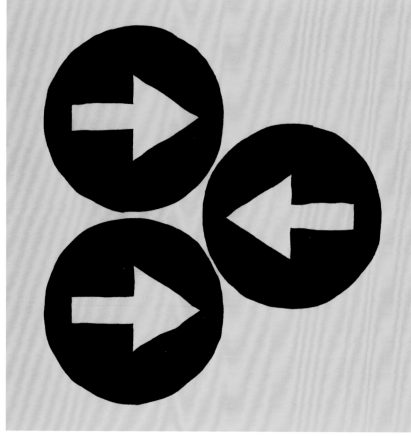

IN CONTEXT

KEY FIGURES
Robert Hooke (1635–1703),
Thomas Young (1773–1829)

BEFORE
1638 Galileo Galilei explores
the bending of wooden beams.

AFTER
1822 French mathematician
Augustin-Louis Cauchy shows
how stress waves move
through an elastic material.

1826 Claude-Louis Navier, a
French engineer and physicist,
develops Young's modulus
into its modern form, the
elastic modulus.

1829 German miner Wilhelm
Albert demonstrates metal
fatigue (weakening of metal
due to stress).

1864 French physicist Jean
Claude St-Venant and German
physicist Gustav Kirchhoff
discover hyperelastic materials.

British physicist and polymath
Robert Hooke made many
crucial contributions in
the Scientific Revolution of the 17th
century, but he became interested
in springs in the 1660s because he
wanted to make a watch. Up to
then, timepieces were typically
pendulum-driven, and pendulum
clocks became erratic when used
on ships. If Hooke could create a
timepiece driven by a spring and
not a pendulum, he could make a
clock that could keep time at sea,
thus solving the key navigational
problem of the age—calculating
a ship's longitude (east–west
distance) required accurate
timekeeping. Using a spring

See also: Pressure 36 ▪ Measuring time 38–39 ▪ Laws of motion 40–45 ▪ Kinetic energy and potential energy 54 ▪ The gas laws 82–85

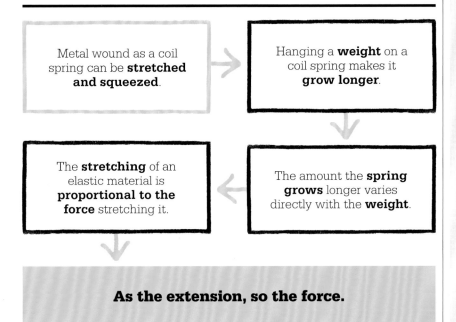

Metal wound as a coil spring can be **stretched and squeezed**.

Hanging a **weight** on a coil spring makes it **grow longer**.

The amount the **spring grows** longer varies directly with the **weight**.

The **stretching** of an elastic material is **proportional to the force** stretching it.

As the extension, so the force.

Robert Hooke

Born on the Isle of Wight in 1635, Robert Hooke went on to make his way through Oxford University, where he gained a passion for science. In 1661, the Royal Society debated an article on the phenomenon of rising water in slender glass pipes; Hooke's explanation was published in a journal. Five years later, the Royal Society hired Hooke as their curator of experiments.

The range of Hooke's scientific achievements is huge. Among his many inventions were the ear trumpet and the spirit level. He also founded the science of meteorology, was the great pioneer of microscope studies (discovering that living things are made from cells), and developed the key law on elasticity, known as Hooke's law. He also collaborated with Robert Boyle on the gas laws, and with Isaac Newton on the laws of gravity.

pendulum also meant that Hooke could make a watch small enough to be put in a pocket.

Spring force
In the 1670s, Hooke heard that Dutch scientist Christiaan Huygens was also developing a spring-driven watch. Anxious not to be beaten, Hooke set to work with master clockmaker Thomas Tompion to make his watch.

As Hooke worked with Tompion, he realized that a coil spring must unwind at a steady rate to keep time. Hooke experimented with stretching and squeezing springs and discovered the simple relationship embodied in the law on elasticity that was later given his name. Hooke's law says that the amount a spring is squeezed or stretched is precisely proportional to the force applied. If you apply twice the force, it stretches twice as far. The relationship can be summarized in a simple equation, $F = kx$, in which F is the force, x is the distance stretched, and k is a constant (a fixed value). This simple law proved to be a key platform for understanding how solids behave.

Hooke wrote his idea down as a Latin anagram, *ceiiinosssttvu*, a common way for scientists at the time to keep their work secret until they were ready to publish it. »

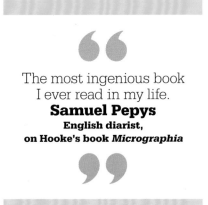

The most ingenious book I ever read in my life.
Samuel Pepys
English diarist,
on Hooke's book *Micrographia*

Key works

1665 *Micrographia*
1678 "Of Spring"
1679 *Collection of Lectures*

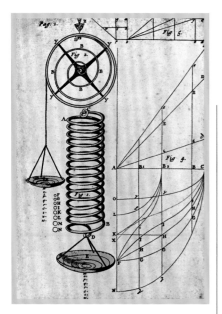

Deciphered, the anagram read *Ut tensio sic vis*, which means "as the extension, so the force"—that is, the extension is proportional to the force. Once the watch was made, Hooke went on to publish his ideas about springs two years later, in his 1678 pamphlet "de Potentia Restitutiva" ("Of Spring"). He began by outlining a simple demonstration for people to try at home—twist wire into a coil, then hang different weights to see how far the coil stretches. He had invented the spring balance.

However, Hooke's paper was of lasting importance. Not only was it a simple observation of how springs behaved, but it also provided a key insight into the strength of materials and the behavior of solids under stress—factors that are central to modern engineering.

Mini springs

In trying to find an explanation for the behavior of springs, Hooke suspected that it was tied to a fundamental property of matter. He speculated that solids were made of vibrating particles that

Hooke's spring balance used the stretching of a spring to show the weight of something. Hooke used this illustration to explain the concept in his "Of Spring" lecture.

constantly collided with each other (anticipating the kinetic theory of gases by more than 160 years). He suggested that squeezing a solid pushed the particles closer and increased collisions making it more resistant; stretching it reduced collisions so that the solid became less able to resist the pressure of the air around itself.

There are clear parallels between Hooke's law, published in 1678, and Boyle's law (1662) on gas pressure, which Robert Boyle called "the spring of the air." Furthermore, Hooke's vision of the role of invisible particles in the strength and elasticity of materials seems remarkably close to our modern understanding. We now know that strength and elasticity do indeed depend on a material's molecular structure and bonding. Metals are hugely resilient, for instance, because of special

metallic bonds between their atoms. Although scientists would not understand this for another 200 years, the Industrial Revolution's engineers soon realized the benefits of Hooke's law when they began to build bridges and other structures with iron in the 1700s.

Engineering math

In 1694, Swiss mathematician Jacob Bernoulli applied the phrase "force per unit area" to the deforming force—the stretching or squeezing force. Force per unit area came to be called "stress" and the amount the material was stretched or squeezed came to be known as the "strain." The direct relationship between stress and strain varies—for example, some materials will deform much more under a certain stress than others. In 1727, another Swiss mathematician, Leonhard Euler, formulated this variation in stress and strain in different materials as the coefficient (a number by which another number is multiplied) "E", and Hooke's equation became $\sigma = E\varepsilon$, in which σ is the stress and ε is the strain.

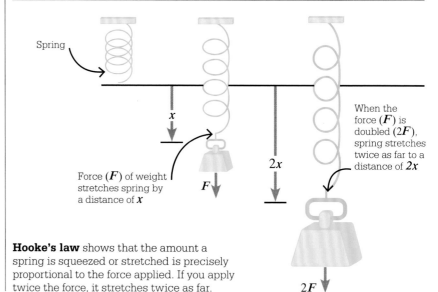

Spring

Force (F) of weight stretches spring by a distance of x

x

F

$2x$

When the force (F) is doubled ($2F$), spring stretches twice as far to a distance of $2x$

$2F$

Hooke's law shows that the amount a spring is squeezed or stretched is precisely proportional to the force applied. If you apply twice the force, it stretches twice as far.

Tensile strength

When materials are stretched beyond their elastic limit, they will not return to their original size, even when the stress is removed. If they are stretched even further, they may eventually snap. The maximum stress that a material can take in tension—being pulled longer—before it snaps is known as its tensile strength, and is crucial in deciding the suitability of a material for a particular task.

Some of the first tests of tensile strength were carried out by Leonardo da Vinci, who wrote in 1500 about "Testing the strength of iron wires of various lengths." We now know that structural steel has a high tensile strength of over 400 MPa (megapascals).

A pascal is the measurement unit for pressure: 1 Pa is defined as 1 N (newton) per square meter. Pascals are named after mathematician and physicist Blaise Pascal.

Structural steel is often used for today's suspension bridges, such as the George Washington Bridge in New Jersey (see left). Carbon nanotubes can be over a hundred times as strong as structural steel (63,000 MPa).

Young's measure

During experiments carried out in 1782, Italian scientist Giordano Riccati had discovered that steel was about twice as resistant to stretching and squeezing as brass. Riccati's experiments were very similar in concept to the work of Euler and also, 25 years later, the work of Thomas Young.

Young was, like Robert Hooke, a British polymath. He earned his living as a physician but his scientific achievements were wide ranging and his work on stress and strain in materials was a keystone for 19th-century engineering. In 1807, Young revealed the mechanical property that was Euler's coefficient "E". In his remarkable series of lectures, during the same year, entitled "Natural Philosophy and the Mechanical Arts," Young introduced the concept of a "modulus" or measure to describe the elasticity of a material.

Stress and strain

Young was interested in what he called the "passive strength" of a material, by which he meant the elasticity, and he tested the strength of various materials to derive his measures. Young's modulus is a measure of a chosen material's ability to resist stretching or squeezing in one direction. It is the ratio of the stress to the strain. A material such as rubber has a low Young's modulus—less than 0.1 Pa (pascals)—so it will stretch a lot with very little stress. Carbon fiber has a modulus of around 40 Pa, which means it is 400 or more times more resistant to stretching than rubber.

Elastic limit

Young realized that the linear relationship (in which one quantity increases directly proportionally to another) between a material's stress and strain works over a limited range. This varies between materials, but in any material subjected to too much stress, a nonlinear (disproportionate) relationship between stress and strain will eventually develop. If the stress continues, the material will reach its elastic limit (the point at which it stops returning to its original length after the stress is removed). Young's modulus only applies when the relationship between a material's stress and strain is linear. Young's contributions about the strength of materials, and also their resistance to stress, were of huge value to engineers. Young's modulus and his equations opened the door to the development of a whole series of calculation systems that allow engineers to work out the stresses and strains on proposed structures precisely before they are built. These calculation systems are fundamental to constructing everything from sports cars to suspension bridges. Total collapses of these structures are rare. ∎

A permanent alteration of form limits the strength of materials with regard to practical purposes.
Thomas Young

THE MINUTE PARTS OF MATTER ARE IN RAPID MOTION

FLUIDS

A fluid is defined as a phase of matter that has no fixed shape, yields easily to external pressure, deforms to the shape of its container, and flows from one point to another. Liquids and gases are among the most common types. All fluids can be compressed to some degree, but a great deal of pressure is required to compress a liquid by just a small amount. Gases are more easily compressed since there is more space between their atoms and molecules.

One of the greatest contributors to the field of fluid dynamics—the study of how forces affect the motion of fluids—was Swiss mathematician and physicist Daniel Bernoulli,

See also: Pressure 36 ▪ Laws of motion 40–45 ▪ Kinetic energy and potential energy 54 ▪ The gas laws 82–85 ▪ The development of statistical mechanics 104–111

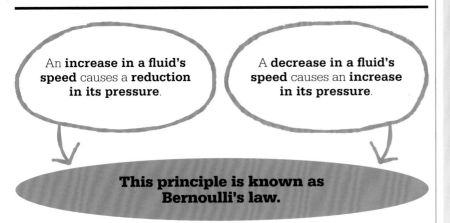

An **increase in a fluid's speed** causes a **reduction in its pressure**.

A **decrease in a fluid's speed** causes an **increase in its pressure**.

This principle is known as Bernoulli's law.

Daniel Bernoulli

Born in 1700 in Groningen in the Netherlands into a family of prominent mathematicians, Bernoulli studied medicine at the University of Basel in Switzerland, Heidelberg University in Germany, and the University of Strasbourg (at the time also in Germany). He gained a doctorate in anatomy and botany in 1721.

Bernoulli's 1724 paper on differential equations and the physics of flowing water earned him a post at the Academy of Sciences in St. Petersburg, Russia, where he taught and produced important mathematical work. *Hydrodynamica* was published after his return to the University of Basel. He worked with Leonhard Euler on the flow of fluids, especially blood in the circulatory system, and also worked on the conservation of energy in fluids. He was elected to the Royal Society of London in 1750, and died in 1782 in Basel, Switzerland, aged 82.

Key work

1738 *Hydrodynamica* (*Hydrodynamics*)

whose 1738 *Hydrodynamica* (*Hydrodynamics*) laid the basis for the kinetic theory of gases. His principle states that an increase in the speed of motion of a fluid occurs simultaneously with a reduction in its pressure or potential energy.

From bath tubs to barrels

Bernoulli's principle built on the discoveries of earlier scientists. The first major work on fluids was *On Floating Bodies* by the ancient Greek philosopher Archimedes. This 3rd-century BCE text states that a body immersed in a liquid receives a buoyant force equal to the weight of the fluid it displaces, a fact that Archimedes is said to have realized when taking a bath, leading to the famous cry *"Eureka!"* ("I found it!").

Centuries later, in 1643, Italian mathematician and inventor Evangelista Torricelli formulated Torricelli's law. This principle of fluid dynamics explains that the velocity of flow (*v*) of a fluid leaving a hole in a container, where *h* is the depth of fluid above the hole, is the same as the velocity that a droplet of fluid would acquire falling freely from a height *h*. If *h* increases, so do the velocity of the falling droplet

and the fluid leaving the hole. The velocity at which the fluid leaves the hole is proportional to the height of fluid above the hole. So, $v = \sqrt{2gh}$ where g is the acceleration due to gravity. »

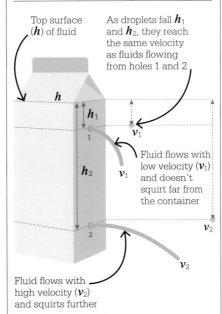

Top surface (*h*) of fluid

As droplets fall h_1 and h_2, they reach the same velocity as fluids flowing from holes 1 and 2

h

h_1

1

v_1

h_2

v_1

Fluid flows with low velocity (v_1) and doesn't squirt far from the container

2

v_2

v_2

Fluid flows with high velocity (v_2) and squirts further

Under Torricelli's law, fluid squirts from holes 1 and 2—placed at a depth of h_1 and h_2 from the top of the fluid—in a container. As *h* increases, the velocity of the fluid also increases. The same applies for droplets in free fall.

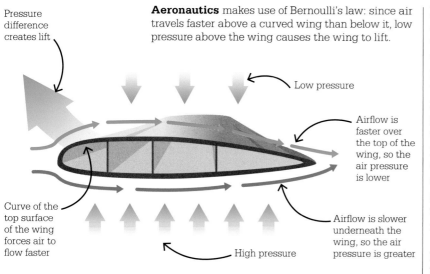

Pressure difference creates lift

Aeronautics makes use of Bernoulli's law: since air travels faster above a curved wing than below it, low pressure above the wing causes the wing to lift.

Low pressure

Airflow is faster over the top of the wing, so the air pressure is lower

Curve of the top surface of the wing forces air to flow faster

High pressure

Airflow is slower underneath the wing, so the air pressure is greater

Another breakthrough came in 1647, when French scientist Blaise Pascal proved that for an incompressible fluid inside a container, any change in pressure is transmitted equally to every part of that fluid. This is the principle behind the hydraulic press and the hydraulic jack.

Pascal also proved that hydrostatic pressure (the pressure of a fluid due to the force of gravity) depends not on the weight of fluid above it, but on the height between that point and the top of the liquid. In Pascal's famous (if apocryphal) barrel experiment, he is said to have inserted a long, narrow, water-filled pipe into a water-filled barrel. When the pipe was raised above the barrel, the increased hydrostatic pressure burst the barrel.

Viscosity and flow

In the 1680s, Isaac Newton studied fluid viscosity—how easily fluids flow. Almost all fluids are viscous, exerting some resistance to deformation. Viscosity is a measure of a fluid's internal resistance to flow: fluids with low viscosity possess low resistance and flow easily, while high-viscosity fluids resist deformation and do not flow easily. According to Newton's viscosity law, a fluid's viscosity is its "shear stress" divided by its "shear rate." While not all liquids follow this law, those that do are called Newtonian liquids. Shear stress and rate can be pictured as a fluid sandwiched between two plates. One is fixed below the fluid, and the other drifts slowly on the surface of the fluid. The fluid is subject to shear stress (the force moving the upper plate, divided by the plate's area). The shear rate is the velocity of the moving plate divided by the distance between the plates.

Later studies showed that there are also different kinds of fluid flow. A flow is described as "turbulent" when it exhibits recirculation, eddying, and apparent randomness. Flows that lack these characteristics are described as "laminar."

Bernoulli's law

Bernoulli studied pressure, density, and velocity in static and moving fluids. He was familiar both with Newton's *Principia* and with Robert Boyle's discovery that the pressure of a given mass of gas, at a constant temperature, increases as the volume of the container holding it decreases. Bernoulli argued that gases are made up of huge numbers of molecules moving randomly in all directions, and that their impact on a surface causes pressure. He wrote that what is experienced as heat is the kinetic energy of their motion, and that—given the random motion of molecules—motion and pressure increase as temperatures rise.

By drawing these conclusions, Bernoulli laid the basis of the kinetic theory of gases. It was not widely accepted at the time of publication in 1738, since the principle of the conservation of energy would not be proven for more than a century. Bernoulli discovered that as fluids flow more quickly, they produce less pressure and, conversely, when fluids flow more slowly, they produce greater pressure. This became known as Bernoulli's law, which now has many applications, such as the lift generated by airflow in aeronautics.

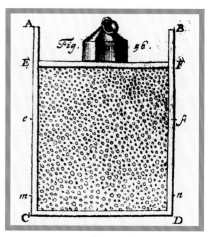

A diagram in Bernoulli's *Hydrodynamica* depicts air molecules colliding with the walls of a container, creating enough pressure to support a weight resting on a movable surface.

Nature always tends to act in the simplest ways.
Daniel Bernoulli

Kinetic theory

While Bernoulli and other scientists laid the foundations for the kinetic theory of gases, Scottish scientist James Clerk Maxwell attempted to quantify the nature of molecular motion within them. He explained the macroscopic qualities of gases: their pressure, temperature, viscosity, and thermal conductivity. Along with Austrian physicist Ludwig Boltzmann, Maxwell developed a statistical means of describing this theory.

In the mid-19th century, most scientists assumed that all gas molecules travel at the same speed, but Maxwell disagreed. In his 1859 paper "Illustration of the Dynamic Theory of Gases," he produced an equation to describe a distribution curve, now known as the Maxwell–Boltzmann distribution, that showed the range of different speeds of gas molecules. He also calculated the mean free path (the average distance traveled by gas molecules between collisions) and the number of collisions at a given temperature, and found that the higher the temperature, the faster the molecular movement, and the greater the number of collisions. He concluded that the temperature of a gas is a measure of its mean kinetic energy. Maxwell also confirmed Italian scientist Amedeo Avogadro's law of 1811, which states that equal volumes of two gases, at equal temperatures and pressures, contain equal numbers of molecules.

Superfluid discoveries

Findings in the 20th century revealed how fluids behave at very cold temperatures. In 1938, Canadian physicists John F. Allen and Don Misener and Russian physicist Pyotr Kapitsa discovered that a helium isotope behaved strangely when cooled to a temperature close to absolute zero. Below its boiling point of −452.1 °F (−268.94 °C) it behaved as a normal colorless liquid, but below −455.75 °F (−270.97 °C), it

As the atoms are cooled, they start piling into the lowest possible energy state.
Lene Hau
Danish physicist, on superfluids

exhibited zero viscosity, flowing without losing kinetic energy. At such low temperatures, atoms almost stop moving. The scientists had discovered a "superfluid."

When superfluids are stirred, vortices form that can rotate indefinitely. They have greater thermal conductivity than any known substance—hundreds of times greater than copper, which itself has high thermal conductivity. Superfluids called "Bose–Einstein condensates" have been used experimentally as coolants, and in 1998 Danish physicist Lene Hau used them to slow the speed of light to 10 mph (17 km/h). Such "slow light" optical switches could cut power requirements dramatically. ∎

Applied fluid dynamics

Predicting how fluids behave is fundamental to many modern technological processes. For instance, factory-based food production systems are designed to convey ingredients and final foodstuffs—from binding syrups to soups—through pipes and ducts. Integral to this process is computational fluid dynamics (CFD), a branch of fluid dynamics that can maximize efficiency, cut costs, and maintain quality. CFD has its roots in the work of French engineer Claude-Louis Navier. Building on earlier work by Swiss physicist Leonhard Euler, in 1822 Navier published equations that applied Isaac Newton's second law of motion to fluids. Later known as Navier–Stokes equations after further contributions from Anglo-Irish physicist George Stokes in the mid-19th century, they were able to explain, for example, the movement of water in channels.

CFD is a branch of fluid dynamics that uses flow modeling and other tools to analyze problems and predict flows. It can take into account variables such as changes in a fluid's viscosity due to temperature, altered flow speeds caused by phase change (such as melting, freezing, and boiling), and it can even predict the effects of turbulent flow in parts of a pipe system.

SEARCHING OUT THE FIRE-SECRET
HEAT AND TRANSFERS

IN CONTEXT

KEY FIGURES
Joseph Black (1728–1799),
James Watt (1736–1819)

BEFORE
1593 Galileo Galilei creates
the thermoscope for showing
changes in hotness.

1654 Ferdinando II de Medici,
Grand Duke of Tuscany, makes
the first sealed thermometer.

1714 Daniel Fahrenheit makes
the first mercury thermometer.

1724 Fahrenheit establishes
a temperature scale.

1742 Anders Celsius invents
the centigrade scale.

AFTER
1777 Carl Scheele identifies
radiant heat.

c. 1780 Jan Ingenhousz
clarifies the idea of heat
conduction.

In the early 1600s, thermoscopes began to appear across Europe. These liquid-filled glass tubes (shown opposite) were the first ever instruments for measuring how hot things are. In 1714, German-born Dutch scientist and instrument-maker Daniel Fahrenheit created the first modern thermometer, filled with mercury—he established his famous temperature scale in 1724. Swedish scientist Anders Celsius invented the more convenient centigrade scale in 1742.

The introduction of the first successful steam engine in 1712 by British inventor Thomas Newcomen sparked huge interest in heat. In a 1761 lecture, Scottish chemist Joseph Black spoke of experiments he had done on melting. These showed that the temperature did not change when ice melted to water, yet melting the ice required the same heat as it took to warm water from melting point to 140°F (60°C). Black realized that heat must be absorbed when ice melted, and he

When ice melts to water, there is **no temperature change**.

It takes the same amount of heat to **melt water** as to **raise its temperature** to 140°F (60°C).

Water must **absorb heat** when it melts—it becomes latent.

Heat and temperature must be different.

See also: The gas laws 82–85 ▪ Internal energy and the first law of thermodynamics 86–89 ▪ Heat engines 90–93
▪ Entropy and the second law of thermodynamics 94–99 ▪ Thermal radiation 112–117

> The heat which disappears
> in the conversion of water
> into vapor is not lost,
> but is retained in vapor.
> **Joseph Black**

called the heat absorbed "latent heat." The latent (hidden) heat is the energy required to change a material to another state. Black had made a crucial distinction between heat, which we now know is a form of energy, and temperature, which is a measurement of energy.

James Watt also discovered the concept of latent heat in 1764. Watt was conducting experiments on steam engines and noticed that adding a little boiling water to a lot

Galilean thermoscopes are tubes filled with a liquid (often ethanol), holding liquid-filled "floats." Heat causes the density of all the liquids to change, making the floats rise or fall.

of cold water barely affects the cold water's temperature, but bubbling a little steam through the water brings it quickly to a boil.

How heat moves

In 1777, an apothecary in Sweden, Carl Scheele, made a few simple yet crucial observations—such as the fact that on a cold day you can feel the heat of a glowing fire a few feet away while still seeing your breath on the cold air. This is radiant heat, and is infrared radiation (emitted from a source, such as fire or the sun), which travels like light—radiation is quite different from convective heat. Convection is how heat moves through a liquid or gas; heat causes the molecules and atoms to spread out—for example, when air above a stove is warmed it will then rise.

Meanwhile, around 1780, Dutch scientist Jan Ingenhousz pinned down a third kind of heat transfer: conduction. This is when atoms in a hot part of a solid vibrate a lot, collide with their neighboring atoms, and by doing so transfer energy (heat). Ingenhousz coated wires of different metals with wax, heated one end of each wire, and noted how quickly the wax melted for each metal. ∎

James Watt

Scottish engineer James Watt was one of the pivotal figures in the history of the steam engine. The son of a ship's instrument-maker, Watt became highly skilled in making instruments, in his father's workshop and in London as an apprentice. He then returned to Glasgow to make instruments for the university.

In 1764, Watt was asked to repair a model Newcomen steam engine. Before making practical adjustments to the model, Watt conducted some scientific experiments, during which he discovered latent heat.

Watt noticed that the engine wasted a lot of steam, and came up with a revolutionary improvement—introducing a second cylinder with one running hot and the other cold. This change transformed the steam engine from a pump of limited use to the universal source of power that drove the Industrial Revolution.

Key inventions

1775 Watt steam engine
1779 Copying machine
1782 Horsepower

ELASTICAL POWER IN THE AIR

THE GAS LAWS

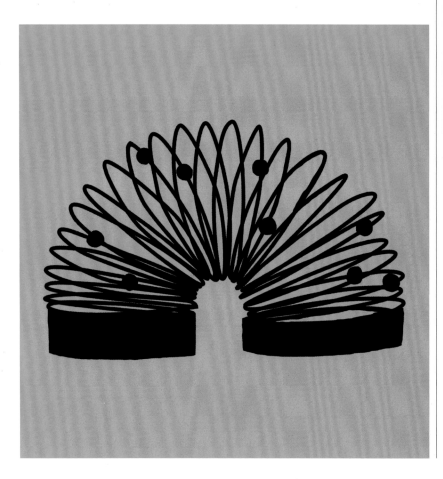

IN CONTEXT

KEY FIGURES
Robert Boyle (1627–1691),
Jacques Charles (1746–1823),
Joseph Gay-Lussac
(1778–1850)

BEFORE
1618 Isaac Beeckman
suggests that, like water, air
exerts pressure.

1643 Italian physicist
Evangelista Torricelli makes
the first barometer and
measures air pressure.

1646 French mathematician
Blaise Pascal shows that air
pressure varies with height.

AFTER
1820 British scientist John
Herapath introduces the
kinetic theory of gases.

1859 Rudolf Clausius shows
that pressure is related to the
speed of gas molecules.

The fact that gases are
so transparent and
seemingly insubstantial
meant that it took natural
philosophers a long time to
appreciate that they have any
physical properties at all.

However, during the 17th
and 18th centuries, European
scientists gradually realized that,
like liquids and solids, gases do
indeed have physical properties.
These scientists discovered the
crucial relationship between the
temperature, pressure, and volume
of gases. Over a period of 150 years,
studies carried out by three
individuals—Robert Boyle in
Britain and Frenchmen Jacques

See also: Pressure 36 ▪ Models of matter 68–71 ▪ Fluids 76–79 ▪ Heat engines 90–93 ▪ Changes of state and making bonds 100–103 ▪ The development of statistical mechanics 104–111

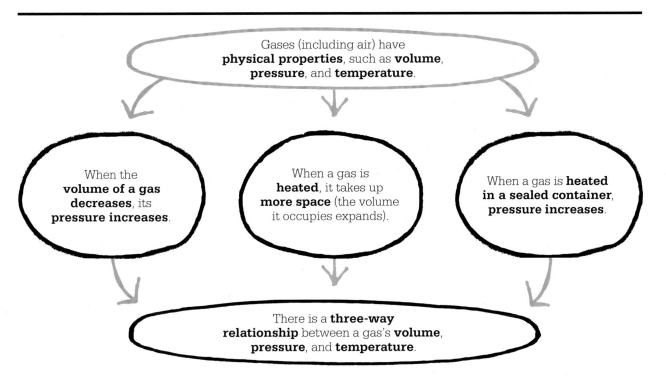

Gases (including air) have **physical properties**, such as **volume**, **pressure**, and **temperature**.

When the **volume of a gas decreases**, its **pressure increases**.

When a gas is **heated**, it takes up **more space** (the volume it occupies expands).

When a gas is **heated in a sealed container**, **pressure increases**.

There is a **three-way relationship** between a gas's **volume**, **pressure**, and **temperature**.

Charles and Joseph Gay-Lussac—finally produced the laws that explain the behavior of gases.

The pressure of air

Early in the 17th century, Dutch scientist Isaac Beeckman suggested that, just like water, air exerts pressure. The great Italian scientist Galileo Galilei disagreed, but Galileo's young protégé Evangelista Torricelli not only proved Beeckman right but showed how to measure pressure by inventing the world's first barometer.

Galileo had observed that a siphon could never raise water above 33 ft (10 m). At the time, vacuums were thought to "suck" liquids, and Galileo wrongly thought this was the maximum weight of water that a vacuum above it could draw up. In 1643, Torricelli showed that the limit was instead the maximum weight of water the pressure of air outside could support.

To prove his point, Torricelli filled a tube closed at one end with mercury, a far denser liquid than water, then turned it upside down. The mercury dropped to about 30 in (76 cm) below the closed end, then stopped falling. He concluded that this was the maximum height the pressure of air outside could support. The height of mercury in the tube would vary slightly in response to changes in air pressure, which is why this is described as the first barometer.

Boyle's "spring of the air"

Torricelli's groundbreaking invention paved the way for the discovery of the first of the gas laws, known as Boyle's law, after Robert Boyle. The youngest son of Richard Boyle, 1st Earl of Cork and once the richest man in Ireland, Robert Boyle used his inherited wealth to set up in »

Evangelista Torricelli used a column of mercury to measure air pressure. He deduced that air pressing down on the mercury in the cistern balanced the column in the tube.

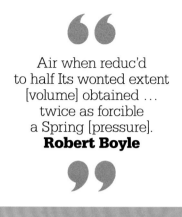

> Air when reduc'd
> to half Its wonted extent
> [volume] obtained …
> twice as forcible
> a Spring [pressure].
> **Robert Boyle**

Oxford his own private scientific research laboratory, the first ever. Boyle was a pioneering advocate of experimental science and it was here that he conducted crucial experiments on air pressure that he described in his book known in short as *Touching the Spring and Weight of the Air* (published in 1662). "Spring" was his word for pressure—he viewed squeezed air acting as if it had springs that recoil when pushed.

Inspired by Torricelli's barometer, Boyle poured mercury into a J-shaped glass tube sealed at the lower end. He could see that the volume of air trapped in the lower tip of the J varied according to how much mercury he added. In other words, there was a clear relationship between how much mercury the air could support and its volume.

Boyle argued that the volume (*v*) of a gas and its pressure (*p*) vary in inverse proportion, provided the temperature stays the same. Mathematically this is expressed as $pv = k$, a constant (an unchanging number). In other words, if you decrease the volume of a gas, its pressure increases. Some people credit the crucial

discovery to Boyle's friend Richard Townley, and a friend of Townley, physician Henry Power. Boyle himself called the idea "Townley's hypothesis," but it was Boyle who made the idea well known.

Charles's hot air discovery
Just over a century later, French scientist and balloon pioneer Jacques Charles added in a third element to the relationship between volume and pressure—temperature. Charles was the first person to experiment with balloons filled with hydrogen rather than hot air, and on August 27, 1783, in Paris, he sent up the first large hydrogen balloon.

In 1787, Charles conducted an experiment with a container of gas in which the volume could vary freely. He then heated up the gas and measured the volume as the temperature rose. He saw that for each degree temperature rise, the gas expanded by $1/273$ of its volume at 32°F (0°C). It contracted at the

same rate as it cooled. If plotted on a graph, it showed the volume would shrink to zero at −460°F, now known as absolute zero, and the zero point on the Kelvin scale. Charles had discovered a law that describes how volume varies with temperature provided the pressure stays steady.

Charles never wrote down his ideas. Instead they were described and clarified in the early 1800s in a paper written by fellow French scientist Joseph Gay-Lussac, at much the same time as English scientist John Dalton showed that the rule applied universally to all gases.

A third dimension
Gay-Lussac added a third gas law to those of Boyle and Charles. Known as Gay-Lussac's law, it showed that if the mass and volume of a gas are held constant then pressure rises in line with the temperature. As was soon clear,

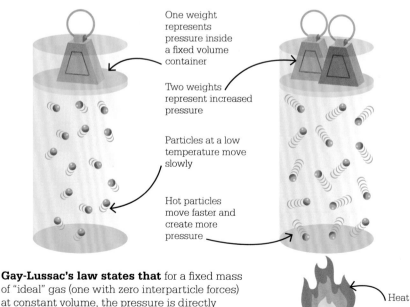

One weight represents pressure inside a fixed volume container

Two weights represent increased pressure

Particles at a low temperature move slowly

Hot particles move faster and create more pressure

Heat

Gay-Lussac's law states that for a fixed mass of "ideal" gas (one with zero interparticle forces) at constant volume, the pressure is directly proportional to the absolute temperature. As more heat is applied, the particles move faster and pressure inside the container rises.

Joseph Gay-Lussac employed atmospheric balloons for several experiments. In this 1804 ascent with Jean-Baptiste Biot, he studied how Earth's electromagnetic intensity varied with altitude.

there is a simple three-way relationship between the volume, pressure, and temperature of gases. This relationship applies to ideal gases (gases with zero interparticle forces), although it is approximately true for all gases.

How gases combine

Gay-Lussac went on to make another important contribution to our understanding of gases. In 1808, he realized that when gases combine, they do so in simple proportions by volume—and that when two gases react, the volume of gases produced depends on the original volumes. So, two volumes of hydrogen join to one volume of oxygen in a ratio of 2:1 to create two volumes of water vapor.

Two years later, Italian scientist Amedeo Avogadro explained this discovery by tying it in with the rapidly emerging ideas about atoms and other particles. He theorized that at a given temperature and pressure, equal volumes of all gases have the same number of "molecules." Indeed, the number of

molecules varies exactly with the volume. This is called Avogadro's hypothesis, and it explained Gay-Lussac's discovery that gases combine in particular proportions.

Crucially, Avogadro's hypothesis indicated that oxygen by itself existed as double-atom molecules, which split to combine with two atoms of hydrogen in water vapor—this must be so if there are to be as many water molecules as there were molecules of hydrogen and oxygen.

This work was important to the development of atomic theory and the relationship between atoms and molecules. It was also vital to the kinetic theory of gases developed by James Clerk Maxwell and others. This states that gas particles move randomly and produce heat when they collide. It helps explain the relationship between pressure, volume, and temperature. ∎

Compounds of gaseous substances with each other are always formed in very simple ratios, so that representing one of the terms by unity, the other is 1, 2, or at most 3.
Joseph Gay-Lussac

Joseph Gay-Lussac

French chemist and physicist Joseph Gay was the son of a wealthy lawyer. The family owned so much of the village of Lussac in southwestern France that, in 1803, both father and son incorporated Lussac into their names. Joseph studied chemistry in Paris before working as a researcher in Claude-Louis Berthollet's laboratory. By age 24, he had already discovered the gas law named after him.

Gay-Lussac was also a balloon pioneer and in 1804 he ascended in a balloon to more than 22,965 ft (7,000 m) with fellow French physicist Jean-Baptiste Biot to sample air at different heights. From these experiments, he showed that the composition of the atmosphere does not change with height and decreasing pressure. In addition to his work on gases, Gay-Lussac also discovered two new elements, boron and iodine.

Key works

1802 "On the Expansion of Gases and Vapors," *Annals of Chemistry*
1828 *Lessons of Physics*

THE ENERGY OF THE UNIVERSE IS CONSTANT

INTERNAL ENERGY AND THE FIRST LAW OF THERMODYNAMICS

IN CONTEXT

KEY FIGURE
William Rankine
(1820–1872)

BEFORE
1749 Émilie du Châtelet
implicitly introduces the idea
of energy and its conservation.

1798 Benjamin Thompson
develops the idea that heat
is a form of kinetic energy.

1824 French scientist Sadi
Carnot concludes there are no
reversible processes in nature.

1840s James Joule, Hermann
von Helmholtz, and Julius von
Mayer introduce the theory of
conservation of energy.

AFTER
1854 William Rankine
introduces potential energy.

1854 Rudolf Clausius publishes
his statement of the second
law of thermodynamics.

In the late 18th century, scientists had begun to understand that heat was different from temperature. Joseph Black and James Watt had shown that heat was a quantity (while temperature was a measurement), and the development of steam engines throughout the Industrial Revolution focused scientific interest on how exactly heat gave those engines such power.

At the time, scientists followed the "caloric" theory—the idea that heat was a mysterious fluid or weightless gas called caloric that flowed from hotter to colder bodies. The connection between heat and movement had long

See also: Kinetic energy and potential energy 54 ▪ The conservation of energy 55 ▪ Heat and transfers 80–81
▪ Heat engines 90–93 ▪ Entropy and the second law of thermodynamics 94–99 ▪ Thermal radiation 112–117

Generating electricity

The burning of fossil fuels (coal, oil, and natural gas) to generate electricity is a classic example of a chain of energy conversions. It begins with solar energy from the sun's rays. Plants convert the solar energy into chemical energy as they grow—this then becomes "stored" as chemical potential energy in the chemical bonds made. The stored energy is concentrated as the plants are compressed into coal, oil, and gas.

The fuel is burned, creating heat energy, which heats water to produce steam, and the steam makes turbines turn (converting 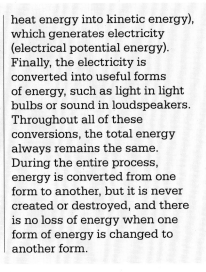 heat energy into kinetic energy), which generates electricity (electrical potential energy). Finally, the electricity is converted into useful forms of energy, such as light in light bulbs or sound in loudspeakers. Throughout all of these conversions, the total energy always remains the same. During the entire process, energy is converted from one form to another, but it is never created or destroyed, and there is no loss of energy when one form of energy is changed to another form.

---

been recognized, but no one fully appreciated how fundamental this link was. In the 1740s, French mathematician Émilie du Châtelet studied the concept of momentum and introduced the idea of mechanical "energy," which is the capacity to make things happen—though she did not name it as such at the time. But it was becoming clear that moving objects had energy, later identified as "kinetic" energy.

Heat is energy

In 1798, American-born physicist Benjamin Thompson, later known as Count Rumford, conducted an experiment in a cannon foundry in Munich. He wanted to measure the heat generated by the friction when the barrels of cannons were bored out. After many hours of continuous friction from a blunt boring machine, heat was still being generated, yet there was no change in the structure of the cannon's metal—so it was clear that nothing physical (and no caloric fluid) was being lost from the metal. It seemed that the heat must be in the movement. In other words, heat is kinetic energy—the energy of movement. But few people accepted this idea and the caloric theory held for another 50 years.

The breakthrough came from several scientists simultaneously in the 1840s, including James Joule in Britain, and Hermann von Helmholtz and Julius von Mayer in Germany. What they all saw was that heat was a form of energy with the capacity to make something

You see, therefore, that living force may be converted into heat, and heat may be converted into living force.
James Joule

happen, just like muscle power. They would come to realize that all forms of energy are interchangeable.

In 1840, Julius von Mayer had been looking at the blood of sailors in the tropics and found that blood returning to the lungs was still rich in oxygen. In colder locations, a person's blood would return to the lungs carrying a lot less oxygen. This meant that, in the tropics, the body needed to burn less oxygen to keep warm. Mayer's conclusion was that heat and all forms of energy (including those in his observations: muscle power, the heat of the body, and the heat of the sun) are interchangeable and they can be changed from one form to another, but never created. The total energy will remain the same. However, Mayer was a medic and physicists paid his work little attention.

Converting energy

Meanwhile, young James Joule began experimenting in a laboratory in the family home in Salford near Manchester. In 1841, he figured out just how much » heat is made by an electric current. He then experimented with ways of converting mechanical movement into heat and developed a famous experiment in which a falling weight turns a paddle wheel in water, heating the water (shown below). By measuring the rise in the water temperature Joule could work out how much heat a certain amount of mechanical work would create. Joule's calculations led him to the belief that no energy is ever lost in this conversion. But like Mayer's research, Joule's ideas were initially largely ignored by the scientific community.

Then, in 1847, Hermann von Helmholtz published a key paper, which drew from his own studies and those of other scientists, including Joule. Helmholtz's paper summarized the theory of the conservation of energy. The same year, Joule presented his work at a meeting of the British Association in Oxford. After the meeting, Joule met William Thomson (who later became Lord Kelvin), and the two of them worked on the theory of gases and on how gases cool as

they expand—the basis of refrigeration. Joule also made the first clear estimate of the average speed of molecules in a gas.

The first law

Throughout the next decade, Helmholtz and Thomson—along with German Rudolf Clausius and Scotsman William Rankine—began to pull their findings together. Thomson first used the phrase "thermo-dynamic" in 1849 to sum up the power of heat. Over the next year, Rankine and Clausius (apparently independently) developed what is now called the first law of thermodynamics. Like Joule, Rankine and Clausius focused on work—the force used to move an object a certain distance. Throughout their studies, they saw a universal connection between heat and work. Significantly, Clausius also began to use the word "energy" to describe the capacity to do work.

In Britain, Thomas Young had coined the word "energy" in 1802 to explain the combined effect of mass and velocity. Around the late 17th century, German polymath

> The commonest objects are by science rendered precious.
> **William Rankine**

Gottfried Leibniz had referred to this as *vis visa* or "living force," which was a term still used by Rankine. But it was only in the 1850s that its full significance emerged, and the word "energy" in its modern sense began to be used regularly.

Clausius and Rankine worked on the concept of energy as a mathematical quantity—in the same way that Newton had revolutionized our understanding of gravity by simply looking at it as a universal mathematical rule, without actually describing how it works. They were finally able to banish the caloric idea of heat as a substance. Heat is energy, a capacity to do work, and heat must therefore conform to another simple mathematical rule: the law of conservation of energy. This law shows that energy cannot be created or destroyed; it can only be transferred from one place to another or converted into other forms of energy. In simple terms, the first law of thermodynamics is the law of conservation of energy applied to heat and work.

Clausius' and Rankine's ideas and research had been inspired by trying to understand theoretically how engines worked. So Clausius looked at the total energy in a closed

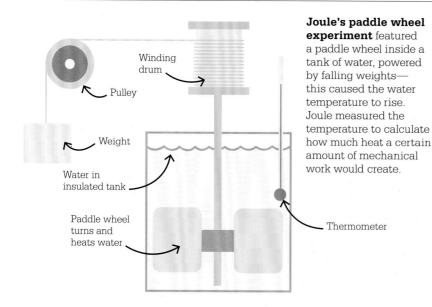

Joule's paddle wheel experiment featured a paddle wheel inside a tank of water, powered by falling weights—this caused the water temperature to rise. Joule measured the temperature to calculate how much heat a certain amount of mechanical work would create.

Winding drum

Pulley

Weight

Water in insulated tank

Paddle wheel turns and heats water

Thermometer

system (a system where matter cannot be moved in or out, but energy can be moved—like in the cylinders of a steam engine) and talked about its "internal energy." You cannot measure the internal energy of the system, but you can measure the energy going in and out. Heat is a transfer of energy into the system and a combination of heat and work is a transfer out.

According to the energy conservation law, any change in internal energy must always be the difference between the energy moving into the system and the energy that is moving out—this also equates to the total difference between heat and work. Put more heat into the same system and you get more work out, and also the other way around—this adheres to the first law of thermodynamics. This must be so because the total energy in the universe (all energy surrounding the system) is constant, so the transfers in and out must match.

Rankine's categories

Rankine was a mechanical engineer, which meant that he liked to put a practical spin on things. So he made a useful division of energy into two kinds: stored energy and work energy. Stored energy is energy held still, ready to move—like a compressed spring or a skier standing at the top of a slope. Today we describe stored energy as potential energy. Work is either the action done to store up the energy or it is the movement when that energy is unleashed. Rankine's categorizing of energy in this way was a simple and lastingly effective way of looking at energy in its resting and moving phases.

By the end of the 1850s, the remarkable work of du Châtelet, Joule, Helmholtz, Mayer, Thomson, Rankine, and Clausius had transformed our understanding of heat. This pioneering group of young scientists had revealed the reciprocal relationship between heat and movement. They had also begun to understand and show the universal importance of this relationship. They summed it up in the term "thermodynamics"—the idea that the total amount of energy in the universe must always be constant and cannot change. ∎

William Rankine

Scotsman William Rankine was born in Edinburgh in 1820. He became a railroad engineer, like his father, but having become fascinated by the science behind the steam engines he was working with, he later switched to studying science instead.

Along with scientists Rudolf Clausius and William Thomson, Rankine became one of the founding fathers of thermodynamics. He helped establish the two key laws of thermodynamics, and defined the idea of potential energy. Rankine and Clausius also independently described the entropy function (the idea that heat is transferred in a disordered way). Rankine wrote a complete theory of the steam engine, and of all heat engines, and contributed to the final move away from the caloric theory of heat as a fluid. He died in Glasgow, at the age of 52, in 1872.

Key works

1853 "On the General Law of Transformation of Energy"
1855 "Outlines of the Science of Energetics"

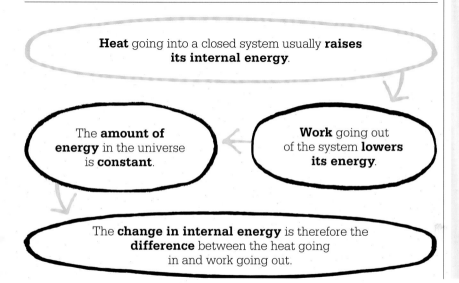

Heat going into a closed system usually **raises its internal energy**.

The **amount of energy** in the universe is **constant**.

Work going out of the system **lowers its energy**.

The **change in internal energy** is therefore the **difference** between the heat going in and work going out.

HEAT CAN BE A CAUSE OF MOTION

HEAT ENGINES

IN CONTEXT

KEY FIGURE
Sadi Carnot (1796–1832)

BEFORE
c. 50 CE Hero of Alexandria builds a small steam-driven engine known as the aeolipile.

1665 Robert Boyle publishes *An Experimental History of Cold*, an attempt to determine the nature of cold.

1712 Thomas Newcomen builds the first successful steam engine.

1769 James Watt creates his improved steam engine.

AFTER
1834 British–American inventor Jacob Perkins makes the first refrigerator.

1859 Belgian engineer Étienne Lenoir develops the first successful internal combustion engine.

It is hard to overestimate the impact of the coming of the steam engine in the 18th century. Steam engines gave people a previously unimaginable source of power. They were practical machines, built by engineers, and were put to use on a grand scale to drive the Industrial Revolution. Scientists became fascinated by how the awesome power of steam engines was created, and their curiosity drove a revolution with heat at its heart.

The idea of steam power is ancient. As long ago as the 3rd century BCE, a Greek inventor in Alexandria called Ctesibius realized that steam jets out

See also: Kinetic energy and potential energy 54 ▪ Heat and transfers 80–81 ▪ Internal energy and the first law of thermodynamics 86–89 ▪ Entropy and the second law of thermodynamics 94–99

> To take away today from England her steam-engines would be to take away at the same time her coal and iron.
> **Sadi Carnot**

powerfully from the spout of a water-filled container heated over a fire. He began to play with the idea of an aeolipile or wind ball, a hollow sphere on a pivot. When the water within boiled, the expanding steam escaped in jets from two curved, directional nozzles, one on each side. The jets set the sphere spinning. About 350 years later, another Alexandrian, named Hero, created a working design for an aeolipile—replicas of which have since been built. It is now known that when liquid water turns to vapor (steam), the bonds holding its molecules to each other break down, causing it to expand.

Hero's device, though, was simply a plaything, and although various inventors experimented with steam, it was another 1,600 years before the first practical steam engine was built. The breakthrough was the discovery of the vacuum and the power of air pressure in the 17th century. In a famous demonstration in 1654, German physicist Otto von Guericke showed that atmospheric pressure was powerful enough to hold together the two halves of a sphere drained

The first successful steam engine was invented by Thomas Newcomen to pump water from mines. It worked by cooling steam in a cylinder to create a partial vacuum and draw up a piston.

of air against the pulling power of eight strong horses. This discovery opened up a new way of using steam, very different from Hero's jets. French inventor Denis Papin realized in the 1670s that if steam trapped in a cylinder cools and condenses, it shrinks dramatically to create a powerful vacuum, strong enough to draw up a heavy piston, a moving component of engines. So, instead of using steam's expansive power, the new discovery utilized the massive contraction when it cools and condenses.

Steam revolution

In 1698, English inventor Thomas Savery built the first large steam engine using Papin's principle. However, Savery's engine used high-pressure steam that made it dangerously explosive and unreliable. A much safer engine, which used low-pressure steam, was built by Devon iron seller Thomas Newcomen in 1712. Although Newcomen's engine was so successful that it was installed in thousands of mines across Britain and Europe by 1755, it was inefficient because the cylinder had to be cooled at every stroke to condense the steam, and this used a huge amount of energy.

In the 1760s, to improve on the Newcomen engine, Scottish engineer James Watt conducted the first scientific experiments on the way heat moves in a steam engine. His experiments led to his discovery, along with his

compatriot Joseph Black, that it is heat, not temperature, that provides the motive power of steam. Watt also realized that the efficiency of steam engines could be hugely improved by using not one cylinder but two—one which was kept hot all the time and a separate cold one to condense the steam. Watt also introduced a crank to convert the up-and-down motion of the piston into the rotary motion needed to drive a wheel. This smoothed the action of the piston strokes to maintain constant power. Watt's innovations were extraordinarily successful and could be said to have launched the age of steam.

Energy and thermodynamics

The efficiency of steam engines intrigued young French military engineer Sadi Carnot. He visited factory after factory, studying not only their steam engines but also »

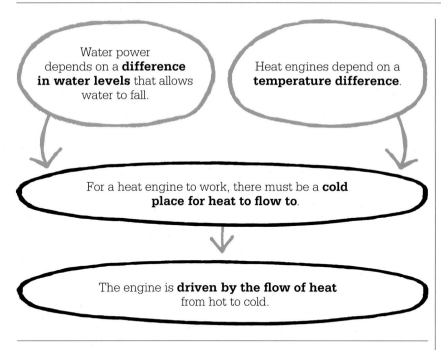

Water power depends on a **difference in water levels** that allows water to fall.

Heat engines depend on a **temperature difference**.

For a heat engine to work, there must be a **cold place for heat to flow to**.

The engine is **driven by the flow of heat** from hot to cold.

those driven by water power. In 1824, he wrote a short book, *Reflections On the Motive Power of Heat*. Carnot realized that heat is the basis of all motion on Earth, driving the winds and ocean currents, earthquakes and other geological shifts, and the body's muscle movements. He saw the cosmos as a giant heat engine made of countless smaller heat engines,

systems driven by heat. This was the first recognition of the true significance of heat in the universe, and provided the launchpad for the science of thermodynamics.

Seeing and comparing water and steam power in factories gave Carnot a key insight into the nature of heat engines. Like most of his contemporaries, he believed in the caloric theory—the false idea that

heat was a fluid. However, this misconception allowed him to see a key analogy between water and steam power. Water power depends on a head of water, a difference in water levels, that allows water to fall. In the same way, Carnot saw that a heat engine depends on a head of heat that allows a "fall of caloric." In other words, for a heat engine to work there must not just be heat, but also a cold place for it to flow to. The engine is driven not by heat but by the flow of heat from hot to cold. So the motive power is the difference between hot and cold, not the heat by itself.

Perfect efficiency

Carnot had a second key insight— that for the generation of maximum power there must be no wasted flow of heat at any place or time. An ideal engine is one in which all the heat flow is converted into useful movement. Any loss of heat that does not generate motive power is a reduction in the heat engine's efficiency.

To model this, Carnot outlined a theoretical ideal heat engine reduced to its basics. Now known

Sadi Carnot

Born in Paris in 1796, Sadi Carnot came from a family of renowned scientists and politicians. His father Lazare was a pioneer in the scientific study of heat, as well as ranking high in the French Revolutionary Army. Sadi followed his father into the military academy. After graduating in 1814, he joined the military engineers as an officer and was sent around France to make a report on its fortifications. Five years later, having become fascinated by steam engines, he retired from the army to pursue his scientific interests.

In 1824, Carnot wrote his groundbreaking *Reflections on the Motive Power of Heat*, which drew attention to the importance of heat engines and introduced the Carnot cycle. Little attention was paid to Carnot's work at the time and before its significance as the starting point of thermodynamics could be appreciated, he died of cholera in 1832.

Key work

1824 *Reflections on the Motive Power of Heat*

> The production of heat alone is not sufficient to give birth to the impelling power: it is necessary that there should also be cold.
> **Sadi Carnot**

Carnot's cycle

⬜ Hot reservoir ⬛ Cold reservoir ⬛ Insulation

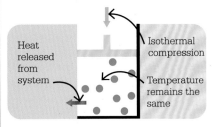

Stage 1: There is a heat transfer from the hot reservoir to the gas in the cylinder. The gas expands, pushing up the piston. This stage is isothermal because there is no temperature change in the system.

Stage 2: The gas, now insulated from the reservoirs, continues to expand as weight is lifted from the piston. The gas cools as it expands, though no heat is lost from the system overall. This expansion is adiabatic.

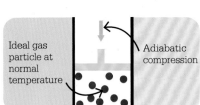

Stage 3: Weight is added above the piston. Since heat is now able to transfer from the cylinder into the cold reservoir, the gas does not increase in temperature, so this stage is isothermal.

Stage 4: More weight is added on the piston, compressing the gas in the cylinder. Since the gas is now insulated from the reservoirs again, the compression causes its temperature to increase adiabatically.

as a Carnot engine, this ideal engine works in a cycle in four stages. First, the gas is heated by conduction from an external source (such as a reservoir of hot water) and expands. Second, the hot gas is kept insulated (inside a cylinder, for example) and as it expands it does work on its surroundings (such as pushing on a piston). As it expands, the gas cools. Third, the surroundings push the piston down, compressing the gas. Heat is transferred from the system to the cold reservoir. Finally, as the system is kept insulated and the piston continues to push down, the gas temperature rises again.

In the first two stages the gas is expanding, and in the second two it is contracting. But the expansion and contraction each go through two phases; isothermal and adiabatic. In the Carnot cycle, isothermal means there is an exchange of heat with the surroundings, but no temperature change in the system. Adiabatic means no heat goes either into or out of the system.

Carnot calculated the efficiency of his ideal heat engine: if the hottest temperature reached is T_H and the coldest is T_C, the fraction of heat energy that comes out as work (the efficiency) can be expressed as $(T_H - T_C)/T_H = 1 - (T_C/T_H)$. Even Carnot's ideal engine is far from 100 percent efficient, but real engines are much less efficient than Carnot's engine. Unlike Carnot's ideal engine, real engines use irreversible processes. Once oil is burned, it stays burned. So the heat available for transfer is continually reduced. And some of the engine's work output is lost as heat through the friction of moving parts. Most motor vehicle engines are barely 25 percent efficient, and even steam turbines are only 60 percent efficient at best, which means a lot of heat is wasted.

Carnot's work on heat was only just beginning when he died of cholera at the age of 36. Unfortunately, his copious notes were burned to destroy the infection, so we will never know how far he got. Two years after his death, Benoît Paul Émile Clapeyron published a summary of Carnot's work using graphs to make the ideas clearer, and updating it to remove the caloric element. As a result, Carnot's pioneering work on heat engines revolutionized our understanding of the key role of heat in the universe and laid the foundations of the science of thermodynamics. ∎

THE ENTROPY
OF THE UNIVERSE
TENDS TO A MAXIMUM

ENTROPY AND THE SECOND
LAW OF THERMODYNAMICS

IN CONTEXT

KEY FIGURE
Rudolf Clausius (1822–1888)

BEFORE
1749 French mathematician and physicist Émilie du Châtelet introduces an early idea of energy and how it is conserved.

1777 In Sweden, pharmacist Carl Scheele discovers how heat can move by radiating through space.

1780 Dutch scientist Jan Ingenhousz discovers that heat can be moved by conduction through materials.

AFTER
1876 American scientist Josiah Gibbs introduces the idea of free energy.

1877 Austrian physicist Ludwig Boltzmann states the relationship between entropy and probability.

I n the mid-1800s, a group of physicists in Britain, Germany, and France revolutionized the understanding of heat. These scientists, including William Thomson and William Rankine in Britain; Hermann von Helmholtz, Julius von Mayer, and Rudolf Clausius in Germany; and Sadi Carnot in France showed that heat and mechanical work are interchangeable. They are both manifestations of what came to be called energy transfers.

In addition, the physicists found that the interchange of heat and mechanical work is entirely balanced: when one form of energy increases, another must decrease. The total energy can never be lost; it simply switches form. This came to be called the law of the conservation of energy and was the first law of thermodynamics. It was later broadened and reframed by Rudolf Clausius as "the energy of the universe is constant."

Heat flow

Scientists quickly realized that there was another fundamental theory of thermodynamics concerning heat flow. In 1824,

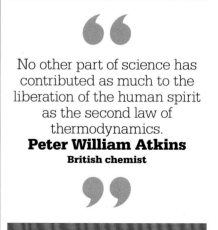

> No other part of science has contributed as much to the liberation of the human spirit as the second law of thermodynamics.
> **Peter William Atkins**
> **British chemist**

French military scientist Sadi Carnot had envisaged an ideal heat engine in which, contrary to what happened in nature, energy changes were reversible: when one form of energy was converted to another, it could be changed back again with no loss of energy. In reality, however, a large portion of the energy used by steam engines was not translated into mechanical movement but lost as heat. Even though engines of the mid-1800s were more efficient than they had been in the 1700s, they were far

---

Rudolf Clausius

The son of a headmaster and pastor, Rudolf Clausius was born in Pomerania, Prussia (now in Poland) in 1822. After studying at the University of Berlin, he became a professor at Berlin's Artillery and Engineering School. In 1855, he became professor of physics at the Swiss Federal Institute of Technology in Zurich. He returned to Germany in 1867.

The publication of his paper "On the Moving Force of Heat" in 1850 marked a key step in the development of thermodynamics. In 1865, he introduced the concept of entropy, leading to his landmark summaries of the laws of thermodynamics: "The energy of the universe is constant" and "The entropy of the universe tends to a maximum." Clausius died in Bonn in 1888.

Key works

1850 "On the Moving Force of Heat"
1856 "On a Modified Form of the Second Fundamental Theorem in the Mechanical Theory of Heat"
1867 *The Mechanical Theory of Heat*

See also: Kinetic energy and potential energy 54–55 ▪ Heat and transfers 80–81 ▪ Internal energy and the first law of thermodynamics 86–89 ▪ Heat engines 90–93 ▪ Thermal radiation 112–117

Clausius realized that in a real heat engine, it is impossible to extract an amount of heat (Q_H) from a hot reservoir and use all the extracted heat to do work (W). Some of the heat (Q_c) must be transferred to a cold reservoir. A perfect heat engine, in which all the extracted heat (Q_H) can be used to do work (W), is impossible according to the second law of thermodynamics.

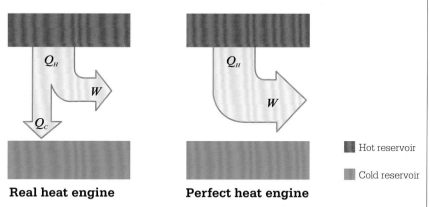

Real heat engine

Perfect heat engine

■ Hot reservoir

■ Cold reservoir

below a 100 percent conversion rate. It was partly the scientists' efforts to understand this energy loss that led them to discover the second law of thermodynamics. Clausius realized, as did Thomson and Rankine, that heat flows only one way: from hot to cold, not from cold to hot.

External help

In 1850, Clausius wrote his first statement of the second law of thermodynamics: "heat cannot by itself flow from a colder body to a warmer one." Clausius was not saying that heat can never flow from cold to hot, but that it needs external help to do so. It needs to do work: the effect of energy. This is how modern refrigerators operate. Like heat engines in reverse, they transfer heat from cold regions inside the device to hot regions outside, making the cold regions even cooler. Such a transfer requires work, which is supplied by an expanding coolant.

Clausius soon realized that the implications of a one-way flow of heat were far more complex than

initially thought. It became clear that heat engines are doomed to inefficiency. However cleverly they are designed, some energy will always leak away as heat, whether as friction, exhaust (gas or steam), or radiation, without doing any useful work.

Work is done by the flow of heat from one place to another. To Clausius and the other scientists researching thermodynamics, it soon became clear that if work is done by heat flow, there must be a concentration of energy stored in one place in order to initiate the flow; one area must be hotter than another. However, if heat is lost every time work is done, then heat gradually spreads out and becomes dissipated. Concentrations of heat become smaller and rarer until no more work can be done. Energy

An eruption of the Sakurajima volcano in Japan transfers thermal energy from Earth's super-hot interior to the cooler exterior, demonstrating the second law of thermodynamics.

supplies available for work are not inexhaustible: in time, they are all reduced to heat and so everything has a limited lifespan.

Energy of the universe

In the early 1850s, Clausius and Thomson independently began to speculate on whether Earth itself was a heat engine with a finite lifespan, and whether this could be true of the entire universe. In 1852, Thomson speculated that there would be a time when the sun's energy would run out. The implications of this to Earth meant that Earth must have a beginning and an end—a new concept. Thomson then attempted to work out how old Earth must be by calculating the time it would have taken to cool to its present temperature, given how long the sun could generate heat as it slowly collapsed under its own gravity.

Thomson's calculation showed Earth to be just a few million years old, which brought him into bitter conflict with geologists and evolutionists, who were convinced that it was considerably older. **»**

The explanation for the discrepancy is that nothing was then known about radioactivity and Einstein's 1905 discovery that matter can be turned into energy. It is the energy of matter that has kept Earth warm for much longer than solar radiation alone. This pushes Earth's history back more than 4 billion years.

Thomson went further still and suggested that in time all the energy in the universe would be dissipated as heat. It would spread out as a uniform "equilibrium" mass of heat, with no concentrations of energy at all. At that point, nothing further would be able to change in the universe and it would effectively be dead. However, Thomson also asserted that the "heat death" theory depended on there being a finite amount of matter in the universe, which he believed was not the case. Because of this, he said, its dynamic processes would carry on. Cosmologists now know much more about the universe than Thomson could and no longer accept the heat death theory, although the universe's ultimate fate remains unknown.

Stating the second law

In 1865, Clausius introduced the word "entropy" (coined from the Greek for "intrinsic" and "direction") to sum up the one-way flow of heat. The concept of entropy brought together the work that Clausius, Thomson, and Rankine had been doing for the previous 15 years as they developed what was to become the second law of thermodynamics. Yet entropy came to mean much more than just one-way flow. As Clausius's ideas took shape, entropy developed into a mathematical measure of just how much energy dissipated.

Clausius argued that because a concentration of energy is needed to hold the shape and order of the universe, dissipation leads to a

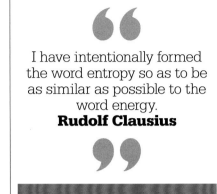

> I have intentionally formed the word entropy so as to be as similar as possible to the word energy.
> **Rudolf Clausius**

random mess of low-level energy. As a result, entropy is now considered a measure of the degree of dissipation, or more precisely the degree of randomness. But Clausius and his peers were talking specifically about heat. In fact, Clausius defined entropy as a measure of the heat that a body transfers per unit of temperature. When a body contains a lot of heat but its temperature is low, the heat must be dissipated.

The fate of all things

Clausius summed up his version of the second law of thermodynamics as "the entropy of the universe tends to a maximum." Because this wording is vague, many people now imagine it applies to everything. It has become a metaphor for the fate of all things—which will ultimately be consumed by chaos.

In 1854, however, Clausius was specifically talking about heat and energy. His definition contained the first mathematical formulation of entropy, though at the time he

Crab Nebula is a supernova, an exploded star. According to the heat death theory, the heat released into space by such explosions will eventually lead to a thermal equilibrium.

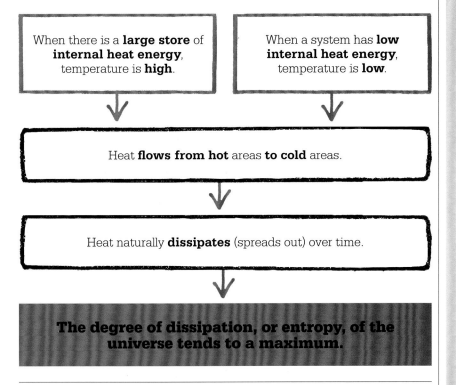

When there is a **large store** of **internal heat energy**, temperature is **high**.

When a system has **low internal heat energy**, temperature is **low**.

↓ ↓

Heat **flows from hot** areas **to cold** areas.

↓

Heat naturally **dissipates** (spreads out) over time.

↓

The degree of dissipation, or entropy, of the universe tends to a maximum.

---

called it "equivalence value," with one equation for **S** (entropy) for open energy systems and another for closed systems. An energy system is a region where energy flows—it could be a car engine or the entire atmosphere. An open system can exchange both energy and matter with its surroundings; a closed system can only exchange energy (as heat or work).

Within a finite period of time to come the earth again must be unfit for human habitation.
William Thomson

Thomson came up with a way of describing the second law of thermodynamics in relation to the limits of heat engines. This became the basis of what is now known as the Kelvin-Planck statement of the law (Lord Kelvin was the title Thomson took when given a peerage in 1892). German physicist Max Planck refined Kelvin's idea so that it read: "it is impossible to devise a cyclically operating heat engine, the effect of which is to absorb energy in the form of heat from a single thermal reservoir and to deliver an equivalent amount of work." In other words, it is impossible to make a heat engine that is 100 percent efficient. It is not easy to see that this is what Clausius was also saying, which has caused confusion ever since. Essentially, these ideas are all based on the same thermodynamics law: the inevitability of heat loss when heat flows one way. ∎

Time's arrow

The significance of the discovery of the second law of thermodynamics is often overlooked, because other scientists quickly built on the work of Clausius and his peers. In fact, the second law of thermodynamics is as crucial to physics as Newton's discovery of the laws of motion, and it played a key part in modifying the Newtonian vision of the universe that had prevailed until then.

In Newton's universe, all actions happen equally in every direction, so time has no direction—like an eternal mechanism that can be run backward or forward. Clausius's second law of thermodynamics overturned this view. If heat flows one way, so must time. Things decay, run down, come to an end, and time's arrow points in one direction only—toward the end. The implications of this discovery shook many people of religious faith who believed that the universe was everlasting.

Once a candle has been burned, the wax that was burned cannot be restored. The thermodynamic arrow of time points in one direction: toward the end.

THE FLUID AND ITS VAPOR BECOME ONE

CHANGES OF STATE AND MAKING BONDS

IN CONTEXT

KEY FIGURE
**Johannes Diderik
van der Waals** (1837–1923)

BEFORE
c. 75 BCE The Roman thinker
Lucretius suggests liquids are
made from smooth round
atoms but solids are bound
together by hooked atoms.

1704 Isaac Newton theorizes
that atoms are held together
by an invisible force of
attraction.

1869 Irish chemist and
physicist Thomas Andrews
discovers the continuity
between the two fluid states
of matter—liquid and gas.

AFTER
1898 Scottish chemist James
Dewar liquefies hydrogen.

1908 Dutch physicist Heike
Kamerlingh Onnes liquefies
helium.

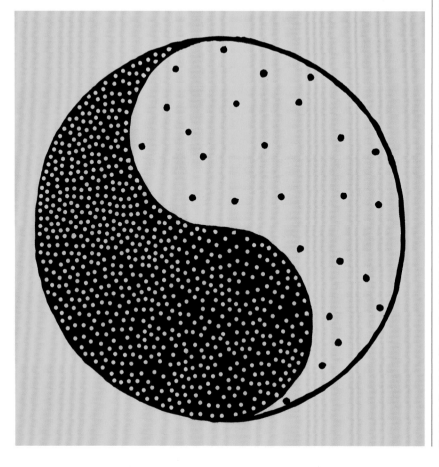

It has long been known that the
same substance can exist in
at least three phases—solid,
liquid, and gas. Water, for example,
can be ice, liquid water, and
vapor. But what was going on
in the changes between these
phases seemed for much of the
19th century to present an obstacle
to the gas laws that had been
established in the late 18th century.

A particular focus was the two
fluid states—liquid and gas. In
both states, the substance flows to
take up the shape of any container,
and it cannot hold its own shape
like a solid. Scientists had shown
that if a gas is compressed more
and more, its pressure does not

See also: Models of matter 68–71 ▪ Fluids 76–79 ▪ Heat and transfers 80–81 ▪ The gas laws 82–85 ▪ Entropy and the second law of thermodynamics 94–99 ▪ The development of statistical mechanics 104–111

> How am I to name this point at which the fluid and its vapor become one according to a law of continuity?
> **Michael Faraday**
> **In a letter to fellow scientist, William Whewell (1844)**

increase indefinitely, and eventually it turns to liquid. Similarly, if a liquid is heated, a little evaporates at first, and then eventually all of it evaporates. The boiling point of water—the maximum temperature that water can reach—is easily measured, and this measurably rises under pressure, which is the principle behind a pressure cooker.

The points of change

Scientists wanted to go beyond these observations to know what happens within a substance when liquid changes to gas. In 1822, French engineer and physicist Baron Charles Cagniard de la Tour experimented with a "steam digester," a pressurized device that generated steam from water heated beyond its normal boiling point. He partially filled the digester cylinder with water and dropped a flint ball into it. Rolling the cylinder like a log, he could hear the ball splashing as it met the water. The cylinder was then heated to a temperature, estimated by de la Tour to be 683.6°F

(362°C), at which point no splashing was heard. The boundary between the gas and the liquid was gone.

It was known that keeping a liquid under pressure can stop it all turning to gas, but de la Tour's experiments revealed that there is a temperature at which a liquid will always turn to gas, no matter how much pressure it is subjected to. At this temperature, there is no distinction between the liquid phase and the gas phase—both become equally dense. Decreasing the temperature then restores the differences between the phases.

The point where liquid and gas are in equilibrium remained a vague concept until the 1860s, when physicist Thomas Andrews investigated the phenomenon. He studied the relationship between

temperature, pressure, and volume, and how it might affect the phases of a substance. In 1869, he described experiments in which he trapped carbon dioxide above mercury in a glass tube. By pushing the mercury up, he could raise the pressure of the gas until it turned liquid. Yet it would never liquefy above 91.26°F (32.92°C), no matter how much pressure he applied. He called this temperature the "critical point" of carbon dioxide. Andrews further observed, "We have seen that the gaseous and liquid phases are essentially only distinct stages of one and the same state of matter and that they are able to go into each other by a continuous change."

The idea of the continuity between the liquid and gas phases was a key insight, highlighting »

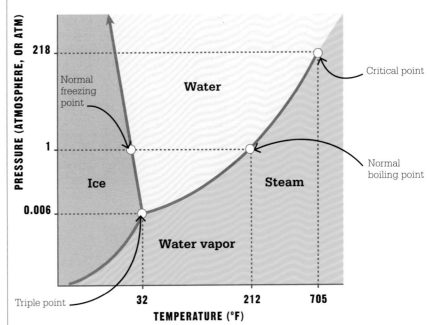

A phase diagram plots the temperature and pressure at which a substance—in this case, water—is solid, liquid, or gas. At the "triple point," a substance can exist simultaneously as a solid, a liquid, and a gas. At the critical point, a liquid and its gas become identical.

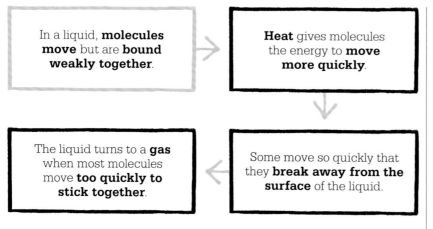

In a liquid, **molecules move** but are **bound weakly together**.

Heat gives molecules the energy to **move more quickly**.

Some move so quickly that they **break away from the surface** of the liquid.

The liquid turns to a **gas** when most molecules move **too quickly to stick together**.

their fundamental similarity. What had yet to be discovered, though, were the forces that lay behind these different phases of matter and how they interacted.

Molecular ties

Early in the 19th century, British scientist Thomas Young suggested that the surface of a liquid is held together by intermolecular bonds. It is this "surface tension" that pulls water into drops and forms a curve on the top of a glass of water, as the molecules are drawn together. This work was taken further by the Dutch physicist Johannes Diderik van der Waals, who looked at what happened when the surface tension broke and allowed the molecules to break away, turning liquid water into water vapor.

Van der Waals proposed that the change in state is part of a continuum and not a distinct break between liquid and gas. There is a transition layer in which the water is neither exclusively liquid nor gas. He found that as temperature rises, surface tension diminishes, and at the critical temperature surface tension vanishes altogether, allowing the transition layer to become infinitely thick. Van der Waals then gradually

developed a key "equation of state" to mathematically describe the behavior of gases and their condensation to liquid, which could be applied to different substances.

Van der Waals' work helped establish the reality of molecules and identify intermolecular bonds. These are far weaker than the bonds between atoms, which are based on powerful electrostatic forces. Molecules of the same substance are held together differently, in both the liquid and gas phases. For instance, the bonds that bind water molecules are not the same as the bonds that tie together oxygen and hydrogen atoms within each water molecule.

When a liquid is turned to a gas, the forces between the molecules need to be overcome to allow those molecules to move freely. Heat provides the energy to make the molecules vibrate. Once the vibrations are powerful enough, the molecules break free of the forces that bind them and become gas.

Forces of attraction

Three key forces of intermolecular attraction—dipole–dipole, London dispersion, and hydrogen bond— have become known collectively as Van der Waals forces. Dipole–dipole

forces occur in "polar" molecules, where electrons are shared unequally between the atoms of the molecule. In hydrochloric acid, for example, the chlorine atom has an extra electron, taken from the hydrogen atom. This gives the chlorine part of the molecule a slight negative charge, unlike the hydrogen part. The result is that in liquid hydrochloric acid solution, negative sides of some molecules are attracted to the positive sides of others—this binds them together.

London dispersion force (named after German–American scientist Fritz London, who first recognized it in 1930) occurs between nonpolar molecules. For example, in chlorine gas the two atoms in each molecule are equally charged on either side. But, the electrons in an atom are constantly moving. This means that one side of the molecule may become briefly negative while the other becomes briefly positive, so bonds within molecules are constantly formed and reformed.

The third force, the hydrogen bond, is a special kind of dipole–dipole bond that occurs within hydrogen. It is the interaction between a hydrogen atom and an atom of either oxygen, fluorine, or nitrogen. It is especially strong for an intermolecular bond because

I was quite convinced of the real existence of molecules.
Johannes Diderik van der Waals

oxygen, fluorine, and nitrogen atoms are strong attractors of electrons, while hydrogen is prone to losing them. So, a molecule that combines them becomes strongly polar, creating, for instance, the robust hydrogen bonds that hold water (H_2O) together.

Dispersion bonds are the weakest of the van der Waals forces. Some elements that are bound together by them, such as chlorine and fluorine, remain as a gas unless cooled to an extremely low temperature ($-29.2°F$ [$-34°C$] and $-306.4°F$ [$-188°C$] respectively), when the bonds become strong enough to enter the liquid phase. Hydrogen bonds are the strongest, which is why water has an unusually high boiling point for a substance made of oxygen and hydrogen.

Critical discoveries

In showing that the attraction forces between gas molecules were not zero but could be forced, under pressure, into state-changing bonds, van der Waals laid the foundations for the understanding of how liquids change into gases

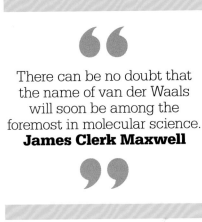

> There can be no doubt that the name of van der Waals will soon be among the foremost in molecular science.
> **James Clerk Maxwell**

and back again. His "equation of state" enabled critical points to be found for a range of substances, making it possible to liquefy gases such as oxygen, nitrogen, and helium. It also led to the discovery of superconductors—substances that lose all electrical resistance when cooled to ultra-low temperatures. ∎

At a liquid oxygen plant, oxygen gas is extracted from air in separation columns and cooled by passing through heat exchangers to its liquefying temperature of $-297.4°F$ ($-183°C$).

Johannes Diderik van der Waals

Born the son of a carpenter in the Dutch city of Leiden in 1837, Johannes Diderik van der Waals lacked sufficient schooling to enter higher education. He became a teacher of mathematics and physics and studied part-time at Leiden University, finally achieving his doctorate—in molecular attraction—in 1873.

Van der Waals was hailed immediately as one of the leading physicists of the day, and in 1876 became Professor of Physics at the University of Amsterdam. He remained there for the rest of his career, until succeeded as professor by his son, also called Johannes. In 1910, Van der Waals was awarded the Nobel Prize in Physics "for his work on the equation of state for gases and liquids." He died in Amsterdam in 1923.

Key works

1873 *On the Continuity of the Gaseous and Liquid State*
1880 *Law of Corresponding States*
1890 *Theory of Binary Solutions*

COLLIDING BILLIARD BALLS IN A BOX

THE DEVELOPMENT OF STATISTICAL MECHANICS

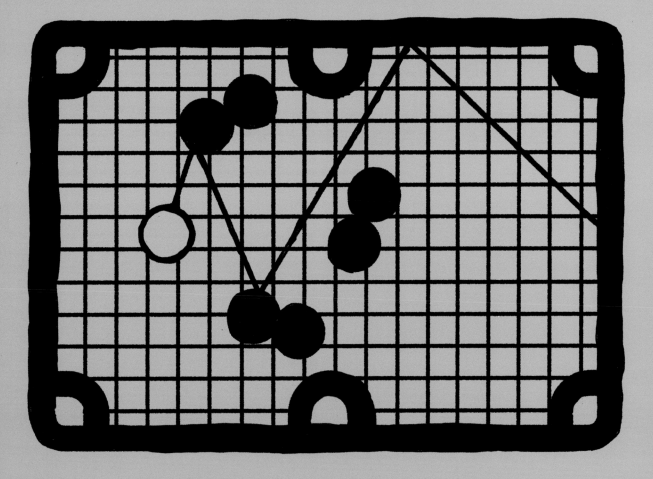

IN CONTEXT

KEY FIGURE
Ludwig Boltzmann
(1844–1906)

BEFORE
1738 Daniel Bernoulli makes the first statistical analysis of particle movement.

1821 John Herapath gives the first clear statement of kinetic theory.

1845 John Waterston calculates the average speed of gas molecules.

1859 James Clerk Maxwell lays out his kinetic theory.

AFTER
1902 Willard Gibbs publishes the first major textbook on statistical mechanics.

1905 Marian von Smoluchowski and Albert Einstein demonstrate Brownian motion as statistical mechanics in action.

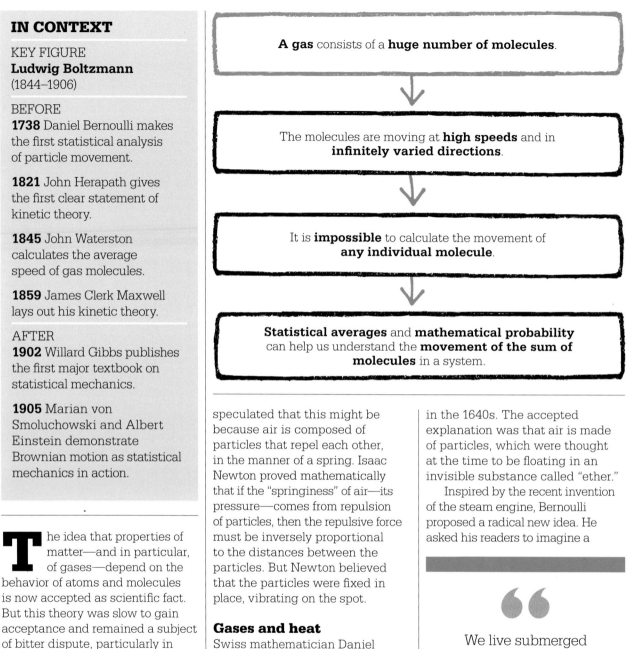

A **gas** consists of a **huge number of molecules**.

The molecules are moving at **high speeds** and in **infinitely varied directions**.

It is **impossible** to calculate the movement of **any individual molecule**.

Statistical averages and **mathematical probability** can help us understand the **movement of the sum of molecules** in a system.

T he idea that properties of matter—and in particular, of gases—depend on the behavior of atoms and molecules is now accepted as scientific fact. But this theory was slow to gain acceptance and remained a subject of bitter dispute, particularly in the 19th century. Several pioneers faced at best neglect and at worst derision, and it was a long time before the "kinetic theory"—the idea that heat is the rapid movement of molecules—was truly accepted.

In the 17th century, Robert Boyle showed that air is elastic, and expands and contracts. He speculated that this might be because air is composed of particles that repel each other, in the manner of a spring. Isaac Newton proved mathematically that if the "springiness" of air—its pressure—comes from repulsion of particles, then the repulsive force must be inversely proportional to the distances between the particles. But Newton believed that the particles were fixed in place, vibrating on the spot.

Gases and heat

Swiss mathematician Daniel Bernoulli made the first serious proposal of the kinetic (movement) theory of gases in 1738. Prior to that date, scientists already knew that air exerted pressure—for instance, enough pressure to hold up a heavy column of mercury, which had been demonstrated by Evangelista Torricelli's barometer in the 1640s. The accepted explanation was that air is made of particles, which were thought at the time to be floating in an invisible substance called "ether."

Inspired by the recent invention of the steam engine, Bernoulli proposed a radical new idea. He asked his readers to imagine a

> We live submerged at the bottom of an ocean of the element air.
> **Evangelista Torricelli**

See also: Kinetic energy and potential energy 54 ▪ Fluids 76–79 ▪ Heat engines 90–93 ▪ Entropy and the second law of thermodynamics 94–99

A well-constructed theory is in some respects undoubtedly an artistic production. A fine example is the famous kinetic theory.
Ernest Rutherford

piston within a cylinder, which contained tiny, round particles zooming this way and that. Bernoulli argued that as the particles collided with the piston, they created pressure. If the air was heated, the particles would speed up, striking the piston more often and pushing it up through the cylinder. His proposal summed up the kinetic theory of gases and heat, but his ideas were forgotten because first the theory that combustible materials contain a fire element called phlogiston, then the caloric theory—that heat is a kind of fluid—held sway for the next 130 years, until Ludwig Boltzmann's statistical analysis in 1868 banished it for good.

Heat and motion

There were other unrecognized pioneers, too, such as Russian polymath Mikhail Lomonosov, who in 1745 argued that heat is a measure of the motion of particles—that is, the kinetic theory of heat. He went on to say that absolute zero would be reached when particles stopped moving, more than a century before William Thomson (later Lord Kelvin,

who would be forever remembered in the Kelvin temperature scale) drew the same conclusion in 1848.

It was in 1821 that British physicist John Herapath gave the first clear statement of kinetic theory. Heat was still seen as a fluid and gases were considered to be made of repelling particles, as Newton had suggested. But Herapath rejected this idea, suggesting instead that gases are made of "mutually impinging atoms." If such particles were infinitely small, he reasoned, collisions would increase as a gas was compressed, so pressure would rise and heat would be generated. Unfortunately, Herapath's work was rejected by the Royal Society in London as overly conceptual and unproven.

In 1845, the Royal Society also rejected a major paper on kinetic theory by Scotsman John Waterston, which used statistical rules to explain how energy is distributed among gas atoms and molecules. Waterston understood that molecules do not all move at the same speed, but travel at a range of different speeds around a statistical average. Like Herapath before him, Waterston's important contribution was neglected, and the only copy of his groundbreaking work was lost by the Royal Society. It was rediscovered in 1891, but by this time Waterston had gone missing and was presumed to have drowned in a canal near his Edinburgh home.

Messy universe

Waterston's work was especially significant because this was the first time that physics rejected the perfect clockwork of the Newtonian universe. Instead, Waterston was »

Brownian motion

In 1827, Scottish botanist Robert Brown described the random movement of pollen grains suspended in water. Although he was not the first to notice this phenomenon, he was the first to study it in detail. More investigation showed that the tiny to-and-fro movements of pollen grains got faster as the temperature of the liquid increased.

The existence of atoms and molecules was still a matter of heated debate at the start of the 20th century, but in 1905 Einstein argued that Brownian motion could be explained by invisible atoms and molecules bombarding the tiny but visible particles suspended in a liquid, causing them to vibrate back and forth. A year later, Polish physicist Marian Smoluchowski published a similar theory, and in 1908 Frenchman Jean Baptiste Perrin conducted experiments that confirmed this theory.

Brownian motion of particles in a fluid results from collisions with fast-moving molecules of the fluid. It was eventually explained using statistical mechanics.

looking at values whose range was so messy that they could only be looked at in terms of statistical averages and probabilities, not certainties. Although Waterston's work was originally rejected, the idea of understanding gas and heat in terms of high-speed movements of minute particles was at last beginning to take hold. The work of British physicists James Joule and William Thomson, German physicist Rudolf Clausius, and others was showing that heat and mechanical movement are interchangeable forms of energy—rendering redundant the idea that heat is some kind of "caloric fluid."

Molecular movement

Joule had calculated the very high speeds of gas molecules with some accuracy in 1847, but assumed that they all moved at the same speed. Ten years later, Clausius furthered understanding with his proposition of a "mean free path." As he saw it, molecules repeatedly collide and bounce off each other in different directions. The mean free path is the average distance each molecule travels before bumping into another. Clausius calculated this to be barely a millionth of a millimeter at ambient

> Available energy is the main object at stake in the struggle for existence and the evolution of the world.
> **Ludwig Boltzmann**

A gas molecule collides repeatedly with other molecules, causing it to change direction. The molecule shown here has 25 such collisions, and the average distance it travels between each collision is what Rudolf Clausius called its "mean free path." Compare the shortest distance between point A and point B with the distance actually traveled.

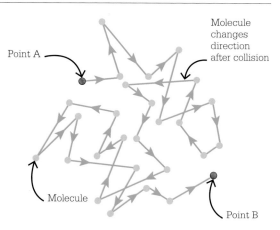

Point A

Molecule changes direction after collision

Molecule

Point B

temperature, meaning that each molecule collides with another more than 8 billion times a second. It is the sheer tininess and frequency of these collisions that make a gas appear to be fluid and smooth, rather than a raging sea.

Within a few years, James Clerk Maxwell provided such a solid exposition of kinetic theory that it at last became more widely accepted. Significantly, in 1859, Maxwell introduced the first-ever statistical law in physics, the Maxwell distribution, which shows the probable proportion of molecules moving at a particular velocity in an ideal gas. Maxwell also established that the rate of molecular collisions corresponds to temperature—the more frequent the collisions are, the higher the temperature. In 1873, Maxwell estimated that there are 19 trillion molecules in a cubic centimeter of gas in ideal conditions—not that far from the modern estimate of 26.9 trillion.

Maxwell also compared molecular analysis to the science of population statistics, which divided people according to factors such as education, hair color, and build, and analyzed them in order to determine average characteristics. Maxwell observed that the vast population of atoms in just a cubic centimeter

of gas is in fact far less varied than this, making the statistical task of analyzing them far simpler.

Boltzmann's breakthrough

The key figure in the development of the statistical analysis of moving molecules was Austrian physicist Ludwig Boltzmann. In major papers in 1868 and 1877, Boltzmann developed Maxwell's statistical approach into an entire branch of science—statistical mechanics. Remarkably, this new discipline allowed the properties of gases and heat to be explained and predicted in simple mechanical terms, such as mass, momentum, and velocity. These particles, although tiny, behaved according to Newton's laws of motion, and the variety in their movement is simply down to chance. Heat, which had previously been thought of as a mysterious and intangible fluid known as "caloric," could now be understood as the high-speed movement of particles—an entirely mechanical phenomenon.

Boltzmann faced a particular challenge in testing his theory: molecules are so innumerable and so tiny that making individual calculations would be impossible. More significantly, their movements vary hugely in speed, and infinitely in direction. Boltzmann realized that the only way to investigate the idea in a rigorous and practical way was to employ the math of statistics and probability. He was forced to forego the certainties and precision of Newton's clockwork world, and enter into the far messier world of statistics and averages.

Micro and macrostates

The law of conservation of energy states that the total energy (E) in an isolated volume of gas must be constant. However, the energy of individual molecules can vary. So, the energy of each molecule cannot be E divided by the total number of molecules (E/N), 19 trillion, for instance—as it would be if they all had the same energy. Instead, Boltzmann looked at the range of possible energies that individual molecules might have, considering factors including their position and velocity. Boltzmann called this range of energies a "microstate."

As the atoms within each molecule interact, the microstate changes many trillions of times every second, but the overall condition of the gas—its pressure, temperature, and volume, which

Boltzmann called its "macrostate"—remains stable. Boltzmann realized that a macrostate can be calculated by averaging the microstates.

To average the microstates that make up the macrostate, Boltzmann had to assume that all microstates are equally probable. He justified this assumption with what came to be known as the "ergodic hypothesis"—that over a very long period of time, any dynamic system will, on average, spend the same amount of time in each microstate. This idea of things averaging out was vital to Boltzmann's thinking.

Statistical thermodynamics

Boltzmann's statistical approach has had huge ramifications. It has become the primary means of understanding heat and energy, and made thermodynamics—the study of the relationships between heat and other forms of energy—a central pillar of physics. His approach also became a hugely valuable way of examining the subatomic world, paving the way for the development of quantum science, and it now underpins much of modern technology.

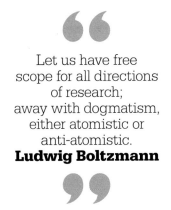

Let us have free scope for all directions of research; away with dogmatism, either atomistic or anti-atomistic.
Ludwig Boltzmann

Scientists now understand that the subatomic world can be explored through probabilities and averages, not just as a way of understanding or measuring it, but as a glimpse of its very reality—the apparently solid world we live in is essentially a sea of subatomic probabilities. Back in the 1870s, however, Boltzmann faced dogged opposition to his ideas when he laid out the mathematical basics of thermodynamics. He wrote two key papers on the second law of thermodynamics (previously »

Ludwig Boltzmann

Born in Vienna in 1844, at the height of the Austro-Hungarian Empire, Ludwig Boltzmann studied physics at the University of Vienna and wrote his PhD thesis on the kinetic theory of gases. At the age of 25, he became a professor at the University of Graz, and then took teaching posts at Vienna and Munich before returning to Graz. In 1900, he moved to the University of Leipzig to escape his bitter long-term rival, Ernst Mach.

It was at Graz that Boltzmann completed his work on statistical mechanics. He established the

kinetic theory of gases and the mathematical basis of thermodynamics in relation to the probable movements of atoms. His ideas brought him opponents, and he suffered from bouts of depression. He committed suicide in 1906.

Key works

1871 Paper on Maxwell–Boltzmann distribution
1877 Paper on second law of thermodynamics and probability ("On the Relationship between the Second Fundamental")

developed by Rudolf Clausius, William Thomson, and William Rankine), which shows that heat flows only in one direction, from hot to cold, not from cold to hot. Boltzmann explained that the law could be understood precisely by applying both the basic laws of mechanics (that is, Newton's laws of motion) and the theory of probability to the movement of atoms.

In other words, the second law of thermodynamics is a statistical law. It states that a system tends toward equilibrium, or maximum entropy—the state of a physical system at greatest disorder—because this is by far the most probable outcome of atomic motion; things average out over time. By 1871, Boltzmann had also developed Maxwell's distribution law of 1859 into a rule that defines the distribution of speeds of molecules for a gas at a certain temperature. The resulting Maxwell–Boltzmann distribution is central to the kinetic theory of gases. It can be used to show the average speed of the molecules and also to show the most probable speed. The distribution highlights the "equipartition" of energy—which shows that the energy of moving atoms averages out as the same in any direction.

Atomic denial

Boltzmann's approach was such a novel concept that he faced fierce opposition from some of his contemporaries. Many considered his ideas to be fanciful, and it is possible that hostility to his work contributed to his eventual suicide. One reason for such opposition was that many scientists at the time were unconvinced of the existence of atoms. Some, including Austrian physicist Ernst Mach—a fierce rival to Boltzmann, and known for his work on shock waves—believed that scientists should only accept what they could directly observe, and atoms could not be seen at the time.

The St. Louis World's Fair in 1904 was the venue for a lecture by Boltzmann on applied mathematics. His American tour also included visits to Stanford and Berkeley universities.

The majority of scientists only accepted the existence of atoms after contributions from Albert Einstein and Polish physicist Marian Smoluchowski. Working independently, they explored Brownian motion, the unexplained random flitting to and fro of tiny, suspended particles in a fluid. In 1905, Einstein, and the following year, Smoluchowski, both showed that it could be explained via statistical mechanics as the result of collisions of the particles with the fast-moving molecules of the fluid itself.

Wider acceptance

Although a brilliant lecturer who was much loved by his students, Boltzmann did not achieve wider popularity for his work, perhaps because he did not promote it. The widespread acceptance of his theoretical approach was due in some part to American physicist

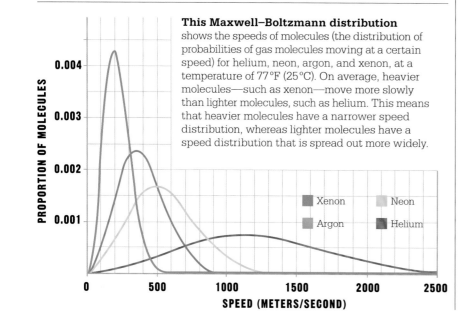

This Maxwell–Boltzmann distribution shows the speeds of molecules (the distribution of probabilities of gas molecules moving at a certain speed) for helium, neon, argon, and xenon, at a temperature of 77°F (25°C). On average, heavier molecules—such as xenon—move more slowly than lighter molecules, such as helium. This means that heavier molecules have a narrower speed distribution, whereas lighter molecules have a speed distribution that is spread out more widely.

Xenon   Neon

Argon   Helium

PROPORTION OF MOLECULES

0.004

0.003

0.002

0.001

0   500   1000   1500   2000   2500

SPEED (METERS/SECOND)

Willard Gibbs, who wrote the first major textbook on the subject, *Statistical Mechanics*, in 1902.

It was Gibbs who coined the phrase "statistical mechanics" to encapsulate the study of the mechanical movement of particles. He also introduced the idea of an "ensemble," a set of comparable microstates that combine to form a similar macrostate. This became a core idea in thermodynamics, and it also has applications in other areas of science, ranging from the study of neural pathways to weather forecasting.

Final vindication

Thanks to Gibbs, Boltzmann was invited on a lecture tour in the US in 1904. By that time, the hostility to his life's work was beginning to take its toll. He had a medical history of bipolar disorder, and in 1906 he hanged himself while on a family vacation in Trieste, Italy. In a bitter twist of fate, his death came in the same year that the work of Einstein and Smoluchowski was gaining acceptance, vindicating Boltzmann. His overarching idea was that

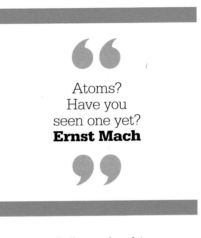

> 66
> Atoms?
> Have you
> seen one yet?
> **Ernst Mach**
> 99

matter, and all complex things, are subject to probability and entropy. It is impossible to overestimate the immense shift in outlook his ideas created among physicists. The certainties of Newtonian physics had been replaced by a view of the universe in which there is only a fizzing sea of probabilities, and the only certainties are decay and disorder. ■

A tornado is a chaotic system that can be analyzed using statistical mechanics. Projecting the distribution of atmospheric molecules can help gauge its temperature and intensity.

Weather forecasts

The methods developed in statistical mechanics to analyze and predict mass movements of particles have been used in many situations beyond thermodynamics.

One real-world application, for instance, is the calculation of "ensemble" weather forecasts. More conventional methods of numerical weather forecasting involve collecting data from weather stations and instruments around the world and using it to simulate future weather conditions. In contrast, ensemble forecasting is based on a large number of possible future weather predictions, rather than on a single predicted outcome. The probability that a single forecast will be wrong is relatively high, but forecasters can have a strong degree of confidence that the weather will fall within a given range of an ensemble forecast.

The idea was proposed by American mathematician Edward Lorenz in a 1963 paper, which also outlined "chaos theory." Known for the so-called "butterfly effect," his theory explored how events occur in a chaotic system such as Earth's atmosphere. Lorenz famously suggested that a butterfly flapping its wings can set off a chain of events that eventually triggers a hurricane.

The power of a statistical approach is immense, and allowing uncertainty to play a part has enabled weather forecasting to become far more reliable. Forecasters can confidently predict weather locally for weeks in advance—within a given range.

FETCHING SOME GOLD FROM THE SUN

THERMAL RADIATION

IN CONTEXT

KEY FIGURE
Gustav Kirchhoff
(1824–1887)

BEFORE
1798 Benjamin Thompson
(Count Rumford) suggests
that heat is related to motion.

1844 James Joule argues
that heat is a form of energy
and that other forms of energy
can be converted to heat.

1848 William Thomson (Lord
Kelvin) defines absolute zero.

AFTER
1900 German physicist Max
Planck proposes a new theory
for blackbody radiation, and
introduces the idea of the
quantum of energy.

1905 Albert Einstein uses
Planck's idea of blackbody
radiation to solve the problem
of the photoelectric effect.

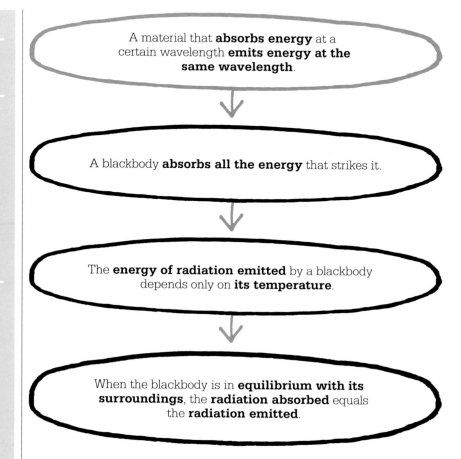

A material that **absorbs energy** at a certain wavelength **emits energy at the same wavelength**.

A blackbody **absorbs all the energy** that strikes it.

The **energy of radiation emitted** by a blackbody depends only on **its temperature**.

When the blackbody is in **equilibrium with its surroundings**, the **radiation absorbed** equals the **radiation emitted**.

H eat energy can be transferred from one place to another in one of three ways: by conduction in solids, by convection in liquids and gases, and by radiation. This radiation, known as thermal—or heat—radiation, does not require physical contact. Along with radio waves, visible light, and X-rays, thermal radiation is a form of electromagnetic radiation that travels in waves through space.

James Clerk Maxwell was the first to propose the existence of electromagnetic waves in 1865. He predicted that there would be a whole range, or spectrum, of electromagnetic waves, and later experiments showed that his theory was correct. Everything that has a temperature above absolute zero (equal to –459.67 °F or –273.15 °C) emits radiation. All of the objects in the universe are exchanging electromagnetic radiation with each other all of the time. This constant flow of energy from one object to another prevents anything from cooling to absolute zero, the theoretical minimum of temperature at which an object would transmit no energy at all.

Heat and light
German-born British astronomer William Herschel was one of the first scientists to observe a connection between heat and light. In 1800, he used a prism to split light into a spectrum, and measured the temperature at different points within that spectrum. He noticed that the temperature increased as he moved his thermometer from the violet part of the spectrum to the red part of the spectrum.

To his surprise, Herschel found that the temperature also increased beyond the red end of the spectrum, where no light was visible at all. He had discovered infrared radiation—a type of energy that is invisible to the eye, but which can be detected as heat. For example, modern-day toasters use infrared radiation to transmit heat energy to the bread.

See also: The conservation of energy 55 ▪ Heat and transfers 80–81 ▪ Internal energy and the first law of thermodynamics 86–89 ▪ Heat engines 90–93 ▪ Electromagnetic waves 192–195 ▪ Energy quanta 208–211

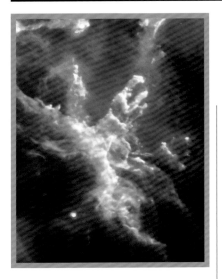

The amount of thermal radiation given off by an object depends on its temperature. The hotter the object, the more energy it emits. If an object is hot enough, a large portion of the radiation it emits can be seen as visible light. For example, a metal rod heated to a high enough temperature will begin to glow first a dull red, then yellow, then brilliant white. A metal rod glows red when it reaches a

Intensely cold gas and dust in the Eagle nebula are rendered in reds (–442°F or –263°C) and blues (–388°F or –205°C) by the Herschel Space Observatory far-infrared telescope.

temperature of more than 1,292°F (700°C). Objects with equal radiative properties emit light of the same color when they reach the same temperature.

Absorption equals emission

In 1858, Scottish physicist Balfour Stewart presented a paper titled "An Account of Some Experiments on Radiant Heat." While investigating the absorption and emission of heat in thin plates of different materials, he found that at all temperatures, the wavelengths of absorbed and emitted radiation are equal. A material that tends to absorb energy at a certain wavelength tends also to emit energy at that same wavelength. Stewart noted that "the absorption of a plate equals its radiation [emission], and that for every description [wavelength] of heat."

Bodies can be imagined which … completely absorb all incident rays, and neither reflect nor transmit any. I shall call such bodies … blackbodies.
Gustav Kirchhoff

Two years after Stewart's paper was published, German physicist Gustav Kirchhoff, unaware of the Scotsman's work, published similar conclusions. At the time, the academic community judged that Kirchhoff's work was more rigorous than Stewart's investigations, and found more immediate applications to other fields, such as astronomy. Despite his discovery being the earlier »

Gustav Kirchhoff

Born in 1824, Kirchhoff was educated in Königsberg, Prussia (modern-day Kaliningrad, Russia). He proved his mathematical skill in 1845 as a student, by extending Ohm's law of electrical current into a formula that allowed the calculation of currents, voltages, and resistances in electrical circuits. In 1857, he discovered that the velocity of electricity in a highly conductive wire was almost exactly equal to the velocity of light, but he dismissed this as a coincidence, rather than inferring that light was an electromagnetic phenomenon. In 1860, he showed

that each chemical element has a unique, characteristic spectrum. He then worked with Robert Bunsen in 1861 to identify the elements in the sun's atmosphere by examining its spectrum.

Although poor health in later life prevented Kirchhoff from laboratory work, he continued to teach. He died in Berlin in 1887.

Key work

1876 *Vorlesungen über mathematische Physik* (*Lectures on mathematical physics*)

> Since we can produce all types of light by means of hot bodies, we can ascribe, to the radiation in thermal equilibrium with hot bodies, the temperature of these bodies.
> **Wilhelm Wien**

of the two, Stewart's contribution to the theory of thermal radiation was largely forgotten.

Blackbody radiation

Kirchhoff's findings can be explained as follows. Imagine an object that perfectly absorbs all of the electromagnetic radiation that strikes it. Since no radiation is reflected from it, all of the energy that it emits depends solely on its temperature, and not its chemical composition or physical shape. In 1862, Kirchhoff coined the term "blackbodies" to describe these hypothetical objects. Perfect blackbodies do not exist in nature.

An ideal blackbody absorbs and emits energy with 100 percent efficiency. Most of its energy output is concentrated around a peak frequency—denoted λ_{MAX}, where λ is the wavelength of the radiation emitted—which increases as the temperature increases. When plotted on a graph, the spread of energy-emitting wavelengths around the object's peak frequency takes a distinctive profile known as a "blackbody curve." The blackbody curve of the sun, for example, has its peak at the center of the visible light range. Since perfect blackbodies do not exist, to help explain his theory Kirchhoff conjectured a hollow container with a single, tiny hole. Radiation can only enter the container through the hole, and is then absorbed inside the cavity, so the hole acts as a perfect absorber. Some radiation will be emitted through the hole and through the surface of the cavity. Kirchhoff proved that the radiation inside the cavity depends only on the temperature of the object, and not on its shape, size, or the material it is made from.

Law of thermal radiation

Kirchhoff's 1860 law of thermal radiation states that when an object is in thermodynamic equilibrium— at the same temperature as the objects around it—the amount of radiation that is absorbed by the surface is equal to the amount emitted, at any temperature and wavelength. Hence, the efficiency with which an object absorbs radiation at a given wavelength is the same as the efficiency with which it emits energy at that wavelength. This can be expressed more concisely as: absorptivity equals emissivity.

In 1893, German physicist Wilhelm Wien discovered the mathematical relationship between temperature change and the shape of the blackbody curve. He found that when the wavelength at which the maximum amount of radiation is emitted was multiplied by the temperature of the blackbody, the resulting value is always a constant.

This finding meant that the peak wavelength could be calculated for any temperature,

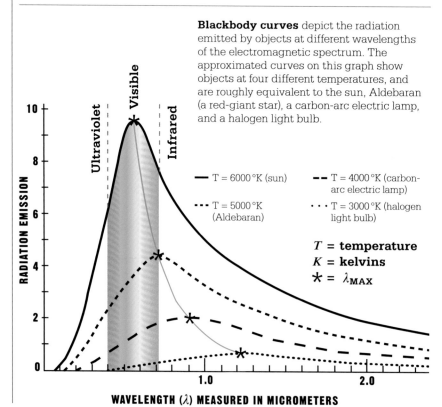

Blackbody curves depict the radiation emitted by objects at different wavelengths of the electromagnetic spectrum. The approximated curves on this graph show objects at four different temperatures, and are roughly equivalent to the sun, Aldebaran (a red-giant star), a carbon-arc electric lamp, and a halogen light bulb.

— T = 6000 °K (sun)
···· T = 5000 °K (Aldebaran)
– – T = 4000 °K (carbon-arc electric lamp)
··· T = 3000 °K (halogen light bulb)

T = **temperature**
K = **kelvins**
✳ = λ_{MAX}

RADIATION EMISSION

Ultraviolet · Visible · Infrared

WAVELENGTH (λ) MEASURED IN MICROMETERS

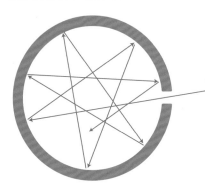

Kirchhoff envisaged a blackbody as a container with a small hole. Most of the radiation that enters the enclosure will be trapped. The amount of radiation emitted depends on the surroundings.

and it explained why objects change color as they get hotter. As the temperature increases the peak wavelength decreases, moving from longer infrared waves to shorter ultraviolet waves. By 1899, however, careful experiments showed that Wien's predictions were not accurate for wavelengths in the infrared range.

Ultraviolet catastrophe

In 1900, British physicists Lord Rayleigh and Sir James Jeans published a formula that seemed to explain what had been observed at the infrared end of the spectrum. However, their findings were soon called into question. According to their theory, there was effectively no upper limit to the higher frequencies of ultraviolet energy that would be generated by the blackbody radiation, meaning that an infinite number of highly energetic waves would be produced. If this was the case, opening the oven door to check on a cake as it bakes would result in instant annihilation in a burst of intense radiation. This came to be known as the "ultraviolet catastrophe," and was obviously incorrect. But explaining why the Rayleigh–Jeans calculations were wrong required bold theoretical physics, the likes of which had never been attempted.

Quantum beginnings

At the same time as the Rayleigh–Jeans findings were announced, Max Planck was working on his own theory of blackbody radiation in Berlin. In October 1900, he proposed an explanation for the blackbody curve that agreed with all the known experimental measurements, yet went beyond the framework of classical physics. His solution was radical, and involved an entirely new way of looking at the world.

Planck found that the ultraviolet catastrophe could be averted by understanding a blackbody's energy emission as occurring not in continuous waves, but in discrete packets, which he called "quanta." On December 19, 1900, Planck presented his findings to a meeting of the German Physical Society in Berlin. This date is generally accepted as marking the birth of quantum mechanics, and a new era in physics. ∎

These laws of light … may have been observed before, but I think they are now for the first time connected with a theory of radiation.
Gustav Kirchhoff

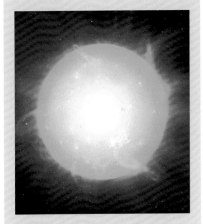

Stellar temperatures

It is possible to calculate the surface temperature of a blackbody by measuring the energy that it emits at specific wavelengths. Since stars, including the sun, produce light spectrums that closely approximate a blackbody spectrum, it is possible to calculate the temperature of a distant star.

A blackbody's temperature is given by the following formula: $T = 2898/\lambda_{MAX}$ where T = the temperature of the blackbody (measured in degrees Kelvin), and λ_{MAX} = the wavelength (λ, measured in micrometers) of the maximum emission of the blackbody.

This formula can be used to calculate the temperature of a star's photosphere—the light-emitting surface—using the wavelength at which it emits the maximum amount of light. Cool stars emit more light from the red and orange end of the spectrum, whereas hotter stars appear blue. For instance, blue supergiants—as depicted in the above artist's impression—are a class of star that can be up to eight times hotter than the sun.

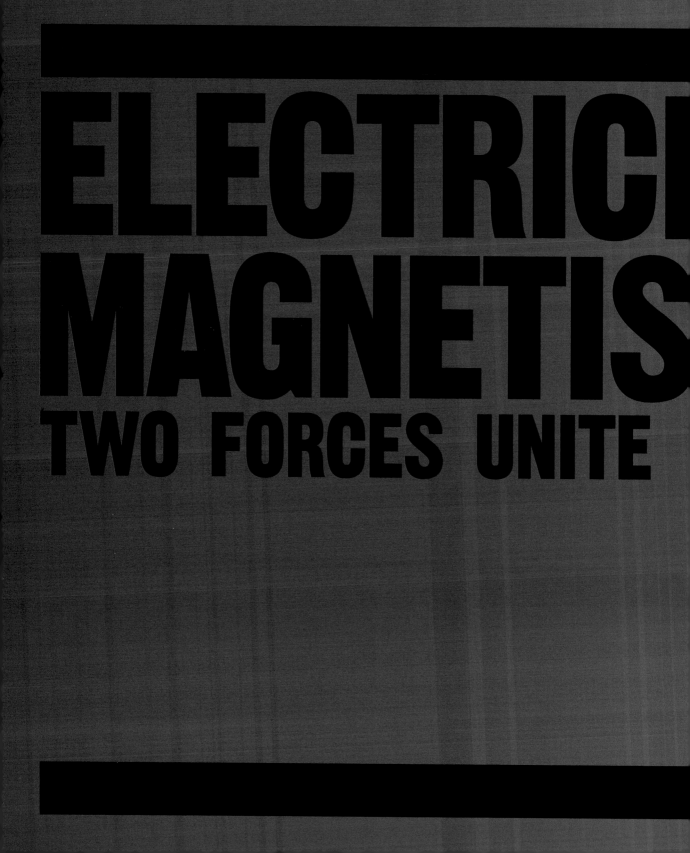

ELECTRICI MAGNETIS
TWO FORCES UNITE

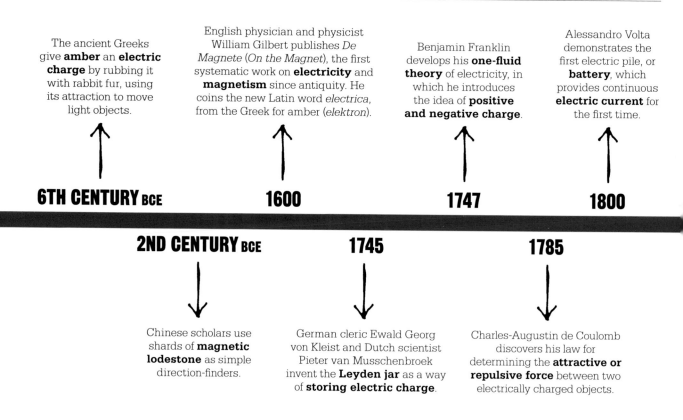

The ancient Greeks give **amber** an **electric charge** by rubbing it with rabbit fur, using its attraction to move light objects.

English physician and physicist William Gilbert publishes *De Magnete (On the Magnet)*, the first systematic work on **electricity** and **magnetism** since antiquity. He coins the new Latin word *electrica*, from the Greek for amber (*elektron*).

Benjamin Franklin develops his **one-fluid theory** of electricity, in which he introduces the idea of **positive and negative charge**.

Alessandro Volta demonstrates the first electric pile, or **battery**, which provides continuous **electric current** for the first time.

6TH CENTURY BCE — **1600** — **1747** — **1800**

2ND CENTURY BCE — **1745** — **1785**

Chinese scholars use shards of **magnetic lodestone** as simple direction-finders.

German cleric Ewald Georg von Kleist and Dutch scientist Pieter van Musschenbroek invent the **Leyden jar** as a way of **storing electric charge**.

Charles-Augustin de Coulomb discovers his law for determining the **attractive or repulsive force** between two electrically charged objects.

In ancient Greece, scholars noticed that some stones from Magnesia, in modern-day Asia Minor, behaved strangely when they were placed near certain metals and iron-rich stones. The stones pulled metals toward them through an unseen attraction. When placed in a particular way, two of these stones were seen to attract each other, but they pushed each other apart when one was flipped around.

Ancient Greek scholars also noted a similar, but subtly different, behavior when amber (fossilized tree sap) was rubbed with animal fur. After rubbing for a little time, the amber would gain a strange ability to make light objects, such as feathers, ground pepper, or hair, dance. The mathematician Thales of Miletus argued that the unseen force producing these phenomena was evidence that the stones and amber had a soul.

The strange forces exhibited by the stones from Magnesia are today known as magnetism, taking its name from the region where it was first found. Forces exhibited by the amber were given the name electricity from the ancient Greek word for amber, *elektron*. Chinese scholars and, later, mariners and other travelers used small shards of magnesia stone placed in water as an early version of the compass, since the stones aligned themselves north–south.

Attraction and repulsion

No new use was found for electricity until the 18th century. By this time, it had been discovered that rubbing other materials together produced behavior similar to that of amber and fur. For example, glass rubbed with silk also made small objects dance. When both amber and glass were rubbed, they pulled together in an attraction, while two lumps of amber or two of glass would push apart. These were identified as two distinct electricities—vitreous electricity for the glass and resinous for the amber.

American polymath Benjamin Franklin chose to identify these two types of electricity with positive or negative numbers, with a magnitude that became known as electric charge. While hiding from the revolutionists to keep his head attached to his body, French physicist and engineer Charles-Augustin de Coulomb carried out a series of experiments.

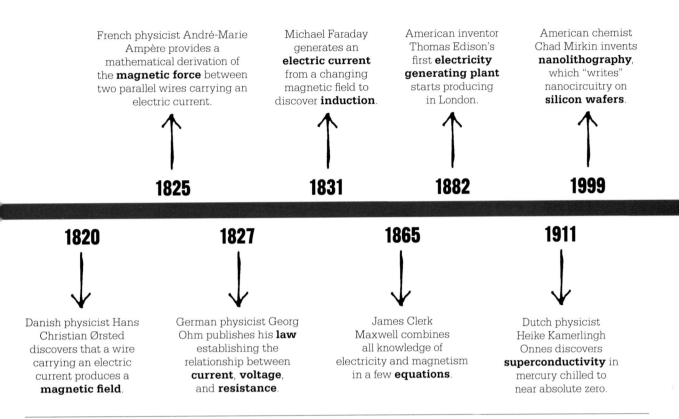

French physicist André-Marie Ampère provides a mathematical derivation of the **magnetic force** between two parallel wires carrying an electric current.

Michael Faraday generates an **electric current** from a changing magnetic field to discover **induction**.

American inventor Thomas Edison's first **electricity generating plant** starts producing in London.

American chemist Chad Mirkin invents **nanolithography**, which "writes" nanocircuitry on **silicon wafers**.

1825 **1831** **1882** **1999**

1820 **1827** **1865** **1911**

Danish physicist Hans Christian Ørsted discovers that a wire carrying an electric current produces a **magnetic field**.

German physicist Georg Ohm publishes his **law** establishing the relationship between **current**, **voltage**, and **resistance**.

James Clerk Maxwell combines all knowledge of electricity and magnetism in a few **equations**.

Dutch physicist Heike Kamerlingh Onnes discovers **superconductivity** in mercury chilled to near absolute zero.

He discovered that the force of attraction or repulsion between electric objects grew weaker as the distance between them increased.

Electricity was also observed to flow. Small sparks would leap from an electrically charged object to one without charge in an attempt to balance out, or neutralize charge. If one object was at a different charge from those surrounding it, then that object was said to have different potential. Any difference in potential can induce a flow of electricity called a current. Electric currents were found to flow easily through most metals, while organic materials seemed far less able to permit a current to flow.

In 1800, Italian physicist Alessandro Volta noticed that differences in the chemical reactivity of metals could lead to an electric potential difference. We now know that chemical reactions and the flow of electricity through a metal are inextricably linked because both result from the movement of subatomic electrons.

A combined force
In mid-19th-century Britain, Michael Faraday and James Clerk Maxwell produced the link between the two apparently distinct forces of electricity and magnetism, giving rise to the combined force of electromagnetism. Faraday created the idea of fields, lines of influence that stretch out from an electric charge or magnet, showing the region where electric and magnetic forces are felt. He also showed that moving magnetic fields can induce an electric current and that electric currents produce magnetic fields.

Maxwell elegantly accommodated Faraday's findings, and those of earlier scientists, in just four equations. In doing so, he discovered that light was a disturbance in electric and magnetic fields. Faraday conducted experiments that demonstrated this, showing that magnetic fields affect the behavior of light.

Physicists' understanding of electromagnetism has revolutionized the modern world through technologies that have been developed to use electricity and magnetism in new and innovative ways. Research into electromagnetism also opened up previously unthought-of areas of study, striking at the heart of fundamental science—guiding us deep inside the atom and far out into the cosmos. ∎

WONDROUS FORCES

MAGNETISM

IN CONTEXT

KEY FIGURE
William Gilbert (1544–1603)

BEFORE
6th century BCE Thales of Miletus states iron is attracted to the "soul" of lodestone.

1086 Astronomer Shen Kuo (Meng Xi Weng) describes a magnetic needle compass.

1269 French scholar Petrus Peregrinus describes magnetic poles and laws of attraction and repulsion.

AFTER
1820 Hans Christian Ørsted discovers that an electric current flowing in a wire deflects a magnetic needle.

1831 Michael Faraday describes invisible "lines of force" around a magnet.

1906 French physicist Pierre-Ernest Weiss advances the theory of magnetic domains to explain ferromagnetism.

The striking properties of the rare, naturally occurring lodestone—an ore of iron called magnetite—fascinated ancient cultures in Greece and China. Early writings from these civilizations describe how lodestone attracts iron, affecting it over a distance without any visible mechanism.

By the 11th century, the Chinese had discovered that a lodestone would orient itself north–south if allowed to move freely (for example, when placed in a vessel floating in a bowl of water). Moreover, an iron needle rubbed with lodestone would inherit its properties and could be used to make a compass. The maritime compass made it possible for ships to navigate away from the shore, and the instrument reached Europe via Chinese seafarers. By the 16th century, the compass was driving the expansion of European empires, as well as being used in land surveying and mining.

Despite centuries of practical application, the underlying physical mechanism of magnetism was poorly understood. The first systematic account on magnetism was Petrus Peregrinus' 13th-century

A compass needle **points approximately north**, but also shows **declination** (deviation away from true north) and **inclination** (tilting toward or away from Earth's surface).

A compass needle shows **exactly the same behavior** when moved over the surface of a **spherical magnetic rock**, or lodestone.

Earth is a giant magnet.

See also: Making magnets 134–135 ▪ The motor effect 136–137 ▪ Induction and the generator effect 138–141 ▪ Magnetic monopoles 159

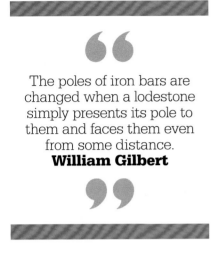

> The poles of iron bars are changed when a lodestone simply presents its pole to them and faces them even from some distance.
> **William Gilbert**

text in which he described polarity (the existence of magnetic north and south poles in pairs). He also found that pieces of a lodestone "inherited" magnetic properties.

Gilbert's little Earths

It was English astronomer William Gilbert's groundbreaking work that dispelled long-held superstitions about magnetism. Gilbert's key innovation was to simulate nature in the laboratory.

Using spheres of lodestone that he called *terella* (Latin for little Earths), he showed that a compass needle was deflected in the same way over different parts of the sphere as it was over the corresponding regions of our planet. He concluded that Earth was itself a giant magnet and published his findings in the groundbreaking *De Magnete* (*On the Magnet*) in 1600.

A new understanding

The magnetism exhibited by magnetite is called ferromagnetism, a property that is also seen with iron, cobalt, nickel, and their alloys.

When a magnet is brought close to an object made from ferromagnetic material, the object itself becomes magnetic. The approaching pole of the magnet induces an opposite pole in the near side of the ferromagnetic object and attracts it. Depending on its exact composition and interaction with the magnet, the ferromagnetic object may become permanently magnetized, retaining this property after the original magnet is removed.

Once physicists connected electricity and magnetism and developed an understanding of atomic structure in the 19th century, a reasonable theory of ferromagnetism began to emerge.

The idea is that the movement of electrons in an atom makes each atom into a miniature magnetic dipole (with north and south poles). In ferromagnetic materials such as iron, groups of neighboring atoms align to form regions called magnetic domains.

These domains are usually arranged in closed loops, but when a piece of iron is magnetized, the domains line up along a single axis, creating north and south poles at opposite ends of the piece. ▪

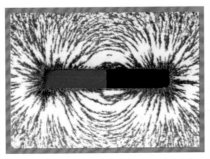

A simple magnet, with a north and south pole, creates lines of force around it. Iron filings scattered around the magnet line up along these lines of force, which is stronger at each pole.

William Gilbert

William Gilbert was born into a prosperous English family in 1544. After graduating from Cambridge, he established himself as a leading physician in London. He met prominent naval officers, including Francis Drake, and cultivated connections at the court of Elizabeth I. Through his connections and on visits to the docks, Gilbert learned of the behavior of compasses at sea and acquired specimens of lodestones. His work with these magnetic rocks informed his masterwork.

In 1600, Gilbert was elected president of the Royal Society of Physicians and appointed personal physician to Elizabeth I. He also invented the electroscope to detect electric charge and distinguished the force of static electricity from that of magnetism. He died in 1603, possibly succumbing to bubonic plague.

Key works

1600 *De Magnete* (*On the Magnet*)
1651 *De Mundo nostro Sublunari Philosophia Nova* (*The New Philosophy*)

THE ATTRACTION OF ELECTRICITY

ELECTRIC CHARGE

For millennia, people have observed electrical effects in nature—for example, lightning strikes, the shocks delivered by electric rays (torpedo fish), and the attractive forces when certain materials touch or are rubbed against each other.

However, it is only in the last few hundred years that we have begun to understand these effects as manifestations of the same underlying phenomenon—electricity. More precisely, these are electrostatic effects, due to electric forces arising from static (stationary) electric charges. Electric current effects, on the other hand, are caused by moving charges.

See also: Laws of gravity 46–51 ▪ Electric potential 128–129 ▪ Electric current and resistance 130–133 ▪ Bioelectricity 156 ▪ Subatomic particles 242–243

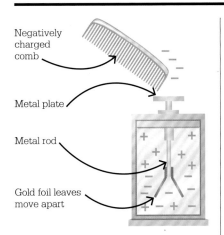

Negatively charged comb

Metal plate

Metal rod

Gold foil leaves move apart

The gold-leaf electroscope detects static electricity through the principle of like charges repelling. When a negatively charged comb is brought near the metal plate, electrons (which carry negative charge) are repelled toward the electroscope's gold foil leaves, which causes the leaves to separate.

The concept of electric charge, and a mathematical description of the forces between charges, emerged in the 18th century. Previously, the ancient Greeks had noted that when a piece of amber (*elektron*), was rubbed with wool, it would attract light objects such as feathers.

In his 1600 book *De Magnete* (*On the Magnet*), William Gilbert named this effect "electricus" and discussed his experiments with the instrument that he had devised to detect the force—the versorium. Gilbert saw that the force had an instantaneous effect over distance and suggested that it must be carried by a fast-moving electrical "fluid" released by the rubbed amber, rather than a slowly diffusing "effluvium," as previously thought.

In 1733, French chemist Charles François du Fay observed that electrical forces could be repulsive as well as attractive, and postulated that there were two types of electrical fluid—vitreous and resinous. Like fluids (such as two vitreous fluids) repelled each other, and unlike fluids attracted.

This theory was simplified by American statesman and polymath Benjamin Franklin in 1747, when he proposed that there was only one type of electrical fluid and different objects could have a surplus or deficiency of this fluid. He labeled a surplus of fluid (charge, in today's terms) to be positive, and a deficit negative, and proposed that the total amount of fluid in the universe was conserved (constant). He also designed (and possibly conducted) an experiment to show that lightning was a flow of electrical fluid, by flying a kite in a storm. Charge is still labeled as positive or negative, although this is simply convention: there is no excess "fluid" in a proton that makes it positive, and nothing lacking in an electron that makes it negative.

Coulomb's law

During the 18th century, scientists suggested mathematical laws that might govern the strength of the »

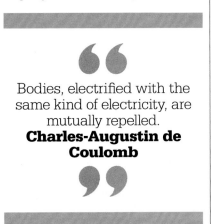

> Bodies, electrified with the same kind of electricity, are mutually repelled.
> **Charles-Augustin de Coulomb**

Electrostatic discharge

An electrostatic discharge occurs when the electric charge carriers (typically electrons) in a charged body or region are rapidly and violently conducted away from it. Lightning is a particularly powerful form of electrostatic discharge that occurs when so much charge has built up between regions of the atmosphere that the intervening space becomes ionized (electrons separate from their atoms) and can conduct a current.

The current is made visible by the electrons heating up the air so much that it emits light. Ionization occurs over short distances, which is why lightning bolts appear forked, changing direction every few feet. Electrostatic discharge happens first at sharp edges—which is why your hair stands on end when statically charged (the hair tips repel each other) and why lightning conductors and static discharge devices on aircraft wings are shaped like spikes.

electrical force, modeled on the inverse square law of gravitation that Newton had established in his hugely influential *Principia* of 1687.

In 1785, French engineer Charles-Augustin de Coulomb developed a torsion balance that was sensitive enough to measure the electrical force between charges. The apparatus consisted of a series of metal spheres connected by a rod, needle, and silk thread. When Coulomb held a charged object next to the external sphere, the charge transferred to an internal sphere and needle. The needle, suspended from a silk thread, moved away from the charged sphere and produced a twisting (torsion) in the silk thread. The degree of torsion could be measured from a scale.

Coulomb published a series of papers detailing his experiments and establishing that the force between two stationary, charged bodies was inversely proportional to the distance between them. He also assumed, but did not prove, that the force was proportional to the product of the charges on the bodies. Today, this law is called Coulomb's law.

The sciences are monuments devoted to the public good; each citizen owes to them a tribute proportional to his talents.
Charles-Augustin de Coulomb

Two **electrically charged** bodies experience a small mutual force.

→

A **torsion balance** can **measure the force** by how much it twists a silk thread.

↓

When the **distance between the charged bodies is doubled**, the amount of torsion (twisting) is **reduced to a quarter** of the original.

↓

The **electrical force** between charged bodies **varies inversely** with the **square of the distance** between them.

Coulomb established that when electric charges attract or repel each other, there is a relationship between the strength of attraction or repulsion and distance. However, it would be more than a century before scientists came closer to understanding the exact nature of electric charge.

Finding the charge carrier
In the 1830s, British scientist Michael Faraday carried out experiments with electrolysis (using electricity to drive chemical reactions) and found that he needed a specific amount of electricity to create a specific amount of a compound or element. Although he was not convinced that matter was made up of atoms (indivisible parts), this result suggested that electricity, at least, might come in "packets." In 1874, Irish physicist George Stoney developed the idea that there was an indivisible packet or unit of electric charge—that is, charge

was quantized—and in 1891, he suggested a name for this unit—the electron.

In 1897, British physicist J.J. Thomson demonstrated that cathode rays—the glowing "rays" of electricity that could be made to travel between two charged plates in a sealed glass tube containing very little gas (a near vacuum)—were in fact made of electrically charged particles. By applying electric and magnetic forces of known strength to the cathode rays, Thomson could bend them by a measurable amount. He could then calculate how much charge a particle must carry per unit mass.

Thomson also deduced that these charge carriers were very much lighter than the smallest atom. They were common in all matter because the behavior of the rays did not vary even if he used plates of different metals. This subatomic particle—the first to be discovered—was then given Stoney's name for the basic unit of charge: the electron. The charge on an electron was assigned as negative. The discovery of the positive charge carrier, the proton, would follow a few years later.

The charge on an electron
Although Thomson had calculated the charge-to-mass ratio for the electron, neither the charge nor the mass were known. From 1909 to 1913, American physicist Robert Millikan performed a series of experiments to find these values. Using special apparatus, he measured the electric field needed to keep a charged droplet of oil suspended in air. From the radius of a droplet, he could work out its weight. When the droplet was still, the upward electric force on it balanced the downward gravitational force, and the charge on the droplet could be calculated.

By repeating the experiment many times, Millikan discovered that droplets all carried charges that were whole-number multiples of a particular smallest number. Millikan reasoned that this smallest number must be the charge on a single electron—called the elementary charge, e—which he computed to be -1.6×10^{-19} C (coulombs), very close to today's accepted value. Nearly a century after Franklin suggested that the total amount of electric "fluid" is constant, Faraday conducted experiments that suggested that charge is conserved—the total amount of charge in the universe remains the same.

Balance of charge
This principle of conservation of charge is a fundamental one in modern physics, although there are circumstances—in high-energy collisions between particles in particle accelerators, for example— where charge is created through a neutral particle splitting into negative and positive particles. However, in this case, the net charge is constant—equal numbers of positive and negative particles carrying equal amounts of negative and positive charge are created.

This balance between charges isn't surprising given the strength of the electric force. The human body has no net charge and contains equal amounts of positive and negative charge, but a hypothetical imbalance of just one percent would make it feel forces of devastating power. Our understanding of electric charge and carriers has not changed dramatically since the discovery of the electron and proton. We also know of other charge carriers, such as the positively charged positron and the negatively charged antiproton that make up an exotic form of matter called antimatter.

In modern terminology, electric charge is a fundamental property of matter that occurs everywhere— in lightning, inside the bodies of electric rays, in stars, and inside us. Static charges create electric fields around them—regions in which other electric charges "feel" a force. Moving charges create electric and magnetic fields, and through the subtle interplay between these fields, give rise to electromagnetic radiation, or light.

Since the development of quantum mechanics and particle physics in the 20th century, we now understand that many of the most familiar properties of matter are fundamentally connected to electromagnetism. Indeed, the electromagnetic force is one of the fundamental forces of nature. ■

Charles-Augustin de Coulomb

Born in 1736 into a relatively wealthy French family, Coulomb graduated as a military engineer. He spent nine years in the French colony of Martinique in the West Indies but was dogged by illness and returned to France in 1773.

While building a wooden fort in Rochefort, southwestern France, he conducted pioneering work on friction and won the Grand Prize of the Académie des Sciences in 1781. He then moved to Paris and devoted most of his time to research. As well as developing the torsion balance, Coulomb wrote memoirs in which he formulated the inverse square law that bears his name. He also consulted on civil engineering projects and oversaw the establishment of secondary schools. He died in Paris in 1806. The SI unit of charge, the coulomb, is named in his honor.

Key works

1784 *Theoretical Research and Experiments on the Torsion Force and Elasticity of Metal Wires*
1785 *Memoirs on Electricity and Magnetism*

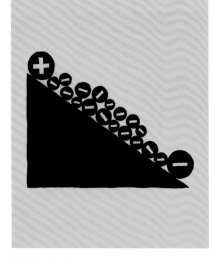

POTENTIAL ENERGY BECOMES PALPABLE MOTION
ELECTRIC POTENTIAL

IN CONTEXT

KEY FIGURE
Alessandro Volta
(1745–1827)

BEFORE
1745 Pieter van Musschenbroek and E. Georg von Kleist invent the Leyden jar, the first practical device that can store electric charge.

1780 Luigi Galvani observes "animal electricity."

AFTER
1813 French mathematician and physicist Siméon-Denis Poisson establishes a general equation for potential.

1828 British mathematician George Green develops Poisson's ideas and introduces the term "potential."

1834 Michael Faraday explains the chemical basis of the voltaic (galvanic) cell.

1836 British chemist John Daniell invents the Daniell cell.

Through the 17th and 18th centuries, an increasing number of investigators began to apply themselves to studying electricity, but it remained an ephemeral phenomenon.

The Leyden jar, invented in 1745 by two Dutch and German chemists working independently, allowed electric charge to be accumulated and stored until it was needed. However, the jar would discharge (unload the charge) rapidly as a spark. It was not until the late 18th century, when Italian chemist Alessandro Volta developed the first electrochemical cell, that scientists had a supply of a moderate flow of electrical charge over time: a current.

Energy and potential
Both the sudden discharge of the Leyden jar and the extended discharge (current) from a battery are caused by a difference in what is called "electric potential" between each device and its surroundings.

Today, electric potential is considered to be a property of the electric field that exists around electric charges. The electric potential at a single point is always

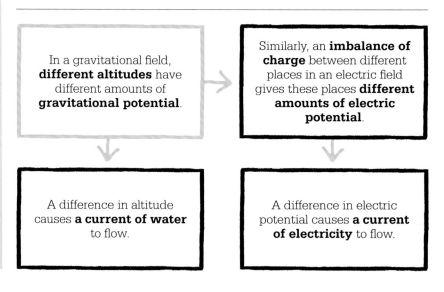

In a gravitational field, **different altitudes** have different amounts of **gravitational potential**.

Similarly, an **imbalance of charge** between different places in an electric field gives these places **different amounts of electric potential**.

A difference in altitude causes **a current of water** to flow.

A difference in electric potential causes **a current of electricity** to flow.

See also: Kinetic energy and potential energy 54 ■ Electric charge 124–127 ■ Electric current and resistance 130–133 ■ Bioelectricity 156

measured relative to that at another point. An imbalance of charge between two points gives rise to a potential difference between them. Potential difference is measured in volts (V) in honor of Volta, and is informally referred to as "voltage." Volta's work paved the way for fundamental breakthroughs in the understanding of electricity.

From animal electricity to batteries

In 1780, Italian physician Luigi Galvani had noticed that when he touched a (dead) frog's leg with two dissimilar metals, or applied an electrical spark to it, the leg twitched. He presumed that the source of this motion was the frog's body and deduced it contained an electrical fluid. Volta performed similar experiments, but without animals, eventually coming up with the theory that the dissimilarity of the metals in the circuit was the source of the electricity.

Volta's simple electrochemical cell consists of two metal pieces (electrodes), separated by a salt solution (an electrolyte). Where each metal meets the electrolyte a chemical reaction takes place, creating "charge carriers" called ions (atoms that have gained or lost electrons, and so are negatively or positively charged). Oppositely charged ions appear at the two electrodes. Because unlike charges attract each other, separating positive and negative charges requires energy (just as holding

The voltaic pile consists of a series of metal disks separated by brine-soaked cloth. A chemical reaction between them creates a potential difference, which drives an electric current.

apart the opposite poles of two magnets does). This energy comes from chemical reactions in the cell. When the cell is connected to an external circuit, the energy that was "stored" in the potential difference appears as the electrical energy that drives the current around the circuit.

Volta made his battery by connecting individual cells made of silver and zinc disks separated by brine-soaked cloth. He demonstrated the resulting voltaic pile to the Royal Society in London in 1800. Voltaic cells only supply current for a short time before the chemical reactions stop. Later developments such as the Daniell cell and the modern dry zinc-carbon or alkaline cell have greatly improved longevity. Like alkaline cells, voltaic cells cannot be recharged once exhausted, and are termed primary cells. Secondary cells, like those found in the lithium polymer batteries of cell phones, can be recharged by applying a potential difference across the electrodes to reverse the chemical reaction. ■

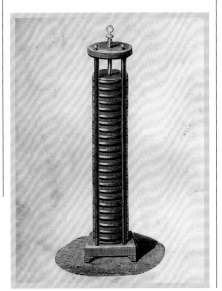

Alessandro Volta

Alessandro Volta was born into an aristocratic family in 1745 in Como, Italy. Volta was just seven years old when his father died. His relatives steered his education toward the Church, but he undertook his own studies in electricity and communicated his ideas to prominent scientists.

Following Volta's early publications on electricity, he was appointed to teach in Como in 1774. The following year, he developed the electrophorus (an instrument for generating electric charge), and in 1776 he discovered methane. Volta became professor of physics at Pavia in 1779. There he engaged in friendly rivalry with Luigi Galvani in Bologna. Volta's doubts about Galvani's ideas of "animal electricity" led to his invention of the voltaic pile. Honored by both Napoleon and the emperor of Austria, Volta was a wealthy man in his later years, and died in 1827.

Key work

1769 *On the Forces of Attraction of Electric Fire*

A TAX ON ELECTRICAL ENERGY

ELECTRIC CURRENT AND RESISTANCE

IN CONTEXT

KEY FIGURE
Georg Simon Ohm
(1789–1854)

BEFORE
1775 Henry Cavendish
anticipates a relationship
between potential difference
and current.

1800 Alessandro Volta invents
the first source of continuous
current, the voltaic pile.

AFTER
1840 British physicist James
Joule studies how resistance
converts electrical energy
into heat.

1845 Gustav Kirchhoff, a
German physicist, proposes
rules that govern current and
potential difference in circuits.

1911 Dutch physicist Heike
Kamerlingh Onnes discovers
superconductivity.

A s early as 1600, scientists
had distinguished
"electric" substances,
such as amber and glass, from
"nonelectric" substances, such as
metals, on the basis that only the
former could hold a charge. In 1729,
British astronomer Stephen Gray
brought a new perspective to
this division of substances by
recognizing that electricity (then,
still thought to be a kind of fluid)
could travel from one electric
substance to another via
a nonelectric substance.

By considering whether
electricity could flow through a
substance rather than whether it
could be stored, Gray established

See also: Electric charge 124–127 ▪ Electric potential 128–129 ▪ Making magnets 134–135 ▪ The motor effect 136–137 ▪ Induction and the generator effect 138–141 ▪ Electromagnetic waves 192–195 ▪ Subatomic particles 242–243

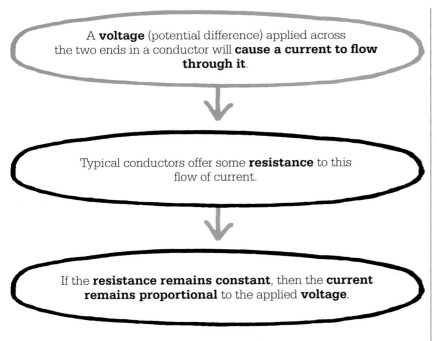

A **voltage** (potential difference) applied across the two ends in a conductor will **cause a current to flow through it**.

⬇

Typical conductors offer some **resistance** to this flow of current.

⬇

If the **resistance remains constant**, then the **current remains proportional** to the applied **voltage**.

the modern distinction between conductors and insulators. It was Alessandro Volta's invention of the electrochemical cell (battery) in 1800 that finally gave scientists a source of continuously flowing electric charge—an electric current—to study conductance and resistance.

Conducting and insulating
As Volta's invention demonstrated, an electric current can only flow if it has a conductive material through which to travel. Metals are generally very good conductors of electricity; ceramics are generally good insulators; other substances, such as a salt solution, water, or graphite, fall somewhere between.

The carriers of electric charge in metals are electrons, which were discovered a century later. Electrons in atoms are found in orbitals at different distances from the nucleus, corresponding to

different energy levels. In metals, there are relatively few electrons in the outermost orbitals, and these electrons easily become "delocalized," moving freely and randomly throughout the metal. Gold, silver, and copper are excellent conductors because their atoms have only one outermost electron, which is easily delocalized. Electrolytes (solutions such as salt water) contain charged ions that can move around fairly easily. In contrast, in insulators, the charge carriers are localized (bound to particular atoms).

Flow of charge
The modern description of current electricity emerged in the late 19th century, when current was finally understood to be a flow of positive or negatively charged particles. In order for a current to flow between two points, they must be connected by a conductor such as a metal

wire, and the points must be at different electric potentials (having an imbalance in charge between the two points). Current flows from higher potential to lower (by scientific convention, from positive to negative).

In metals, the charge carriers are negatively charged, so a current flowing in a metal wire from A to B is equivalent to negatively charged electrons flowing in the opposite direction (toward the higher, or relatively positive potential). Charge carriers in other materials may be positive. For example, salt water contains positively charged sodium ions (among others)—and their movement would be in the same direction as the flow of current. Current is measured in units called amps, short for amperes. A current of 1 amp means about 6 trillion electrons are moving past a particular point every second.

In a copper wire, delocalized electrons move around randomly at more than 1,000 km per second. Because they are moving in random directions, the net (average) velocity is zero, and so there is no net »

The beauty of electricity … is not that the power is mysterious and unexpected … but that it is under law.
Michael Faraday

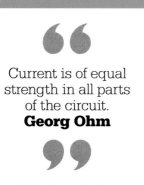

> Current is of equal strength in all parts of the circuit.
> **Georg Ohm**

current. Applying a potential difference across the ends of the wire creates an electric field. This field causes the free, delocalized electrons to experience a net force toward the end at high potential (because they are negatively charged), accelerating them so that they drift through the wire. This drift velocity constitutes the current, and it is very small—typically a fraction of a millimeter per second in a wire.

Although charge carriers in a wire move relatively slowly, they interact with each other via an electric field (due to their charge) and a magnetic field (created by their movement). This interaction is an electromagnetic wave, which travels extremely quickly. The copper wire acts as a "waveguide," and electromagnetic energy travels along the wire at (typically) 80–90 percent of what would be its speed in a vacuum—hence electrons throughout the circuit all begin to drift almost instantaneously, and a current is established.

Electrical resistance

The property of an object to oppose a current is called its resistance. Resistance (and its opposite, conductance) depends not only on an object's intrinsic properties (how the particles that make it up are arranged, and in particular whether the charge carriers are delocalized), but also on extrinsic factors such as its shape, and whether it is subject to high temperature or pressure. A thicker copper wire, for example, is a better conductor than a thinner one of the same length. Such factors are comparable with hydraulic systems. For example, it is harder to push water through a narrow pipe than through a wide one.

Temperature also plays a role in a material's resistance. The resistance of many metals decreases with a decrease in temperature. Some materials exhibit zero resistance when cooled below a specific, very low temperature—a property known as superconductivity.

The resistance of a conductor may vary with the potential difference (also known as the voltage) applied or the current flowing through it. For example, the resistance of a tungsten filament in an incandescent bulb increases with current. The resistance of many conductors remains constant as the current or voltage varies. Such conductors are known as ohmic conductors, named after Georg Ohm, who formulated a law that relates voltage to current.

Ohm's law

The law Ohm established is that current flowing through a conductor is proportional to the voltage across it. Dividing the voltage (measured in volts) by the current (measured in amps) gives a constant number, which is the resistance of the conductor (measured in ohms).

A copper wire is an ohmic conductor—it obeys Ohm's law as long as its temperature does not

Georg Simon Ohm

Ohm was born in Erlangen (now Germany) in 1789. His father, a locksmith, taught him mathematics and science. He was admitted to the University of Erlangen and met mathematician Karl Christian von Langsdorff. In 1806, Ohm's father, worried that his son was wasting his talents, sent him to Switzerland, where he taught mathematics and continued his own studies.

In 1811, Ohm returned to Erlangen and obtained a doctorate. He moved to Cologne to teach in 1817. After hearing of Hans Christian Ørsted's discoveries, he began to experiment with electricity. At first, his publications were not well received, partly because of his mathematical approach, but also because of squabbles over his scientific errors. Later, however, he was awarded the Royal Society's Copley Medal in 1841, and was made Chair of Physics at the University of Munich in 1852, two years before his death.

Key work

1827 *The Galvanic Circuit Investigated Mathematically*

Ohm's law encapsulates the link between voltage (potential difference), current, and resistance. Its formula (see right) can be used to calculate how much current (in amps) passes through a component depending on the voltage (V) of the power source and the resistance (measured in ohms) of items in the circuit.

$$\text{Current (A)} = \frac{\text{Voltage (V)}}{\text{Resistance (Ω)}}$$

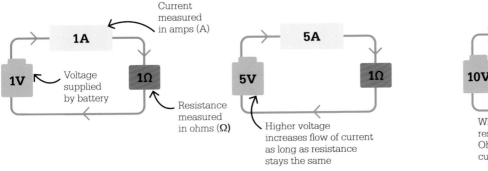

1A

Current measured in amps (A)

1V — Voltage supplied by battery

1Ω

Resistance measured in ohms (Ω)

5A

5V

1Ω

Higher voltage increases flow of current as long as resistance stays the same

5A

10V

1Ω

1Ω

When voltage and resistance are both doubled, Ohm's law means that current stays the same

---

change dramatically. The resistance of ohmic conductors depends on physical factors such as temperature and not on the applied potential difference or the current flowing.

Ohm arrived at his law through a combination of experiments and mathematical theory. In some of his experiments, he made circuits using electrochemical cells to supply the voltage and a torsion balance to measure current. He used wire of different lengths and thicknesses to carry the electricity and noted the difference in current and resistance that occurred as a result. His theoretical work was based on geometrical methods to analyze electrical conductors and circuits.

Ohm also compared the flow of current with Fourier's theory of heat conduction (named after French mathematician Joseph Fourier). In this theory, heat energy is transferred from one particle to

Filament (incandescent) light bulbs work by providing high resistance to current electricity because the wire (the filament) is very narrow. This resistance causes electrical energy to be converted to heat and light.

the next in the direction of a temperature gradient. In describing the flow of electrical current, the potential difference across an electrical conductor is similar to the temperature difference across two ends of a thermal conductor.

Ohm's law is not a universal law, however, and does not hold for all conductors, or under all circumstances. So-called non-ohmic materials include diodes and the tungsten filament in incandescent bulbs. In such cases,

the resistance depends on the applied potential difference (or current flowing).

Joule heating

The higher the current in a metal conductor, the more collisions occur between the electrons and the ionic lattice. These collisions result in the kinetic energy of the electrons being converted into heat. The Joule-Lenz law (named in part after James Prescott Joule who discovered that heat could be generated by electricity in 1840) states that the amount of heat generated by a conductor carrying a current is proportional to its resistance, multiplied by the square of the current.

Joule heating (also called ohmic heating, or resistive heating) has many practical uses. It is responsible for the glow of incandescent lamp filaments, for example. However, Joule heating can also be a significant problem. In electricity transmission grids, for example, it causes major energy losses. These losses are minimized by keeping the current in the grid relatively low, but the potential difference (voltage) relatively high. ∎

EACH METAL HAS A CERTAIN POWER

MAKING MAGNETS

IN CONTEXT

KEY FIGURE
Hans Christian Ørsted
(1777–1851)

BEFORE
1600 English astronomer
William Gilbert realizes that
Earth is a giant magnet.

1800 Alessandro Volta
makes the first battery,
creating a continuous flow
of electric current for the
first time.

AFTER
1820 André-Marie Ampère
develops a mathematical
theory of electromagnetism.

1821 Michael Faraday creates
the first electric motor and
shows electromagnetic
rotation in action.

1876 Alexander Graham Bell,
a Scottish–American physicist,
invents a telephone that uses
electromagnets and a
permanent horseshoe magnet
to transmit sound vibrations.

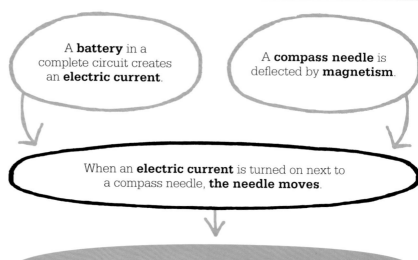

A **battery** in a complete circuit creates an **electric current**.

A **compass needle** is deflected by **magnetism**.

When an **electric current** is turned on next to a compass needle, **the needle moves**.

Electricity produces a magnetic field.

By the end of the 18th century, many magnetic and electrical phenomena had been noticed by scientists. However, most believed electricity and magnetism to be totally distinct forces. It is now known that flowing electrons produce a magnetic field and that spinning magnets cause an electric current to flow in a complete circuit. This relationship between electricity and magnetism is integral to nearly every modern electrical appliance, from headphones to cars, but it was discovered purely by chance.

Ørsted's chance discovery
Alessandro Volta's invention of the voltaic pile (an early battery) in 1800 had already opened up a whole new field of scientific study. For the first time, physicists could produce a steady electric current. In 1820, Danish physicist Hans Christian Ørsted was delivering a

See also: Magnetism 122–123 ▪ Electric charge 124–127 ▪ Induction and the generator effect 138–141 ▪ Force fields and Maxwell's equations 142–147

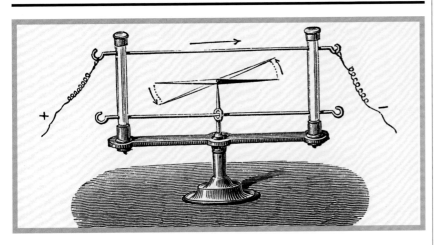

lecture to students at the University of Copenhagen. He noticed that a compass needle was deflected away from magnetic north when he switched an electric current on and off. This was the first time a link had been shown between an electric current and a magnetic field. Ørsted carried out more experiments and found that a current produces a concentric magnetic field around the wire through which it is flowing.

Creating electromagnets

Four years after Ørsted's discovery, British inventor William Sturgeon made a magnet from a horseshoe-shaped piece of iron and wound it with 18 loops of copper wire. He passed an electric current through the wire, magnetizing the horseshoe enough for it to attract other pieces of iron.

In the 1830s, American scientist Joseph Henry developed the electromagnet further, insulating copper wire with silk thread and winding multiple layers around iron cores. One of Henry's magnets lifted a weight of 2,064 lb (936 kg). By the 1850s, small electromagnets were being widely used in receivers in the

By passing an electric current through a wire, Ørsted created a magnetic field around it. This deflected a compass needle.

US's electrical telegraph network. The advantage of an electromagnet is that its magnetic field can be controlled. Whereas a regular magnet's strength is constant, the strength of an electromagnet can be varied by changing the current flowing through its wire coil (called a solenoid). However, electromagnets only work with a continuous supply of electrical energy. ▪

> The agreement of this law with nature will be better seen by the repetition of experiments than by a long explanation.
> **Hans Christian Ørsted**

Hans Christian Ørsted

Born in Rudkøbing, Denmark, in 1777, Ørsted was mostly home-schooled before attending the University of Copenhagen in 1793. After gaining a doctorate in physics and aesthetics, he was awarded a travel bursary and met German experimenter Johann Ritter, who aroused his interest in the possible connection between electricity and magnetism.

In 1806, Ørsted returned to Copenhagen to teach. His 1820 discovery of the link between the two forces brought him international recognition. He was awarded the Royal Society of London's Copley Medal and was later made a member of the Royal Swedish Academy of Science and the American Academy of Arts and Sciences. In 1825, he was the first chemist to produce pure aluminum. He died in Copenhagen in 1851.

Key works

1820 "Experiments on the Effect of a Current of Electricity on the Magnetic Needle"
1821 "Observations on Electro-magnetism"

ELECTRICITY IN MOTION
THE MOTOR EFFECT

IN CONTEXT

KEY FIGURE
André-Marie Ampère
(1775–1836)

BEFORE
1600 William Gilbert conducts the first scientific experiments on electricity and magnetism.

1820 Hans Christian Ørsted proves that an electric current creates a magnetic field.

AFTER
1821 Michael Faraday makes the first electric motor.

1831 Joseph Henry and Faraday use electromagnetic induction to create the first electric generator, converting motion into electricity.

1839 German-born Russian engineer Moritz von Jacobi demonstrates the first practical rotary electrical motor.

1842 Scottish engineer Robert Davidson builds an electric motor to power a locomotive.

Building on Hans Christian Ørsted's discovery of the relationship between electricity and magnetism, French physicist André-Marie Ampère conducted his own experiments.

Ørsted had discovered that a current passing through a wire forms a magnetic field around the wire. Ampère realized that two parallel wires carrying electric currents either attract or repel each other, depending on whether the currents are flowing in the same or opposite directions. If the current flows in the same direction in both, then the wires are attracted; if one flows in the opposite direction, they repel each other.

Ampère's work led to the law bearing his name, which states that the mutual action of two

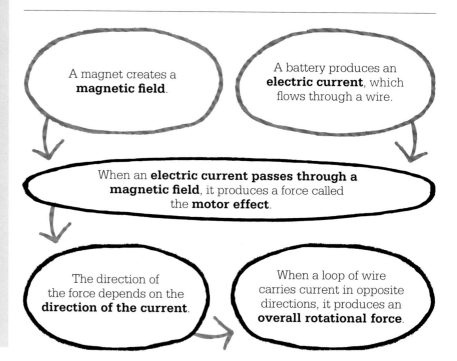

A magnet creates a **magnetic field**.

A battery produces an **electric current**, which flows through a wire.

When an **electric current passes through a magnetic field**, it produces a force called the **motor effect**.

The direction of the force depends on the **direction of the current**.

When a loop of wire carries current in opposite directions, it produces an **overall rotational force**.

See also: Electric potential 128–129 ▪ Making magnets 134–135 ▪ Induction and the generator effect 138–141 ▪ Force fields and Maxwell's equations 142–147 ▪ Generating electricity 148–151

lengths of current-carrying wire is proportional to their lengths and to the magnitudes of their currents. This discovery was the foundation of a new branch of science known as electrodynamics.

Making motors

When a current-carrying wire is placed in a magnetic field, it is subject to a force because the magnetic field interacts with the field created by the current. If the interaction is strong enough, the wire moves. The force is at its greatest when the current is flowing at right angles to the magnetic field lines.

If a loop of wire, with two parallel sides, is placed between the poles of a horseshoe magnet, the interaction of the current on one side causes a downward force, while there is an upward force on the other side. This makes the loop rotate. In other words, electrical potential energy is converted to kinetic (motion) energy, which can do work. However, once the loop has

rotated 180 degrees, the force reverses, so the loop comes to a halt.

French instrument-maker Hippolyte Pixii discovered the solution to this conundrum in 1832 when he attached a metal ring split into two halves to the ends of a coil with an iron core. This device, a commutator, reverses the current in the coil each time it rotates through half a turn—so the loop continues to spin in the same direction.

In the same year, British scientist William Sturgeon invented the first commutator electric motor capable of turning machinery. Five years later, American engineer Thomas Davenport invented a powerful motor that rotated at 600 revolutions per minute and was capable of driving a printing press and machine tools.

An electrodynamic world

Over the years, electrodynamic technology produced more powerful and efficient motors. Torques (turning forces creating rotational motion) were increased by using

The experimental investigation by which Ampère established the law of the mechanical action between electric currents is one of the most brilliant advancements in science.
James Clerk Maxwell

more powerful magnets, increasing the current, or using very thin wire to increase the number of loops. The closer the magnet to the coil, the greater the motor force.

Direct-current (DC) motors are still used for small battery-operated devices; universal motors, which use electromagnets instead of permanent magnets, are used for many household appliances. ▪

André-Marie Ampère

Born to wealthy parents in Lyon, France, in 1775, André-Marie Ampère was encouraged to educate himself at home, in a house with a well-stocked library. Despite a lack of formal education, he took up a teaching position at the new École Polytechnique in Paris in 1804 and was appointed as professor of mathematics there five years later.

After hearing of Ørsted's discovery of electromagnetism, Ampère concentrated his intellectual energies into establishing electrodynamism as a new branch of physics. He also

speculated about the existence of "electrodynamic molecules," anticipating the discovery of electrons. In recognition of his work, the standard unit of electric current—the amp—is named after him. He died in Marseilles in 1836.

Key works

1827 *Memoirs on the Mathematical Theory of Electrodynamics Phenomena, Uniquely Deduced from Experience*

THE DOMINION OF MAGNETIC FORCES

INDUCTION AND THE GENERATOR EFFECT

IN CONTEXT

KEY FIGURE
Michael Faraday (1791–1867)

BEFORE
1820 Hans Christian Ørsted discovers the link between electricity and magnetism.

1821 Michael Faraday invents a device that uses the interaction of electricity and magnetism to produce mechanical motion.

1825 William Sturgeon, a British instrument-maker, builds the first electromagnet.

AFTER
1865 James Clerk Maxwell presents a paper describing electromagnetic waves, including light waves.

1882 The first power stations to use electricity generators are commissioned in London and New York.

E lectromagnetic induction is the production of electromotive force (emf, or a potential difference) across an electrical conductor as the result of a changing magnetic field. Its discovery would transform the world. It is still the foundation of the electricity power industry today, and it made possible the invention of electric generators and transformers, which are at the heart of modern technology.

In 1821, inspired by Hans Christian Ørsted's discovery of the relationship between electricity and magnetism the previous year, British physicist Michael Faraday built two devices that took

See also: Electric potential 128–129 ▪ The motor effect 136–137 ▪ Force fields and Maxwell's equations 142–147 ▪ Generating electricity 148–151

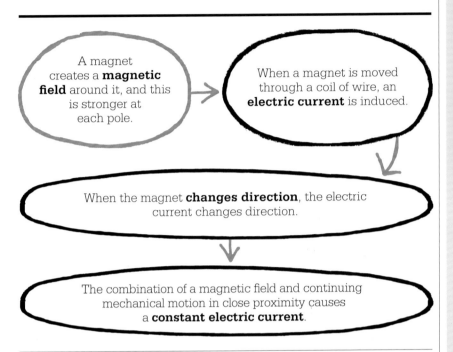

A magnet creates a **magnetic field** around it, and this is stronger at each pole.

When a magnet is moved through a coil of wire, an **electric current** is induced.

When the magnet **changes direction**, the electric current changes direction.

The combination of a magnetic field and continuing mechanical motion in close proximity causes a **constant electric current**.

Michael Faraday

The son of a London blacksmith, Michael Faraday received a very limited formal education. However, when he was 20, he heard renowned chemist Humphry Davy lecturing at London's Royal Institution and sent him his notes. Faraday was invited to become Davy's assistant and traveled around Europe with him from 1813 to 1815.

Famous for inventing the electric motor in 1821, Faraday also devised an early form of Bunsen burner, discovered benzene, and formulated the laws of electrolysis. A pioneer in environmental science, he warned about the dangers of pollution in the Thames River. A man of strong principles, he scorned the pseudoscientific cults of the day. He gave Christmas lectures for the public, refused to offer advice to the government on military matters, and turned down a knighthood. He died in 1867.

Key works

1832 *Experimental Researches in Electricity*
1859 *A Course of Six Lectures on the Various Forces of Matter*

advantage of the so-called motor effect (the creation of a force when current passes through a conductor in a magnetic field). These devices converted mechanical energy into electrical energy.

Faraday conducted many experiments to investigate the interplay of electric currents, magnets, and mechanical motion. These culminated in a series of experiments from July to November 1831 that would have a revolutionary impact.

The induction ring
One of Faraday's first experiments in 1831 was to build an apparatus with two coils of insulated wire wrapped around an iron ring. When a current was passed through one coil, a current was seen to flow temporarily in the other, showing up on a galvanometer, a device that had only recently been invented. This effect became known as

mutual induction, and the apparatus—an induction ring— was the world's first transformer (a device that transfers electrical energy between two conductors).

Faraday also moved a magnet through a coil of wire, making electric current flow in the coil as he did so. However, once the motion of the magnet was stopped, the »

I am busy just now again on electro-magnetism and think I have got hold of a good thing.
Michael Faraday

galvanometer registered no current: the magnet's field only enabled current to flow when the field was increasing or decreasing. When the magnet was moved in the opposite direction, a current was again seen to flow in the coil, this time in the opposite direction. Faraday also found that a current flowed if the coil was moved over a stationary magnet.

The law of induction

Like other physicists of the day, Faraday did not understand the true nature of electricity—that current is a flow of electrons—but he nevertheless realized that when a current flows in one coil, it produces a magnetic field. If the current remains steady, so does the magnetic field, and no potential difference (consequently, no current) is induced in the second coil. However, if the current in the first coil changes, the resulting change in magnetic field will induce a potential difference in the other coil, so a current will flow.

Faraday's conclusion was that no matter how a change in the magnetic environment of a coil is produced, it will cause a current

to be induced. This could come about as a result of a change in the strength of the magnetic field, or by moving the magnet and the coil closer or further from each other, or by rotating the coil or turning the magnet.

An American scientist called Joseph Henry had also discovered electromagnetic induction in 1831, independently of Faraday, but Faraday published first and his findings became known as Faraday's law of induction. It remains the principle behind generators, transformers, and many other devices.

In 1834, Estonian physicist Emil Lenz developed the principle further, stating that the potential difference induced in a conductor by a changing magnetic field opposes the change in that magnetic field. The current resulting from the potential difference generates a magnetic field that will strengthen the original magnetic field if its strength is reducing, and weaken it if its strength is increasing. This principle is known as Lenz's law. One effect of Lenz's law is that some electric current is lost and converted to heat.

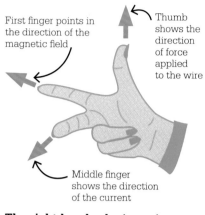

First finger points in the direction of the magnetic field

Thumb shows the direction of force applied to the wire

Middle finger shows the direction of the current

The right-hand rule shows the direction in which a current will flow in a wire when the wire moves in a magnetic field.

Later, in the 1880s, British physicist John Ambrose Fleming described a simple way of working out the direction of induced current flow: the "right-hand rule." This uses the thumb, index finger, and middle finger of the right hand (held perpendicular to one another) to indicate the direction of the current flow from an induced potential difference when a wire moves in a magnetic field (see diagram, above).

Faraday's dynamo

In 1831, the same year as Faraday's experiments on the induction ring, he also created the first electric dynamo. This was a copper disk mounted on brass axes that rotated freely between the two poles of a permanent magnet. He connected the disk to a galvanometer and found that when it rotated, the galvanometer registered a current, which moved out from the center of the disk and flowed through a spring contact into a wire circuit. The apparatus came to be known as a Faraday disk.

The experiment showed that the combination of a magnetic field and continuing mechanical

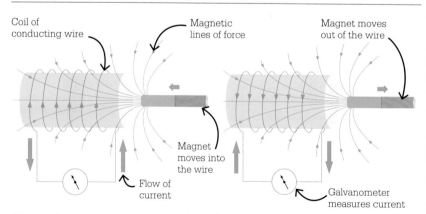

Coil of conducting wire

Magnetic lines of force

Magnet moves out of the wire

Magnet moves into the wire

Flow of current

Galvanometer measures current

When a bar magnet moves in and out of a coil of wire, it produces an electric current. The direction of current changes according to the direction moved by the magnet. The current produced would be greater with more coils or a stronger magnet.

motion in close proximity caused a constant electric current. In a motor, the flow of electrons through a wire in a magnetic field induces a force on the electrons and therefore on the wire, causing it to move. However, in Faraday's disk (and other generators), the law of induction applies—a current is produced as a result of the motion of a conductor (the disk) in a magnetic field. While the characteristic of the motor effect is that electrical energy is changed to mechanical energy, in the generator effect, mechanical energy is converted to electrical energy.

Practical uses

Faraday's discoveries required painstaking experimentation but produced very practical results. They provided the understanding of how to produce electricity on a previously undreamed-of scale.

Although the basic design of the Faraday disk was inefficient, it would soon be taken up by others and developed into practical electricity generators. Within months, French instrument-maker Hippolyte Pixii had built a hand-cranked generator based on Faraday's design. The

current it produced, however, reversed every half-turn and no one had yet discovered a practical way of harnessing this alternating current (AC) electricity to power electronic devices. Pixii's solution was to use a device called a commutator to convert the alternating current to a single-direction current. It was not until the early 1880s that the first large AC generators were constructed by British electrical engineer James Gordon.

The first industrial dynamo was built in 1844 in Birmingham, UK, and was used for electroplating. In 1858, a Kent lighthouse became the first installation to be powered by a steam-powered electrical generator. With the coupling of dynamos to steam-driven turbines, commercial electricity production became possible. The first practical electricity generators went into production in 1870, and by the 1880s areas of New York and London were lit by electricity produced in this way.

A scientific springboard

Faraday's work on the relationship between mechanical movement, magnetism, and electricity was also

Now we find all matter subject to the dominion of Magnetic forces, as they before were known to be to Gravitation, Electricity, cohesion.
Michael Faraday

the springboard for further scientific discovery. In 1861, Scottish physicist James Clerk Maxwell simplified the knowledge to date about electricity and magnetism to 20 equations. Four years later, in a paper presented to the Royal Society in London ("A dynamical theory of the electromagnetic field"), Maxwell unified electric and magnetic fields into one concept—electromagnetic radiation, which moved in waves at close to the speed of light. The paper paved the way for the discovery of radio waves and for Einstein's theories of relativity. ■

I have at last succeeded in illuminating a magnetic curve … and in magnetizing a ray of light.
Michael Faraday

Wireless charging

Many small battery-operated appliances—such as cell phones, electric toothbrushes, and pacemakers—now use induction chargers, which eliminate exposed electrics and reduce reliance on plugs and cables. Two induction coils in close proximity form an electrical transformer, which charges up an appliance's battery. The induction coil in a charging base produces an alternating electromagnetic field, while the receiver coil

inside the device takes power from the field and converts it back into electric current. In small domestic appliances, the coils are small, so they must be in close contact to work.

Inductive charging is also possible for electric vehicles as an alternative to plug-in charging. In this case, larger coils can be used. Robotic, automatic guided vehicles, for example, don't need to be in contact with the charger unit, but can simply pull up close by and charge.

LIGHT ITSELF IS
AN ELECTROMAGNETIC
DISTURBANCE
FORCE FIELDS AND
MAXWELL'S EQUATIONS

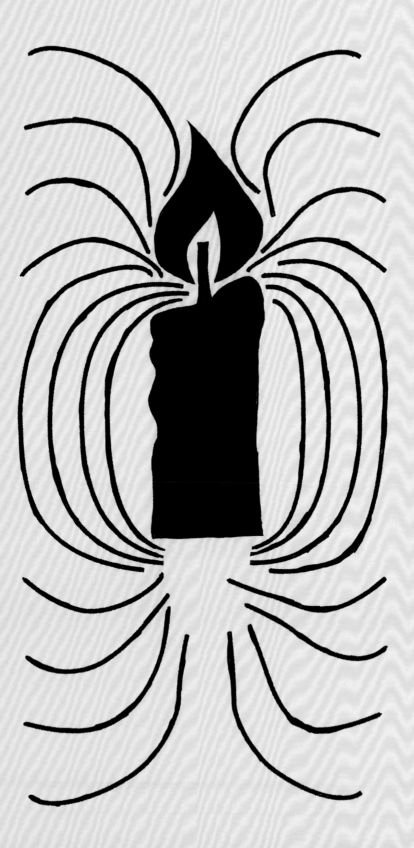

IN CONTEXT

KEY FIGURE
James Clerk Maxwell
(1831–1879)

BEFORE
1820 Hans Christian Ørsted discovers that a current carrying wire deflects the needle of a magnetic compass.

1825 André-Marie Ampère lays the foundations for the study of electromagnetism.

1831 Michael Faraday discovers electromagnetic induction.

AFTER
1892 Dutch physicist Hendrik Lorentz investigates how Maxwell's equations work for different observers, leading to Einstein's theory of special relativity.

1899 Heinrich Hertz discovers radio waves while he is investigating Maxwell's theory of electromagnetism.

I do not perceive in any part of space, whether ... vacant or filled with matter, anything but forces and the lines in which they are exerted.
Michael Faraday

Four equations describe how **electric fields**, **magnetic fields**, **electric charges**, and **currents** are related.

A **single equation** derived from these four describes the **motion of an electromagnetic wave**.

This electromagnetic wave travels at a **constant**, **very high speed**, very close to the observed **speed of light**.

Electromagnetic waves and light are the same phenomenon.

The 19th century witnessed a series of breakthroughs, both experimental and deductive, that would enable the greatest advance in physics since Isaac Newton's laws of motion and gravitation: the theory of electromagnetism. The chief architect of this theory was Scottish physicist James Clerk Maxwell, who formulated a set of equations based upon the work of, among others, Carl Gauss, Michael Faraday, and André-Marie Ampère.

Maxwell's genius was to place the work of his predecessors on a rigorous mathematical footing, recognize symmetries among the equations, and deduce their greater significance in light of experimental results.

Originally published as 20 equations in 1861, Maxwell's theory of electromagnetism describes precisely how electricity and magnetism are intertwined and how this relationship generates wave motion. Although the theory was both fundamental and true, the complexity of the equations (and perhaps, its revolutionary nature) meant that few other physicists understood it immediately.

In 1873, Maxwell condensed the 20 equations into just four, and in 1885, British mathematician Oliver Heaviside developed a much more accessible presentation that allowed a wider community of scientists to appreciate their significance. Even today, Maxwell's equations remain valid and useful at all but the very smallest scales, where quantum effects necessitate their modification.

Lines of force
In a series of experiments in 1831, Michael Faraday discovered the phenomenon of electromagnetic induction—the generation of an electric field by a varying magnetic field. Faraday intuitively proposed a model for induction that turned out to be remarkably close to our current theoretical understanding,

See also: Magnetism 122–123 ▪ Electric charge 124–127 ▪ Making magnets 134–135 ▪ The motor effect 136–137 ▪ Magnetic monopoles 159 ▪ Electromagnetic waves 192–195 ▪ The speed of light 275 ▪ Special relativity 276–279

although his inability to express the model mathematically meant that it was ignored by many of his peers. Ironically, when Maxwell translated Faraday's intuition into equations, he was in turn initially ignored, because of the forbidding depth of his mathematics.

Faraday was keenly aware of a long-standing problem in physics—namely, how a force could be instantaneously transmitted through "empty" space between separated bodies. There is nothing in our everyday experience that suggests a mechanism for this "action at a distance." Inspired by the patterns in iron filings around magnets, Faraday proposed that magnetic effects were carried by invisible lines of force that permeate the space around a magnet. These lines of force point in the direction that a force acts and the density of lines corresponds to the strength of the force.

Faraday's experimental results were first interpreted mathematically by British physicist J.J. Thomson in 1845, but in 1862 Maxwell, who

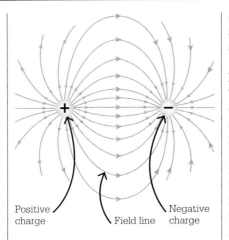

Positive charge / Field line / Negative charge

Electric field lines show the direction of the field between charges. The lines come together at the negative charge, travel away (diverge) from the positive charge, and can never cross.

attended Faraday's lectures in London, turned the descriptive "lines of force" into the mathematical formalism of a field. Any quantity that varies with position can be represented as a field. For example, the temperature in a room can be considered a field, with each point in space defined by three coordinates and associated with a number—the temperature at that point.

Force fields

Taken together, magnetic "field" or "flux" lines describe the region around a magnet in which magnetizable bodies "feel" a force. In this magnetic field, the magnitude of the force at any point in space is related to the density of field lines. Unlike a temperature field, the points on a magnetic field also have a direction, given by the direction of the field line. A magnetic field is therefore a vector field—each spatial point in it has an associated strength and direction, like the velocity field of flowing water.

Similarly, in an electric field, the field line indicates the direction of the force felt by a positive charge, and the concentration of field lines indicates the strength of the field. Like typical fluid flows, electric and magnetic fields may change over time (due, for example, to changing weather patterns), so the vector at each point is time-dependent. »

James Clerk Maxwell

Born in Edinburgh in 1831, James Clerk Maxwell was a precocious child and presented a paper on mathematical curves aged just 14. He studied at Edinburgh and Cambridge universities. In 1856, he was appointed professor at Marischal College, Aberdeen, where he correctly reasoned that the rings of Saturn were made up of many small solid particles.

Maxwell's most productive years were at King's College London from 1860 and then Cambridge from 1871, where he was made the first Professor of Experimental Physics at the new Cavendish Laboratory. He made enormous contributions to the study of electromagnetism, thermodynamics, the kinetic theory of gases, and the theory of optics and color. All of this he accomplished in a short life, before dying of cancer in 1879.

Key works

1861 *On Physical Lines of Force*
1864 *A Dynamical Theory of the Electromagnetic Field*
1870 *Theory of Heat*
1873 *A Treatise on Electricity and Magnetism*

Maxwell's first two equations are statements of Gauss's laws for electric and magnetic fields. Gauss's laws are an application of Gauss's theorem (also known as the divergence theorem), which was first formulated by Joseph-Louis Lagrange in 1762 and rediscovered by Gauss in 1813. In its most general form, it is a statement about vector fields—such as fluid flows—through surfaces.

Gauss formulated the law for electric fields around 1835, but did not publish it in his lifetime. It relates the "divergence" of an electric field at a single point to the presence of a static electric charge. The divergence is zero if there is no charge at that point, positive (field lines flow away) for positive charge, and negative (field lines converge)

for negative charge. Gauss's law for magnetic fields states that the divergence of a magnetic field is zero everywhere; unlike electric fields, there can be no isolated points from which magnetic field lines flow out or in. In other words, magnetic monopoles do not exist and every magnet has both a north and a south pole. As a consequence, magnetic field lines always occur as closed loops, so the line leaving a magnet's north pole returns to the south pole and continues through the magnet to close the loop.

The Faraday and Ampère–Maxwell laws

The third of Maxwell's equations is a rigorous statement of Faraday's law of induction, which the latter had deduced in 1831. Maxwell's

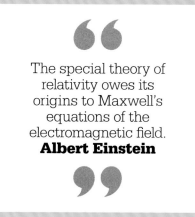

> The special theory of relativity owes its origins to Maxwell's equations of the electromagnetic field.
> **Albert Einstein**

equation relates the rate of change of the magnetic field B with time, to the "curl" of the electric field. The curl describes how electric field lines circulate around a point. Unlike the electric fields created by static point charges, which have divergence but no curl, electric fields that are induced by changing magnetic fields have a circulating character, but no divergence, and can cause current to flow in a coil.

The fourth of Maxwell's equations is a modified version of André-Marie Ampère's circuital law, which was originally formulated in 1826. This states that a constant electric current flowing through a conductor will create a circulating magnetic field around the conductor.

Driven by a sense of symmetry, Maxwell reasoned that just as a changing magnetic field generates an electric field (Faraday's law), so a changing electric field should generate a magnetic field. In order to accommodate this hypothesis, he added the $\partial E/\partial t$ term (which represents the variation of an electric field, E, in time, t) to Ampère's law to make what is now called the Ampère–Maxwell law. Maxwell's addition to the law was not based

Maxwell's equations

Maxwell's four equations contain the variables E and B, representing the electric and magnetic field strengths, which vary with position and time. They may be written as this set of four coupled partial differential equations. They are "differential" because they involve differentiation, a mathematical operation concerned with how things change. They are "partial" because the quantities involved depend on several variables, but each term of the equation only considers a part of the variation, such as dependence on time. They are "coupled" because they involve the same variables and are all simultaneously true.

Name	Equation
Gauss's law for electric fields	$\nabla \bullet E = \rho/\varepsilon_0$
Gauss's law for magnetic fields	$\nabla \bullet B = 0$
Faraday's law	$\nabla \times E = -\partial B/\partial t$
Ampère–Maxwell law	$\nabla \times B = \mu_0 J + \mu_0 \varepsilon_0 \left(\partial E/\partial t \right)$

J = Electric current density (current flowing in a given direction through unit area)

B = Magnetic field E = Electric field

∇ = Differential operator ε_0 = Electric constant

∂t = Partial derivate with respect to time

ρ = Electric charge density (charge per unit volume)

μ_0 = Magnetic constant

on any experimental results but was vindicated by later experiments and advances in theory. The most dramatic consequence of Maxwell's addition to Ampère's law was that it suggested that electric and magnetic fields were commonly associated with a wave.

Electromagnetic waves and light

In 1845, Faraday observed that a magnetic field altered the plane of polarization of light (this is known as the Faraday effect). The phenomenon of polarization had been discovered by Christiaan Huygens back in 1690, but physicists did not understand how it worked. Faraday's discovery did not directly explain polarization, but it did establish a link between light and electromagnetism—a relationship that Maxwell would put on firm mathematical footing a few years later.

From his several equations, Maxwell produced an equation that described wave motion in three-dimensional space. This was his electromagnetic wave equation. The speed of the wave described by the equation is given by the term $1/\sqrt{(\mu_0\varepsilon_0)}$. Maxwell had not only

It turns out that the magnetic and the electric force ... is what ultimately is the deeper thing ... where we can start with to explain many other things.
Richard Feynman

established that electromagnetic phenomena have a wavelike character (having deduced that disturbances in the electromagnetic field propagate as a wave), but that the wave speed, determined theoretically through comparison with the standard form of a wave equation, was very close to the experimentally determined value for the speed of light.

Since nothing else but light was known to travel at anything like the speed of light, Maxwell concluded that light and electromagnetism must be two aspects of the same phenomenon.

Maxwell's legacy

Maxwell's discovery encouraged scientists such as American physicist Albert Michelson to seek a more accurate measurement of the speed of light in the 1880s. Yet Maxwell's theory predicts an entire spectrum of waves, of which visible light is only the most easily sensed by humans. The power and

Heinrich Hertz's experiments at the end of the 19th century proved what Maxwell had predicted and confirmed the existence of electromagnetic waves, including radio waves.

validity of the theory became obvious in 1899 when German physicist Heinrich Hertz—who was determined to test the validity of Maxwell's theory of electromagnetism—discovered radio waves.

Maxwell's four equations today underlie a vast range of technologies, including radar, cellular radio, microwave ovens, and infrared astronomy. Any device that uses electricity or magnets fundamentally depends on them. Classical electromagnetism's impact cannot be overstated—not only does Maxwell's theory contain the fundamentals of Einstein's special theory of relativity, it is also, as the first example of a "field theory," the model for many subsequent theories in physics. ■

MAN WILL IMPRISON THE POWER OF THE SUN

GENERATING ELECTRICITY

IN CONTEXT

KEY FIGURE
Thomas Edison (1847–1931)

BEFORE
1831 Michael Faraday shows that a changing magnetic field interacts with an electric circuit to produce an electromagnetic force.

1832 Hippolyte Pixii develops a prototype DC generator based on Faraday's principles.

1878 German electrical engineer Sigmund Schuckert builds a small steam-driven power station to light a Bavarian palace.

AFTER
1884 Engineer Charles Parsons invents the compound steam turbine, for more efficient power generation.

1954 The first nuclear power station goes online in Russia.

On January 12, 1882, the Edison Electric Light Station at Holborn Viaduct in London began to generate electricity for the first time. This facility, the brainchild of prolific American inventor Thomas Edison, was the world's first coal-fired power station to produce electricity for public use. A few months later, Edison opened a bigger version, at Pearl Street in New York City. The ability to make electricity on a large scale was to be one of the key drivers for the Second Industrial Revolution of 1870–1914.

Until the 1830s, the only way of making electricity was by chemical reactions inside a battery. In 1800,

See also: Magnetism 122–123 ▪ Electric potential 128–129 ▪ Electric current and resistance 130–133 ▪ Making magnets 134–135 ▪ The motor effect 136–137 ▪ Induction and the generator effect 138–141 ▪ Force fields and Maxwell's equations 142–147

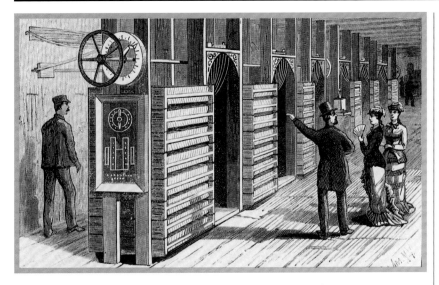

Thomas Edison installed six huge dynamos in the Pearl Street Station plant in Manhattan, New York, in 1882. Each of the dynamos produced enough electricity to power 1,200 lights.

Italian scientist Alessandro Volta converted chemical energy to electricity to produce a steady current in a battery (later known as a Voltaic pile). Although this was the first electric battery, it was impractical, unlike that built by British inventor John Daniell in 1836. The component parts of a Daniell cell were a copper pot filled with copper sulfate solution, in which was immersed an earthenware container filled with zinc sulfate and a zinc electrode. Negatively charged ions migrated to one electrode, and positively charged ions to the other, creating an electric current.

The first dynamo
In the 1820s, British physicist Michael Faraday experimented with magnets and coils of insulated wire. He discovered the relationship between magnetic and electric fields (Faraday's law, or the induction principle) and used this knowledge to build the first electric generator, or dynamo, in 1831. This consisted of a highly conductive copper disk rotating between the poles of a horseshoe magnet, whose magnetic field produced an electric current.

Faraday's explanation of the principles of mechanical electricity generation—and his dynamo—would become the basis for the more powerful generators of the future, but in the early 19th century there wasn't yet the demand for large voltages.

Electricity for telegraphy, arc lighting, and electroplating was supplied by batteries, but this process was very expensive, and scientists from several nations sought alternatives. French inventor Hippolyte Pixii, Belgian electrical engineer Zenobe Gramme, and German inventor Werner von Siemens all worked independently to develop Faraday's induction principle in order to generate electricity more efficiently. In 1871,

Gramme's dynamo was the first electric motor to enter production. It became widely used in manufacturing and farming.

Industry demanded ever more efficient manufacturing processes to increase production. This was the age of prolific invention, and Edison was an outstanding example, turning ideas into commercial gold at his workshops and laboratories. His "light bulb moment" came when he devised a lighting system to replace gas lamps and candles in homes, factories, and public buildings. Edison didn't invent the light bulb, but his 1879 carbon-filament incandescent design was economical, safe, and practical for home use. It ran on low voltage but required a cheap, steady form of electricity to make it work.

Jumbo dynamos
Edison's power plants transformed mechanical energy into electricity. A boiler fired by burning coal converted water to high-pressure steam within a Porter-Allen steam engine. The shaft of the engine was connected directly to the armature (rotating coil) of the »

> We will make electricity so cheap that only the rich will burn candles.
> **Thomas Edison**

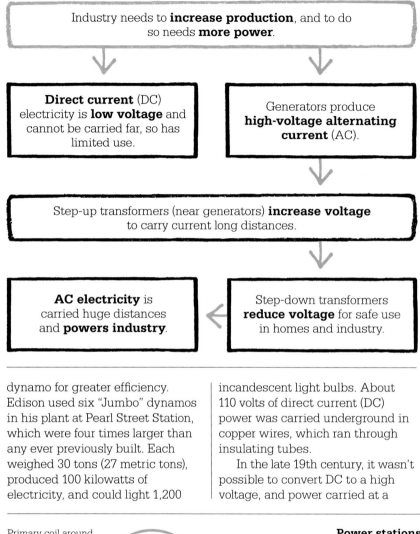

Industry needs to **increase production**, and to do so needs **more power**.

Direct current (DC) electricity is **low voltage** and cannot be carried far, so has limited use.

Generators produce **high-voltage alternating current** (AC).

Step-up transformers (near generators) **increase voltage** to carry current long distances.

AC electricity is carried huge distances and **powers industry**.

Step-down transformers **reduce voltage** for safe use in homes and industry.

dynamo for greater efficiency. Edison used six "Jumbo" dynamos in his plant at Pearl Street Station, which were four times larger than any ever previously built. Each weighed 30 tons (27 metric tons), produced 100 kilowatts of electricity, and could light 1,200 incandescent light bulbs. About 110 volts of direct current (DC) power was carried underground in copper wires, which ran through insulating tubes.

In the late 19th century, it wasn't possible to convert DC to a high voltage, and power carried at a low voltage and high current had a limited range due to resistance in the wires. To get around this problem, Edison proposed a system of local power plants that would provide electricity for local neighborhoods. Because of the problem transmitting over long distances, these generating stations needed to be located within 1 mile (1.6 km) of the user. Edison had built 127 of these stations by 1887, but it was clear that large parts of the United States would not be covered even with thousands of power plants.

The rise of AC electricity

Serbian-American electrical engineer Nikola Tesla had suggested an alternative—using alternating current (AC) generators— in which the polarity of the voltage in a coil reverses as the opposite poles of a rotating magnet pass over it. This regularly reverses the direction of current in the circuit; the faster the magnet turns, the quicker the current flow reverses. When American electronics engineer William Stanley built the first workable transformer in 1886, the voltage of transmission from an

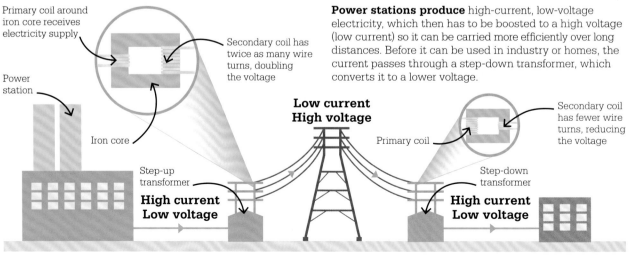

Primary coil around iron core receives electricity supply

Secondary coil has twice as many wire turns, doubling the voltage

Power station

Iron core

Step-up transformer

High current Low voltage

Low current High voltage

Primary coil

Secondary coil has fewer wire turns, reducing the voltage

Step-down transformer

High current Low voltage

Power stations produce high-current, low-voltage electricity, which then has to be boosted to a high voltage (low current) so it can be carried more efficiently over long distances. Before it can be used in industry or homes, the current passes through a step-down transformer, which converts it to a lower voltage.

> The deadly electricity of the alternating current can do no harm unless a man is fool enough to swallow a whole dynamo.
> **George Westinghouse**

AC generator could be "stepped up" and the current simultaneously "stepped down" in smaller-diameter wire. At higher voltages, the same power could be transmitted at a much lower current, and this high voltage could be transmitted over long distances, then "stepped down" again before use in a factory or home.

American entrepreneur George Westinghouse realized the problems of Edison's design, purchased Tesla's many patents, and hired Stanley. In 1893, Westinghouse won a contract to build a dam and hydroelectric power plant at Niagara Falls to harness the immense water power to produce electricity. The installation was soon transmitting AC to power the businesses and homes of Buffalo, New York, and marked the beginning of the end for DC as the default means of transmission in the United States.

New, super-efficient steam turbines allowed generating capacity to expand rapidly. Ever-

larger operating plants employed higher steam pressures for greater efficiency. During the 20th century, electricity generation increased almost 10-fold. High-voltage AC transmission enabled power to be moved from distant power stations to industrial and urban centers hundreds or even thousands of miles away.

At the start of the century, coal was the primary fuel burned to create the steam to drive turbines. It was later supplemented by other fossil fuels (oil and natural gas), wood chippings, and enriched uranium in nuclear power stations.

Sustainable solutions

With fears about rising atmospheric CO_2 levels, sustainable alternatives to fossil fuels have been introduced. Hydroelectric power, which dates back to the 19th century, now produces almost one-fifth of the world's electricity. The power of wind, waves, and tides is now harnessed to drive turbines and provide electricity. Steam originating deep in Earth's crust produces geothermal energy in parts of the world such as Iceland. About 1 percent of global electricity is now generated by solar panels. Scientists are also working to develop hydrogen fuel cells. ∎

Solar panels are made of dozens of photovoltaic cells, which absorb solar radiation, exciting free electrons and generating a flow of electrical current.

Thomas Edison

One of the most prolific inventors of all time, Edison had more than 1,000 patents to his name by the time of his death. He was born in Milan, Ohio, and was mostly home-taught. In an early sign of his entrepreneurial talent, he gained exclusive rights to sell newspapers while working as a telegraph operator with Grand Trunk Railway. His first invention was an electronic vote recorder.

At the age of 29, Edison established an industrial research lab in New Jersey. Some of his most famous patents were for radical improvements to existing devices, such as the telephone, microphone, and light bulb. Others were revolutionary, including the phonograph in 1877. He went on to found the General Electric Company in 1892. Edison was a vegetarian and proud that he never invented weapons.

Key work

2005 *The Papers of Thomas A. Edison: Electrifying New York and Abroad, April 1881–March 1883*

A SMALL STEP IN THE CONTROL OF NATURE

ELECTRONICS

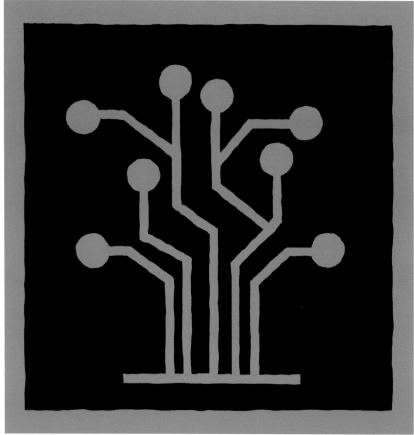

IN CONTEXT

KEY FIGURE
John Bardeen (1908–1991)

BEFORE
1821 German physicist Thomas Seebeck observes a thermoelectric effect in a semiconductor.

1904 British engineer John Ambrose Fleming invents the vacuum tube diode.

1926 Austrian–American engineer Julius Lilienfeld patents the FET (field-effect transistor).

1931 British physicist Alan Herries Wilson establishes the band theory of conduction.

AFTER
1958–59 American engineers Jack Kilby and Robert Noyce develop the integrated circuit.

1971 Intel Corporation releases the first microprocessor.

Electronics encompasses the science, technology, and engineering of components and circuits for the generation, transformation, and control of electrical signals. Electronic circuits contain active components, such as diodes and transistors, that switch, amplify, filter, or otherwise change electrical signals. Passive components, such as cells, lamps, and motors, typically only convert between electrical and other forms of energy, such as heat and light.

The term "electronics" was first applied to the study of the electron, the subatomic particle that carries electricity in solids. The discovery of the electron in 1897 by the

See also: Electric charge 124–127 ▪ Electric current and resistance 130–133 ▪ Quantum applications 226–231
▪ Subatomic particles 242–243

Electronics is the **use of electricity** to produce signals that **carry information** and **control devices**.

→

To produce these signals, the **flow of electric current** must be controlled with great precision.

→

Semiconductors contain a **variable number of charge carriers** (particles that carry electricity).

↓

In most semiconductors, the most important electronic component, **the transistor**, is made of **doped silicon regions**. It can amplify electrical signals or turn them on or off.

←

Doping (adding impurities to) silicon semiconductors **affects the nature of their charge carriers**, enabling current to be controlled precisely.

British physicist J.J. Thomson generated scientific study of how the particle's electric charge could be harnessed. Within 50 years, this research had led to the invention of the transistor, paving the way for the concentration of electrical signals into ever-more compact devices, processed at ever-greater speeds. This led to the spectacular advances in electronic engineering and technology of the digital revolution that began in the late 20th century and continue today.

Valves and currents

The first electronic components were developed from vacuum tubes—glass tubes with the air removed. In 1904, British physicist John Ambrose Fleming developed the vacuum tube into the thermionic diode, which consisted of two electrodes—a cathode and an anode—inside the tube. The metal cathode was heated by an electrical circuit to the point of thermionic emission, whereby its negatively charged electrons gained enough energy to leave the surface and move through the tube. When a voltage was applied across the electrodes, it made the anode relatively positive, so the electrons were attracted to it and a current flowed. When the voltage was reversed, the electrons emitted by the cathode were repelled from the anode and no current flowed. The diode conducted only when the anode was positive relative to the cathode, and so acted as a one-way valve, which could be used to convert AC (alternating current)

The Colossus Mark II computer of 1944 used a vast number of valves to make the complex mathematical calculations needed to decode the Nazi Lorenz cipher systems.

into DC (direct current). This allowed for the detection of AC radio waves, so the valve found wide application as a demodulator (detector) of signals in early AM (amplitude modulation) radio receivers.

In 1906, American inventor Lee de Forest added a third, grid-shaped electrode to Fleming's diode to create a triode. A small, varying voltage applied across the new electrode and the cathode changed the electron flow between the cathode and anode, creating a large voltage variation—in other words, the small input voltage was amplified to create a large output voltage. The triode became a vital component in the development of radio broadcasting and telephony.

Solid-state physics

Although in the following decades valves enabled further advances in technology, such as the television and early computers, they were bulky, fragile, power-hungry, and limited in the frequency of their operation. The British Colossus »

It is at a surface where many of our most interesting and useful phenomena occur.
Walter Brattain

computers of the 1940s, for example, each had up to 2,500 valves, occupied a whole room, and weighed several tons.

All of these limitations were resolved with the transition to electronics based not on vacuum tubes but on the electron properties of semiconducting solids, such as the elements boron, germanium, and silicon. This in turn was born out of an increasing interest from the 1940s onward in solid-state physics—the study of those properties of solids that depend on microscopic structure at the atomic and subatomic scales, including quantum behavior.

A semiconductor is a solid with electrical conductivity varying between that of a conductor and an insulator—so neither high nor low. In effect, it has the potential to control the flow of electric current. Electrons in all solids occupy distinct energy levels grouped into bands, called valence and conduction bands. The valence band contains the highest energy level that electrons occupy when bonding with adjacent atoms. The conduction band has even higher energy levels, and its electrons are not bound to any particular atoms, but have enough energy to move throughout the solid, and so

conduct current. In conductors, the valence and conduction bands overlap, so electrons involved in bonding can also contribute to conduction. In insulators, there is a large band gap, or energy difference, between the valence and conduction bands that keeps most electrons bonding but not conducting. Semiconductors have a small band gap. When given a little extra energy (by applying heat, light, or voltage), their valence electrons can migrate to the conduction band, changing the properties of the material from insulator to conductor.

Control through doping

In 1940, a chance discovery added another dimension to the electrical potential of semiconductors. When testing a crystal of silicon, Russell Ohl, an American electrochemist, found that it produced different electrical effects if probed in different places. When examined, the crystal appeared to have regions that contained distinct impurities. One, phosphorus, had a small excess of electrons; another,

boron, had a small deficiency. It became clear that tiny amounts of impurities in a semiconductor crystal can dramatically change its electrical properties. The controlled introduction of specific impurities to obtain desired properties became known as "doping."

The regions of a crystal can be doped in different ways. In a pure silicon crystal, for example, each atom has four bonding (valence) electrons that it shares with its neighbors. Regions of the crystal can be doped by introducing some atoms of phosphorus (which has five valence electrons) or boron (with three valence electrons). The phosphorus-doped region has extra "free" electrons and is called an n-type semiconductor (n for negative). The boron-doped region, called a p-type semiconductor (p for positive) has fewer electrons, creating charge-carrying "holes." When the two types are joined, it is called a p–n junction. If a voltage is applied to the p-type side, it attracts electrons from the n-type side and current flows. A crystal with one p–n junction

Most semiconductors in transistors are made of silicon (Si), which has been doped with impurities to control the flow of current through it. Adding phosphorus atoms to silicon makes an n-type semiconductor, with negatively charged electrons that are free to move. Adding boron atoms makes a p-type semiconductor, with positively charged "holes" that can move through the silicon.

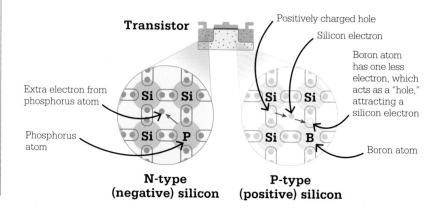

Transistor

Extra electron from phosphorus atom

Phosphorus atom

Positively charged hole

Silicon electron

Boron atom has one less electron, which acts as a "hole," attracting a silicon electron

Boron atom

N-type (negative) silicon

P-type (positive) silicon

John Bardeen

Born in 1908, John Bardeen was just 15 when he began electrical engineering studies at the local University of Wisconsin. After graduating, he joined the Gulf Oil laboratories in 1930 as a geophysicist, developing magnetic and gravitational surveys. In 1936, he earned a PhD in mathematical physics from Princeton University, and later researched solid-state physics at Harvard. During World War II, he worked for the US Navy on torpedoes and mines.

After a productive period at Bell Labs, where Bardeen coinvented the transistor, in 1957 he coauthored the BCS (Bardeen–Cooper–Schrieffer) theory of superconductivity. He shared two Nobel Prizes in Physics—in 1956 (for the transistor) and again in 1972 (for BCS theory). He died in 1991.

Key works

1948 "The transistor, a semi-conductor triode"
1957 "Microscopic theory of superconductivity"
1957 "Theory of superconductivity"

acts as a diode—it conducts a current across the junction in only one direction.

Transistor breakthroughs

After World War II, the search for an effective replacement for the vacuum tube, or valve, continued. In the US, the Bell Telephone Company assembled a team of American physicists, including William Shockley, John Bardeen, and Walter Brattain at its laboratories in New Jersey to develop a semiconductor version of the triode amplifier.

Bardeen was the main theorist, and Brattain the experimenter, in the group. After failed attempts at applying an external electric field to a semiconductor crystal to control its conductivity, a theoretical breakthrough by Bardeen shifted the focus onto the surface of the semiconductor as the key site of changes in conductivity. In 1947, the group began to experiment with electrical contacts on top of a crystal of germanium. A third electrode (the "base") was attached underneath. For the device to work as desired, the two top contacts needed to be very close together,

which was achieved by wrapping gold foil around the corner of a piece of plastic and slitting the foil along the edge to create two tightly spaced contacts. When they were pressed onto the germanium, they formed an amplifier that could boost the signal that was fed to the base. This first working version was simply known as a "point contact" but soon became known as a "transistor."

The point-contact transistor was too delicate for reliable high-volume production, so in 1948 Shockley began work on a new transistor. Based on the p–n semiconductor junction, Shockley's "bipolar" junction transistor was built on the assumption that the positively charged holes created by doping penetrated the body of the semiconductor rather than just floating across its surface. The transistor consisted of a sandwich of p-type material between two n-type layers (npn), or n-type sandwiched between p-type layers (pnp), with the semiconductors separated by p–n junctions. By 1951, Bell was mass-producing the transistor. Although at first mainly used in hearing aids and radios,

the device soon drove spectacular growth in the electronics market and began to replace vacuum tubes in computers.

Processing power

The first transistors were made from germanium, but this was superseded as a base material by the much more abundant and versatile silicon. Its crystals form on their surface a thin insulating layer of oxide. Using a technique called photolithography, this layer can be engineered with microscopic precision to create highly complex patterns of doped regions and other features on the crystal.

The advent of silicon as a base and advances in transistor design drove rapid miniaturization. This led first to integrated circuits (entire circuits on a single crystal) in the late 1960s, and in 1971 to the Intel 4004 microprocessor—an entire CPU (central processing unit) on one 3 × 4mm chip, with more than 2,000 transistors. Since then, the technology has developed with incredible rapidity, to the point where a CPU or GPU (graphics processing unit) chip can contain up to 20 billion transistors. ∎

ANIMAL ELECTRICITY
BIOELECTRICITY

IN CONTEXT

KEY FIGURES
Joseph Erlanger (1874–1965),
Herbert Spencer Gasser
(1888–1963)

BEFORE
1791 Italian physicist Luigi
Galvani publishes his findings
on "animal electricity" in a
frog's leg.

1843 German physician Emil
du Bois-Reymond shows that
electrical conduction travels
along nerves in waves.

AFTER
1944 American physiologists
Joseph Erlanger and Herbert
Gasser receive the Nobel Prize
for Physiology or Medicine for
their work on nerve fibers.

1952 British scientists Alan
Hodgkin and Andrew Huxley
show that nerve cells
communicate with other
cells by means of flows of
ions; this becomes known as
the Hodgkin-Huxley model.

Bioelectricity enables the
entire nervous system of an
animal to function. It allows
the brain to interpret heat, cold,
danger, pain, and hunger, and to
regulate muscle movement,
including heartbeat and breathing.

One of the first scientists to
study bioelectricity was Luigi
Galvani, who coined the term
"animal electricity" in 1791. He
observed muscular contraction in
a dissected frog's leg when a cut
nerve and a muscle were connected
with two pieces of metal. In 1843,
Emil du Bois-Reymond showed that
nerve signals in fish are electrical,
and in 1875, British physiologist

Richard Caton recorded the
electrical fields produced by the
brains of rabbits and monkeys.

The breakthrough in
understanding how the electrical
impulses were produced came
in 1932, when Joseph Erlanger
and Herbert Gasser found that
different fibers in the same nerve
cord possess different functions,
conduct impulses at different rates,
and are stimulated at different
intensities. From the 1930s onward,
Alan Hodgkin and Andrew Huxley
used the giant axon (part of a
nerve cell) of a veined squid to
study how ions (charged atoms
or molecules) move in and out
of nerve cells. They found that
when nerves pass on messages,
sodium, potassium, and chloride
ions create fast-moving pulses of
electrical potential. ∎

Sharks and some other fish have
jelly-filled pores called ampullae of
Lorenzini, which detect changes in
electric fields in the water. These
sensors can detect a change of just
0.01 microvolt.

See also: Magnetism 122–123 ▪ Electric charge 124–127 ▪ Electric potential
128–129 ▪ Electric current and resistance 130–133 ▪ Making magnets 134–135

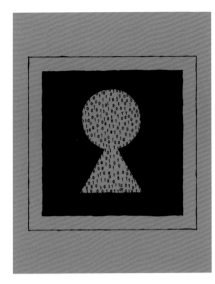

A TOTALLY UNEXPECTED SCIENTIFIC DISCOVERY
STORING DATA

IN CONTEXT

KEY FIGURES
Albert Fert (1938–),
Peter Grünberg (1939–2018)

BEFORE
1856 Scottish physicist
William Thomson (Lord Kelvin)
discovers magnetoresistance.

1928 The quantum-
mechanical idea of electron
spin is postulated by Dutch–
American physicists George
Uhlenbeck and Samuel
Goudsmit.

1957 The first computer
hard-disk drive (HDD) is the
size of two refrigerators, and
can store 3.75 megabytes (MB)
of data.

AFTER
1997 British physicist
Stuart Parkin applies
giant magnetoresistance
to create extremely sensitive
spin-valve technology for
data-reading devices.

Computer hard-disk drives (HDD) store data encoded into bits, written on the disk surface as a series of changes in direction of magnetization. The data is read by detecting these changes as a series of 1s and 0s. The pressure to store more data in less space has driven a constant evolution in HDD technology, but a major problem soon emerged: it was difficult for conventional sensors to read ever-greater amounts of data on ever-smaller disk space.

In 1988, two teams of computer scientists—one led by Albert Fert, the other by Peter Grünberg— independently discovered giant magnetoresistance (GMR). GMR depends on electron spin, a quantum-mechanical property.

Electrons possess either up-spin or down-spin—if an electron's spin is "up," it will move easily through an up-oriented magnetized material, but it will encounter greater resistance through a down-oriented magnet. The study of electron spin became known as "spintronics." By sandwiching a

> We had just participated in the birth of the phenomenon which we called giant magnetoresistance.
> **Albert Fert**

nonmagnetic material between magnetic layers only a few atoms thick, and by applying small magnetic fields, the current flowing through became spin-polarized. The electron spin was either oriented "up" or "down," and if the magnetic field changed, the spin-polarized current was switched on or off, like a valve. This spin-valve could detect minute magnetic impulses when reading HDD data, allowing vastly increased amounts of data to be stored. ∎

See also: Magnetism 122–123 ▪ Nanoelectronics 158 ▪ Quantum numbers 216–217 ▪ Quantum field theory 224–225 ▪ Quantum applications 226–231

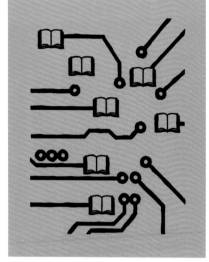

AN ENCYCLOPEDIA ON THE HEAD OF A PIN
NANOELECTRONICS

IN CONTEXT

KEY FIGURE
Gordon Moore (1929–)

BEFORE
Late 1940s The earliest
transistors are made,
measuring ⅓ inch in length.

1958 American electrical
engineer Jack Kilby
demonstrates the first
working integrated circuit.

1959 Richard Feynman
challenges other scientists
to research nanotechnology.

AFTER
1988 Albert Fert and Peter
Grünberg independently
discover the GM (giant
magnetoresistance) effect,
enabling computer hard drives
to store ever-greater amounts
of data.

1999 American electronics
engineer Chad Mirkin invents
dip-pen nanolithography,
which "writes" nanocircuitry
on silicon wafers.

The components that are
integral to virtually every
electronic device, from
smartphones to car ignition systems,
are only a few nanometers (nm)
in scale (1 nm is one-billionth of 1
meter). A handful of tiny integrated
circuits (ICs) can perform functions
that were previously carried out by
thousands of transistors, switching
or amplifying electronic signals.
ICs are collections of components,
such as transistors and diodes, that
are printed onto a silicon wafer, a
semiconducting material.

Reducing the size, weight, and
power consumption of electronic
devices has been a driving trend
since the 1950s. In 1965, American
engineer Gordon Moore forecast the
demand for ever-smaller electronic
components and predicted that the
number of transistors per silicon
chip would double every 18 months.

In 1975, he revised the timescale
to every two years, a maxim that
became known as "Moore's law."
Although the rate of miniaturization
has slowed since 2012, the smallest
modern transistors are 7 nm across,
enabling 20 billion transistor-based
circuits to be integrated into
a single computer microchip.
Photolithography (transferring a
pattern from a photograph onto the
semiconducting material) is used
to fabricate these nanocircuits. ∎

Gordon Moore, who was CEO of the
technology company Intel between
1975 and 1987, is best known for his
observations about the demand for
ever-smaller electronic components.

See also: The motor effect 136–137 ▪ Electronics 152–155 ▪ Storing data 157
▪ Quantum applications 226–231 ▪ Quantum electrodynamics 260

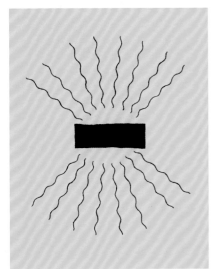

A SINGLE POLE, EITHER NORTH OR SOUTH

MAGNETIC MONOPOLES

IN CONTEXT

KEY FIGURES
Gerard 't Hooft (1946–),
Alexander Polyakov (1945–)

BEFORE
1865 James Clerk Maxwell
unifies electricity and
magnetism in one theory.

1894 Pierre Curie suggests
that magnetic monopoles
could exist.

1931 Paul Dirac proposes that
magnetic monopoles could
explain the quantization of
electric charge.

1974 American physicists
Sheldon Glashow and Howard
Georgi publish the first Grand
Unified Theory (GUT) in
particle physics.

AFTER
1982 Blas Cabrera, a
Spanish physicist at Stanford
University, California, records
an event consistent with a
monopole passing through
a superconducting device.

I n classical magnetism, magnets have two poles that cannot be separated. If a magnet is broken in half, the broken end simply becomes a new pole. However, in particle physics, magnetic monopoles are hypothetical particles with a single pole, either north or south. Theoretically, opposite magnetic monopoles attract, matching monopoles repel, and their trajectories bend in an electric field.

There is no observational or experimental proof that magnetic monopoles exist, but in 1931, British physicist Paul Dirac suggested that they could explain the quantization of electric charge, whereby all electron charges are multiples of 1.6×10^{-19} coulombs.

Gravity, electromagnetism, the weak nuclear force, and the strong nuclear force are the four recognized fundamental forces. In particle physics, several variants of Grand Unified Theory (GUT) propose that in an exceptionally high-energy environment, electromagnetic, weak, and strong

Experimenters will continue to stalk the monopole with unfailing determination and ingenuity.
John Preskill
American theoretical physicist

nuclear forces merge into a single force. In 1974, theoretical physicists Gerard 't Hooft and Alexander Polyakov independently argued that GUT predicts the existence of magnetic monopoles.

In 1982, a detector at Stanford University recorded a particle consistent with a monopole, but such particles have never been found since, despite scientists' efforts to find them using highly sensitive magnetometers. ∎

SOUND AND LIGHT
THE PROPERTIES OF WAVES

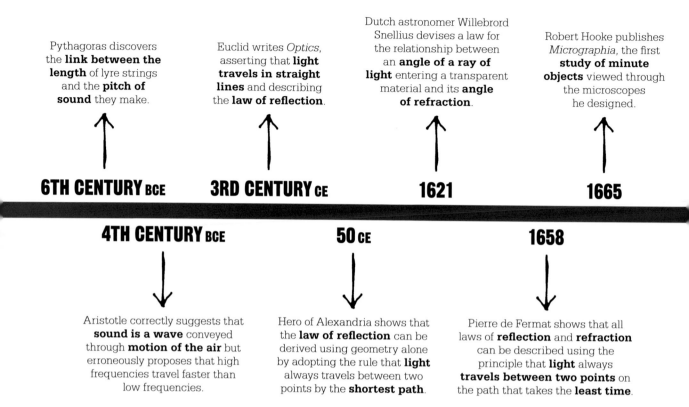

Pythagoras discovers the **link between the length** of lyre strings and the **pitch of sound** they make.

6TH CENTURY BCE

Euclid writes *Optics*, asserting that **light travels in straight lines** and describing the **law of reflection**.

3RD CENTURY CE

Dutch astronomer Willebrord Snellius devises a law for the relationship between an **angle of a ray of light** entering a transparent material and its **angle of refraction**.

1621

Robert Hooke publishes *Micrographia*, the first **study of minute objects** viewed through the microscopes he designed.

1665

4TH CENTURY BCE

Aristotle correctly suggests that **sound is a wave** conveyed through **motion of the air** but erroneously proposes that high frequencies travel faster than low frequencies.

50 CE

Hero of Alexandria shows that the **law of reflection** can be derived using geometry alone by adopting the rule that **light** always travels between two points by the **shortest path**.

1658

Pierre de Fermat shows that all laws of **reflection** and **refraction** can be described using the principle that **light** always **travels between two points** on the path that takes the **least time**.

Hearing and sight are the senses we use most to interpret the world; little wonder that light—a prerequisite for human sight—and sound have fascinated humans since the dawn of civilization.

Music has been a feature of everyday life since Neolithic times, as cave paintings and archaeology testify. The ancient Greeks, whose love of learning infused every element of their culture, sought to work out the principles behind the production of harmonic sounds. Pythagoras, inspired by hearing hammers strike different notes on an anvil, recognized that a relationship existed between sound and the size of the tool or instrument that produced it. In a number of studies, he and his students explored the effects of vibrating lyre strings of different tensions and lengths.

Reflection and refraction
Mirrors and reflections have also long been things of wonder. Again in ancient Greece, scholars proved—using geometry alone—that light always returns from a mirrored surface at the same angle at which it approaches it, a principle known today as the law of reflection.

In the 10th century, Persian mathematician Ibn Sahl noted the relationship between a ray of light's angle of incidence as it entered a transparent material and its angle of refraction within that material. In the 17th century, French mathematician Pierre de Fermat proposed correctly that a single principle pertained in both reflection and refraction: light always chooses to take the shortest path it can between any two points.

The phenomena of reflection and refraction of light allowed pioneers such as Italian physicist and astronomer Galileo Galilei and Robert Hooke in England to create devices that offered new views of the universe, while microscopes unveiled nature. Telescopes revealed hitherto unseen moons orbiting other planets and prompted a reassessment of Earth's place in the cosmos, while microscopes offered a view into an alien world of tiny creatures and the cellular structure of life itself.

The very nature of light was questioned by influential scientists many times over the centuries. Some felt that light was composed

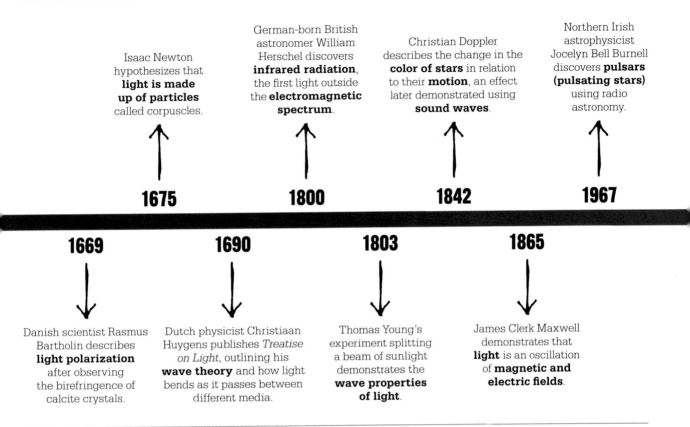

Isaac Newton hypothesizes that **light is made up of particles** called corpuscles.

German-born British astronomer William Herschel discovers **infrared radiation,** the first light outside the **electromagnetic spectrum**.

Christian Doppler describes the change in the **color of stars** in relation to their **motion,** an effect later demonstrated using **sound waves**.

Northern Irish astrophysicist Jocelyn Bell Burnell discovers **pulsars (pulsating stars)** using radio astronomy.

1675　　**1800**　　**1842**　　**1967**

1669　　**1690**　　**1803**　　**1865**

Danish scientist Rasmus Bartholin describes **light polarization** after observing the birefringence of calcite crystals.

Dutch physicist Christiaan Huygens publishes *Treatise on Light*, outlining his **wave theory** and how light bends as it passes between different media.

Thomas Young's experiment splitting a beam of sunlight demonstrates the **wave properties of light**.

James Clerk Maxwell demonstrates that **light** is an oscillation of **magnetic and electric fields**.

of small particles which move through the air from a source to an observer, possibly after reflection from an object. Others saw light as a wave, citing behavior such as diffraction (spreading of light as it passed through narrow gaps). English physicists Isaac Newton and Robert Hooke opposed each other; Newton favored particles and Hooke opted for waves.

British polymath and physician Thomas Young provided an answer. In an experiment outlined to the Royal Society in 1803, he split a beam of sunlight, dividing it by means of a thin card, so that it diffracted and produced a pattern on a screen. The pattern was not that of two bright patches from two rays of light, but a series of bright and dark repeating lines. The results could only be explained if

light were acting as a wave, and the waves each side of the thin card interfered with one another.

Wave theory took off, but only when it became clear that light traveled in transverse waves, unlike sound, which travels in longitudinal waves. Soon, physicists examining the properties of light also noticed that light waves could be forced to oscillate in particular orientations known as polarization. By 1821, French physicist Augustin-Jean Fresnel had produced a complete theory of light in terms of waves.

The Doppler effect

While considering the light from pairs of stars orbiting each other, Austrian physicist Christian Doppler noticed that most pairs showed one star red and the other blue. In 1842, he proposed that this was due to

their relative motion as one moved away and the other toward Earth. The theory was proven for sound and deemed correct for light; the color of light depends on its wavelength, which is shorter as a star approaches and longer when it is moving away.

In the 19th century, physicists also discovered new light invisible to the human eye—first infrared and then ultraviolet. In 1865, Scottish physicist James Clerk Maxwell interpreted light as an electromagnetic wave, prompting questions about how far the electromagnetic spectrum might extend. Soon physicists discovered light with more extreme frequencies, such as X-rays, gamma rays, radio waves, and microwaves—and their multiple uses. Light that is invisible to humans is now an essential part of modern life. ■

THERE IS GEOMETRY IN THE HUMMING OF THE STRINGS

MUSIC

H umans have been making music since prehistoric times, but it was not until the golden age of ancient Greece (500–300 BCE) that some of the physical principles behind the production of harmonic sounds, in particular frequency and pitch, were established. The earliest scientific attempt to understand these fundamental aspects of music is generally attributed to the Greek philosopher Pythagoras.

The sound of music

Pitch is determined by the frequency of sound waves (the number of waves that pass a fixed point each second). A combination

See also: The scientific method 20–23 ▪ The language of physics 24–31 ▪ Pressure 36 ▪ Harmonic motion 52–53 ▪ Electromagnetic waves 192–195 ▪ Piezoelectricity and ultrasound 200–201 ▪ The heavens 270–271

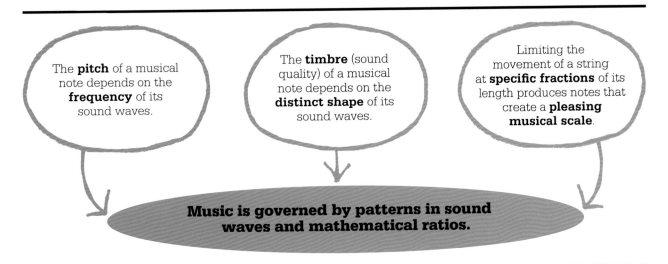

The **pitch** of a musical note depends on the **frequency** of its sound waves.

The **timbre** (sound quality) of a musical note depends on the **distinct shape** of its sound waves.

Limiting the movement of a string at **specific fractions** of its length produces notes that create a **pleasing musical scale**.

Music is governed by patterns in sound waves and mathematical ratios.

of physics and biology enables its perception by the human ear. Today, we know that sound waves are longitudinal—they are created by air being displaced back and forth in directions parallel to that in which the wave is propagating (moving through the medium that carries it). We perceive sound when these oscillations cause our eardrums to vibrate.

While the frequency or pitch of a sound dictates the note that we hear, music also has another quality known as "timbre." This is the subtle variation in the rise and fall of the waves produced by a particular musical instrument—features of the wave's "shape" that are distinct from its simple frequency and wavelength (the distance between two successive peaks and troughs in a wave). No nondigital musical instrument can produce sound waves that vary consistently and entirely smoothly, and timbre is the reason why a note sung by a human voice can sound very different when it is played on a stringed, wind, or brass instrument,

or even sung by another singer. Musicians can also alter the timbre of an instrument by using different playing techniques. A violinist, for example, can change the way the strings vibrate by using the bow in different ways.

Pythagorean findings

According to legend, Pythagoras formulated his ideas about pitch while listening to the melodic sound of hammers striking anvils as he passed a busy blacksmith's workshop. Hearing a note that sounded different from the others, he is said to have rushed inside

to test the hammers and anvils, and in doing so discovered a relationship between the size of the hammer used and the pitch of the sound it produced.

Like many other stories about Pythagoras, this episode is certainly invented (there is no relationship between pitch and the size of the hammer striking an anvil) but it is true that the philosopher and his followers found fundamental links between the physics of musical instruments and the notes they produced. They realized that there was a relationship between the size and »

Even if two notes have the same pitch, their sound depends on the shape of their waves. A tuning fork produces a pure sound with just one pitch. A violin has a jagged waveform with pitches called overtones on top of its fundamental pitch.

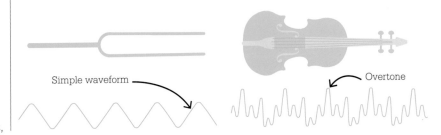

Simple waveform

Overtone

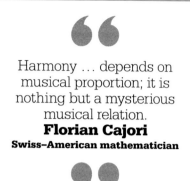

Harmony … depends on musical proportion; it is nothing but a mysterious musical relation.
Florian Cajori
Swiss–American mathematician

sound of musical instruments. In particular, they found similar relationships among the sounds produced by strings of different lengths, organ pipes of varying heights, and wind instruments of different diameters and lengths.

The Pythagoreans' discoveries had more to do with systematic experiments involving vibrating strings than hammers and anvils. The observation that shorter strings vibrate more quickly and produce higher notes was nothing new—it had been the basis of stringed instruments for at least 2,000 years. However, vibrating identical strings in different ways produced more interesting results.

Pythagoras and his students tested strings from lyres (an early musical instrument) with varying lengths and tensions. Plucking a string in the middle of its length, for example, creates a "standing wave" in which the middle oscillates back and forth while the ends remain fixed in place. In effect, the string produces a wave with a wavelength that is twice its own length and a frequency determined by the wavelength and the tension in the string. This is known as the fundamental tone or

"first harmonic". Standing waves of shorter wavelengths, known as "higher harmonics," can be created by "stopping" a string (holding or limiting its movement at another point on its length). The "second harmonic" is produced by stopping the string precisely halfway along its length. This results in a wave whose entire wavelength matches the string length—in other words, the wavelength is half and the frequency double that of the fundamental tone. To human hearing, this creates a note with many of the same characteristics as the fundamental tone but with a higher pitch—in musical terms it is an octave higher. A stop one-third of the way down the string creates the third harmonic, with a wavelength of two-thirds the length of the string and a frequency three times that of the fundamental tone.

The perfect fifth

The difference between the second and third harmonics was important. Equivalent to a 3:2 ratio between the frequencies of vibrating waves, it separated pitches (musical notes) that blended together pleasingly, but were also musically more distinct from each other than harmonic notes separated by a whole octave.

Experimentation soon allowed the Pythagoreans to construct an entire musical system around this relationship. By building and correctly "tuning" other strings to vibrate at frequencies set by simple numerical relationships, they constructed a musical bridge, or progression, between the

The lyre of ancient Greece
originally had four strings, but had acquired as many as 12 by the 5th century BCE. Players plucked the strings with a plectrum.

fundamental tone and the octave through a series of seven steps without striking discordant notes. The 3:2 ratio defined the fifth of these steps, and became known as the "perfect fifth."

A set of seven musical notes (equivalent to the white keys, designated A–G, of an octave on a modern piano) proved somewhat limiting, however, and smaller tuning divisions called semitones were later introduced. These led to a versatile system of twelve notes (both the white and black keys on a modern piano). While the seven white keys alone can only make a pleasing progression (known as the "diatonic scale") when played in order up or down from one "C" to the next, the additional black keys ("sharps" and "flats") allow such a progression to be followed from any point.

"Pythagorean tuning" based on the perfect fifth was used to find the desirable pitch of notes on Western musical instruments for many centuries, until changes in musical taste in the Renaissance era led to more complex tunings. Musical cultures outside Europe, such as those of India and China,

A medieval woodcut depicts the musical investigations of Pythagoras and his follower Philolaus, including the various sounds produced by different sized wind instruments.

followed their own traditions, although they, too, recognized the pleasing effects when notes of certain pitches were played in sequence or together.

Music of the spheres?

For Pythagorean philosophers, the realization that music was shaped by mathematics revealed a profound truth about the universe as a whole. It inspired them to look for mathematical patterns elsewhere, including in the heavens. Studies of the cyclic patterns in which the planets and stars moved across the sky led to a theory of cosmic harmony that later became known as the "music of the spheres."

Several followers of Pythagoras also attempted to explain the nature of musical notes by considering the physics involved. The Greek philosopher Archytas (c. 428–347 BCE) suggested that oscillating strings created sounds that moved at different speeds, which humans heard as different pitches. Although incorrect, this theory was adopted and repeated in the teachings of the hugely influential philosophers Plato

and Aristotle and passed into the canon of Western pre-Renaissance musical theory.

Another long-standing musical misunderstanding passed down from the Pythagoreans was their claim that the pitch of a string had a proportional relationship to both its length and to the tension at which it was strung. When Italian lute-player and musical theorist Vincenzo Galilei (father of Galileo) investigated these supposed laws in the mid-16th century, he found that while the claimed relationship between length and pitch was correct, the law of tension was more complex—the pitch varied in proportion to the square root of the tension applied.

Galilei's discovery led to wider questions about the supposed superiority of ancient Greek knowledge, while his experimental method—carrying out practical tests and mathematical analysis rather than taking the claims of authority for granted—became a major influence on his son. ∎

> ❝
> They saw in numbers the attributes and ratios of the musical scales.
> **Aristotle**
> **on the Pythagoreans**
> ❞

Pythagoras

Little is known about the early life of Pythagoras, but later Greek historians agreed that he was born on the Aegean island of Samos, a major trading center, in around 570 BCE. Some legends tell of youthful travels in the Near East, and studying under Egyptian or Persian priests and philosophers, as well as Greek scholars. They also say that on returning to Samos, he rose rapidly in public life.

At around the age of 40, Pythagoras moved to Croton, a Greek city in southern Italy, where he founded a school of philosophy that attracted many followers. Surviving writings from Pythagoras's pupils suggest that his teachings included not only mathematics and music but also ethics, politics, metaphysics (philosophical inquiry into the nature of reality itself), and mysticism.

Pythagoras acquired great political influence over the leaders of Croton and may have died during a civil uprising that was triggered by his rejection of calls for a democratic constitution. His death is placed at around 495 BCE.

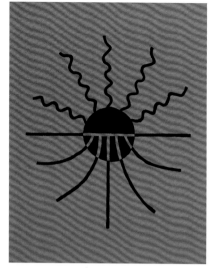

LIGHT FOLLOWS THE PATH OF LEAST TIME
REFLECTION AND REFRACTION

IN CONTEXT

KEY FIGURE
Pierre de Fermat (1607–1665)

BEFORE
c. 160 CE Ptolemy puts forward
a theory that the eye emits
rays which return information
to the observer.

c. 990 Persian mathematician
Ibn Sahl develops a law of
refraction after studying the
bending of light.

1010 Arabic scholar Ibn
al-Haytham's *Book of Optics*
proposes that vision is the
result of rays entering the eye.

AFTER
1807 Thomas Young coins
the term "refractive index" to
describe the ratio of the speed
of light in a vacuum to its
speed in a refracting material.

1821 Frenchman Augustin
Fresnel outlines a complete
theory of light and describes
refraction and reflection in
terms of waves.

Reflection and refraction
are the two fundamental
behaviors of light.
Reflection is the tendency for
light to bounce off a surface in a
direction related to the angle at
which it approaches that surface.
Early studies led the Greek
mathematician Euclid to note
that light is reflected off a mirror
at an "angle of reflection" equal to
its "angle of incidence"—the angle
at which it approaches in relation to
a line perpendicular to the mirror's
surface, known as the "normal." The
angle between the incoming ray
and the normal is the same as
that between the normal and the
reflected ray. In the 1st century CE,
the mathematician Hero of
Alexandria showed how this
path always involves the light
ray traveling over the shortest
distance (and spending the
least time doing so).

Refraction is the way rays of light
change direction when passing
from one transparent material to

Light approaching the boundary to another material can either be
reflected off at the same angle to the "normal" line, perpendicular to the
surface, or refracted at an angle that relates to the angle of approach and
the relative speed of light in the two materials. Whether light is reflected
or refracted, it always follows the shortest and simplest path.

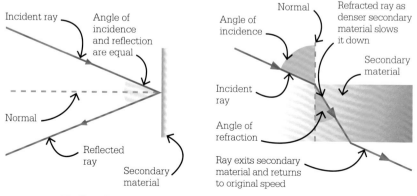

Reflection

Refraction

See also: Energy and motion 56–57 ▪ Focusing light 170–175 ▪ Lumpy and wavelike light 176–179 ▪ Diffraction and interference 180–183 ▪ Polarization 184–187 ▪ The speed of light 275 ▪ Dark matter 302–305

another. It was 10th-century Persian mathematician Ibn Sahl who first found a law linking a ray's angle of incidence at the boundary of two materials with its angle of refraction in the second material, as well as the properties of the two materials. This was rediscovered in Europe in the early 17th century, most notably by Dutch astronomer Willebrord Snellius, and is known as Snell's law.

Shortest path and time

In 1658, French mathematician Pierre de Fermat realized that both reflection and refraction could be described using the same basic principle—an extension of that put forward by Hero of Alexandria. Fermat's principle states that the path light will take between any two points is that which can be crossed in the least time.

Fermat created his principle by considering Dutch physicist Christiaan Huygens' earlier concept describing the motion of light in the form of waves and how it applied to cases with the

A landscape is reflected in a lake on a still day. The angle of reflection equals the angle of incidence—the angle at which sunlight reflected from the landscape hits the water.

smallest imaginable wavelength. This is often seen as justification for the concept of a light "ray"— the widely used idea of a beam of light that travels through space along the path of least time. When light is moving through a single unchanging medium, this means that it will travel in straight lines (except when the space through which it travels is distorted). However, when light reflects from a boundary or passes into a transparent medium in which its speed is slower or faster, this "principle of least time" dictates the path it will take.

By assuming that the speed of light was finite, and that light moved more slowly in denser transparent materials, Fermat was able to derive Snell's law from his principle, describing correctly how light will bend toward the normal

on entering a denser material, and away from it when entering one that is less dense.

Fermat's discovery was important in its own right, but also widely regarded as the first example of what is known as a family of "variational principles" in physics, describing the tendency of processes to always follow the most efficient pathway. ▪

Pierre de Fermat

Born in 1607, the son of a merchant from southwestern France, Pierre de Fermat trained and worked as a lawyer, but is mostly remembered for mathematics. He found a means of calculating the slopes and turning points of curves a generation before Isaac Newton and Gottfried Leibniz used calculus to do so.

Fermat's "principle of least time" is regarded as a key step toward the more universal "principle of least action"—an observation that many physical phenomena behave in ways that minimize (or sometimes maximize)

the energy required. This was important for understanding not only light and large-scale motion, but also the behavior of atoms on the quantum level. Fermat died in 1665; three centuries later, his famous Fermat's last theorem about the behavior of numbers at powers higher than 2 was finally proved in 1995.

Key work

1636 "Method of Finding the Maxima, Minima and Linear Tangents of Curves"

A NEW VISIBLE WORLD

WORLD

FOCUSING LIGHT

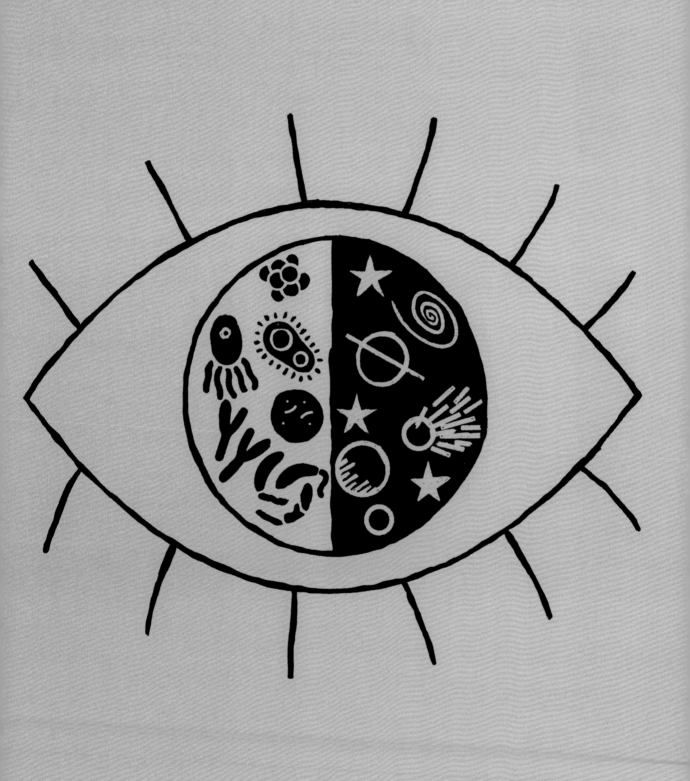

IN CONTEXT

KEY FIGURE
Antonie van Leeuwenhoek
(1632–1723)

BEFORE
5th century BCE In ancient Greece, convex disks of crystal or glass are used to ignite fires by focusing sunlight on a single point.

c. 1000 CE The first lenses, glass "reading stones" with a flat base and convex upper surface, are used in Europe.

AFTER
1893 German scientist August Köhler devises an illumination system for the microscope.

1931 In Germany, Ernst Ruska and Max Knoll build the first electron microscope, which uses the quantum properties of electrons to view objects.

1979 The Multiple Mirror Telescope at Mount Hopkins, Arizona, is built.

> Where the telescope ends, the microscope begins. Which of the two has the grander view?
> **Victor Hugo**
> *Les Misérables*

Just as rays of light can be reflected in different directions by plane mirrors (with a flat surface), or refracted from one angle to another as they cross a flat boundary between different transparent materials, so curved surfaces can be used to bend light rays onto converging paths—bringing them together at a point called a focus. The bending and focusing of light rays using lenses or mirrors is the key to many optical instruments, including the groundbreaking single-lens microscope designed by Antonie van Leeuwenhoek in the 1670s.

Magnifying principles

The first form of optical instrument was the lens, used in Europe from the 11th century. The refraction process as light enters and leaves a convex lens (with an outward-curving surface) bends the spreading light rays from a source onto more closely parallel paths, creating an image that occupies a larger part of the eye's field of vision. These magnifying lenses, however, had limitations. Studying objects that were very close to the lens, using larger lenses to achieve a greater field of view (area of the object that could be magnified) involved bending light rays that would be diverging strongly between opposite sides of the lens. This required a more powerful lens (sharply curved and thicker at the center), which due to the limitations of early glass manufacture was more likely to produce distortions. High magnifications therefore seemed impossible to achieve. These problems held up the development of optical instruments for centuries.

The first telescopes

In the 17th century, scientists realized that by using a combination of multiple lenses instead of a single lens in an instrument they could improve magnification significantly,

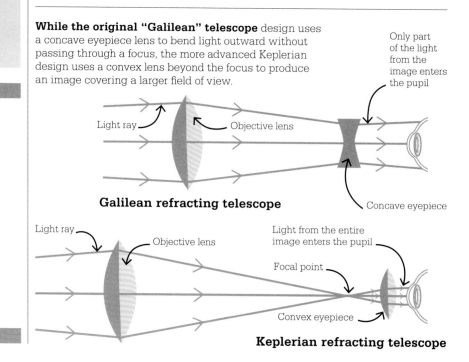

While the original "Galilean" telescope design uses a concave eyepiece lens to bend light outward without passing through a focus, the more advanced Keplerian design uses a convex lens beyond the focus to produce an image covering a larger field of view.

Light ray — Objective lens — Only part of the light from the image enters the pupil

Galilean refracting telescope — Concave eyepiece

Light ray — Objective lens — Light from the entire image enters the pupil — Focal point — Convex eyepiece

Keplerian refracting telescope

See also: Reflection and refraction 168–169 ▪ Lumpy and wavelike light 176–179 ▪ Diffraction and interference 180–183
▪ Polarization 184–187 ▪ Seeing beyond light 202–203

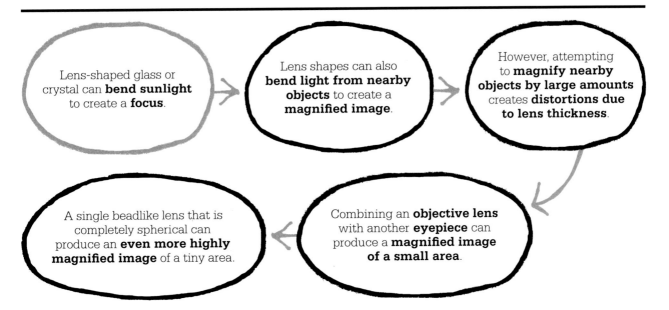

Lens-shaped glass or crystal can **bend sunlight** to create a **focus**.

Lens shapes can also **bend light from nearby objects** to create a **magnified image**.

However, attempting to **magnify nearby objects by large amounts** creates **distortions due to lens thickness**.

A single beadlike lens that is completely spherical can produce an **even more highly magnified image** of a tiny area.

Combining an **objective lens** with another **eyepiece** can produce a **magnified image of a small area**.

creating optical tools that could magnify not only nearby objects but also those very far away.

The first compound optical instrument (one using multiple lenses) was the telescope, usually said to have been invented by Dutch lens-maker Hans Lippershey around 1608. Lippershey's device, consisting of two lenses mounted at either end of a tube, produced images of distant objects magnified by a factor of three. Just as a magnifying glass bends diverging rays from nearby objects onto more parallel paths, so the front, or "objective." lens of a telescope gathers near-parallel rays from more distant objects and bends them onto a converging path. However before the rays can meet, a concave (inward-curving) "eyepiece" lens bends them back onto diverging paths—creating an image in the observer's eye that appears larger and (because the objective lens gathered much more light than a human pupil) also brighter.

In 1609, Italian physicist Galileo Galilei built his own telescope based on this model. His rigorous approach allowed him to improve on the original, producing devices that could magnify by a factor of 30. These telescopes allowed him to make important astronomical discoveries, but the images were still blurry, and the field of view tiny.

[T]he effect of my instrument is such that it makes an object fifty miles off appear as large as if it were only five miles away.
Galileo Galilei

In 1611, German astronomer Johannes Kepler came up with a better design. In the Keplerian telescope, the rays are allowed to meet, and the eyepiece lens is convex rather than concave. Here, the lens is placed beyond the focus point, at a distance where rays have begun to diverge again. The eyepiece therefore acts more like a normal magnifying glass, creating an image with a larger field of view and potentially offering far higher magnification. Keplerian telescopes form images that are upside-down and back-to-front, but this was not a significant problem for astronomical observations, or even for practiced terrestrial users.

Tackling aberration
The key to increasing a telescope's magnifying power lay in the strength and separation of its lenses, but stronger lenses brought with them problems of their own. Astronomers identified two major problems—colorful »

Ten times more powerful than the Hubble Space Telescope, the Giant Magellan Telescope in Chile, due to be completed in 2025, will capture light from the outer reaches of the universe.

fringes around objects known as "chromatic aberration" (caused by different colors of light refracting at different angles) and blurred images due to "spherical aberration" (difficulties in making a lens of ideal curvature).

It seemed that the only practical route to high magnification with minimal aberration was to make the objective lens larger and thinner, and place it much further away from the eyepiece. In the late 17th century, this led to the building of extraordinary "aerial telescopes" focusing light over distances of more than 330 ft (100 meters).

A more practical solution was discovered by British lawyer Chester Moore Hall in around 1730. He realized that nesting an objective lens made of weakly refractive glass next to a concave second lens with much stronger refraction would create a "doublet" that focused all colors at a single distance, avoiding blurry fringes. This "achromatic" lens became widespread after Hall's technique was discovered and marketed by optician John Dollond in the 1750s. Later developments of the same idea allowed telescope makers to eliminate spherical aberration, too.

Mirrored telescopes

From the 1660s, astronomers began using reflector telescopes. These use a concave, curved primary mirror to gather parallel light rays and bounce them onto converging paths. Various designs were proposed in the early 17th century, but the first practical example was built by Isaac Newton in 1668. This used a spherically curved primary lens, with a small secondary mirror suspended at a diagonal angle in front of it, which intercepted the converging rays and sent them toward an eyepiece on the side of the telescope tube (crossing over at a focus point along the way).

Using reflection rather than refraction sidestepped the issue of chromatic aberration, but early reflectors were far from perfect. The "speculum metal" (a highly polished copper and tin alloy), used for mirrors of the time, produced faint images tinted with brassy colors. Silver-coated glass, a better reflector, was pioneered by French physicist Léon Foucault in the 1850s.

Compound microscopes

Just as scientists combined lenses to magnify distant objects, they did the same to improve the power of the magnifying glass to look at minuscule objects that were close. Credit for inventing the "compound microscope" is disputed, but Dutch inventor Cornelis Drebbel demonstrated such an instrument in London in 1621. In order to minimize thickness and reduce aberration,

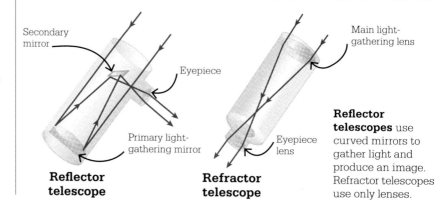

Reflector telescope

Refractor telescope

Reflector telescopes use curved mirrors to gather light and produce an image. Refractor telescopes use only lenses.

> My method for seeing the very smallest animalcules and minute eels, I do not impart to others; nor how to see very many animalcules at one time.
> **Antonie van Leeuwenhoek**

the objective lens of a compound microscope has a very small diameter (so light rays passing through different points on the lens are only slightly divergent). The eyepiece lens, which is often larger than the objective, does most of the work of producing the magnified image.

The most successful early compound microscope was designed by Robert Hooke in the 1660s. He mounted a third convex lens between the objective lens and the eyepiece, bending the light more sharply and increasing the magnification, at a cost of greater aberration. He also addressed another important issue—the tiny field of view on which the microscope focused meant that images tended to be dim (simply because there are fewer light rays reflecting from a smaller area). To correct this, he added an artificial source of illumination— a candle whose light was focused onto the object by a spherical glass bulb filled with water.

Hooke's large-format drawings of plant and insect structures, created by using the microscope and published in the influential 1665 book *Micrographia*, caused a sensation, but his work was soon surpassed by that of Dutch scientist Antonie van Leeuwenhoek.

Simple ingenuity

Van Leeuwenhoek's single-lens microscope was far more powerful than compound microscopes of the time. The lens itself was a tiny glass bead—a polished sphere able to bend light rays from a small area very sharply to create a highly magnified image. A metal frame, holding the tiny lens in place, was held close to the eye, in order for the sky to act as a light source behind it. A pin on the frame held the sample in place, and three screws could move the pin in three dimensions, allowing adjustments to the focus and the area on which the light focused by the lens fell.

Van Leeuwenhoek kept his lens-making technique a secret, but it achieved a magnification of at least 270 times, revealing features far smaller than Hooke had seen. With this powerful tool, he was able to discover the first bacteria, human spermatozoa, and the internal structure of the "cells" already identified by Hooke, for which he is known as the founding father of microbiology. ∎

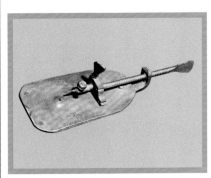

Van Leeuwenhoek made his microscopes himself. They comprised a tiny spherical lens held between two plates with a hole in each side and a pin to hold the specimen.

Antonie van Leeuwenhoek

Born in the Dutch town of Delft in 1632, van Leeuwenhoek became an apprentice in a linen shop at the age of six. On marrying in 1654, he set up his own draper's business. Wishing to study the quality of fibers up close, but unhappy with the power of available magnifying glasses, he studied optics and began to make his own microscopes.

Van Leeuwenhoek soon started to use his microscopes for scientific studies, and a physician friend, Regnier de Graaf, brought his work to the attention of London's Royal Society. Van Leeuwenhoek's studies of the microscopic world, including his discovery of single-celled organisms, amazed its members. Elected to the Royal Society in 1680, he was visited by leading scientists of the day until his death in 1723.

Key works

375 letters in *Philosophical Transactions of the Royal Society*
27 letters in *Memoirs of the Paris Academy of Sciences*

LIGHT IS A WAVE

LUMPY AND WAVELIKE LIGHT

IN CONTEXT

KEY FIGURE
Thomas Young (1773–1829)

BEFORE
5th century BCE The Greek philosopher Empedocles claims that "light corpuscles" emerge from the human eye to illuminate the world around it.

c. 60 BCE Lucretius, a Roman philosopher, proposes that light is a form of particle emitted by luminous objects.

c. 1020 In his *Book of Optics*, Arab polymath Ibn al-Haytham theorizes that objects are lit by the reflection of light from the sun.

AFTER
1969 At Bell Labs, in the US, Willard Boyle and George E. Smith invent the charge-coupled device (CCD), which generates digital images by collecting photons.

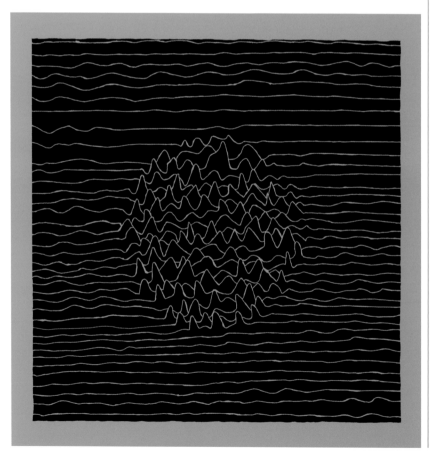

Theories about the nature of light had occupied the minds of philosophers and scientists since ancient times, but it was the development of optical instruments such as the telescope and microscope in the 1600s that led to important breakthroughs, such as confirmation that Earth was not the center of the solar system and the discovery of a microscopic world.

In around 1630, French scientist and philosopher René Descartes, seeking to explain the phenomenon of light refraction (how light bends when it passes between media), proposed that light was a traveling disturbance—a wave moving at

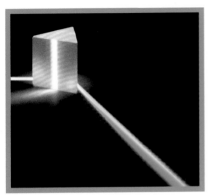

White light splits into the rainbow components of the visible spectrum when it passes through a prism. The precise color depends on the length of the wave—red has the longest.

infinite speed through a medium that filled empty space. (He called this medium the plenum.) Around 1665, English scientist Robert Hooke was the first to make a link between the diffraction of light (its ability to spread out after passing through narrow openings) and the similar behavior of water waves. This led him to suggest that not only was light a wave, but it was a transverse wave, like water—one in which the direction of disturbance is at right angles to the direction of motion, or "propagation."

The color spectrum
Hooke's ideas were largely overlooked thanks to the influence within the scientific community of his great rival, Isaac Newton. In around 1670, Newton began a series of experiments on light and optics in which he showed that color is an intrinsic property of light. Until then, most researchers had assumed that light acquired colors through its interaction with different matter, but Newton's

experiments—in which he split white light into its color components using a prism and then recreated a white beam using lenses—revealed the truth.

Newton also investigated reflection. He showed that beams of light always reflect in straight lines and cast sharp-edged shadows. In his view, wavelike light would show more signs of bending and spreading, so he concluded that light must be corpuscular—made up of tiny, lumpy particles. Newton published his findings in 1675 and developed them further in his book *Opticks* in 1704. His premise dominated theories about light for more than a century.

Light bends
Newton's work had some failings, particularly when it came to refraction. In 1690, Dutch scientist and inventor Christiaan Huygens published *Treatise on Light*, in which he explained how light

One may conceive light to spread successively, by spherical waves.
Christiaan Huygens

bends when it passes between different media (such as water and air) in terms of wave behavior. Huygens rejected the corpuscular model on the grounds that two beams could collide without scattering in unexpected directions. He suggested that light was a disturbance moving at very high (though finite) speed through what he called the "luminiferous ether"—a light-carrying medium. »

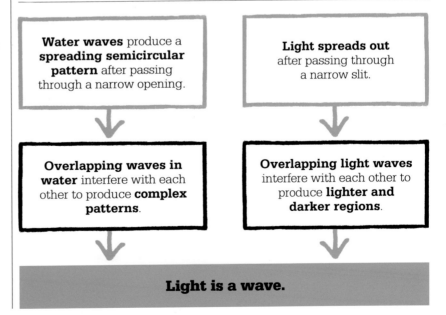

Water waves produce a **spreading semicircular pattern** after passing through a narrow opening.

Light spreads out after passing through a narrow slit.

Overlapping waves in water interfere with each other to produce **complex patterns**.

Overlapping light waves interfere with each other to produce **lighter and darker regions**.

Light is a wave.

> The experiments I am about to relate … may be repeated with great ease, whenever the sun shines, and without any other apparatus than is at hand to every one.
> **Thomas Young**

Huygens developed a useful principle in which each point on the advancing "wave front" of a light beam is treated as a source of smaller "wavelets" spreading in all directions. The overall course of the beam can be predicted by finding the direction in which the wavelets align and reinforce each other.

Young's experiments

For most of the 18th century, the corpuscular model dominated thinking on light. That changed in the early 19th century with the work of British polymath Thomas Young. As a medical student in the 1790s, Young investigated the properties of sound waves and became interested in general wave phenomena. Similarities between sound and light convinced him that light might also form waves, and in 1799 he wrote a letter setting out his argument to the Royal Society in London.

Faced with fierce skepticism from the followers of Newton's model, Young set about devising experiments that would prove the wave behavior of light beyond doubt. With a shrewd grasp of the power of analogy, he built a ripple tank—a shallow bath of water with a paddle at one end to generate periodic waves. Placing obstacles, such as a barrier with one or more openings, in the bath allowed him to study wave behavior. He showed how after passing through a narrow slit in a barrier, parallel straight waves would produce a spreading semicircular pattern— an effect similar to the diffraction of light through narrow openings.

A barrier with two slits created a pair of spreading wave patterns that overlapped. Waves from the two slits could freely pass through each other, and where they crossed, the wave height was determined by the phases of the overlapping waves—an effect known as interference. When two wave peaks or two troughs overlapped, they increased the height or depth of the wave; when a peak and a trough met, they cancelled each other out.

Young's next step, outlined in a lecture to the Royal Society in 1803, was to demonstrate that light behaved in a similar way to waves. To do this, he split a narrow beam of sunlight into two beams using a piece of thin card and compared the way they illuminated a screen to how a single beam illuminated it. A pattern of light and dark areas appeared on the screen when the dividing card was in place, but disappeared when the card was removed. Young had elegantly shown that light was behaving in the same way as water waves— as the two separated beams diffracted and overlapped, areas of greater and lesser intensity interacted to produce an interference pattern that the corpuscular theory could not explain. (The experiment was later refined with two narrow slits in an opaque glass surface, for which it became known as the "double-slit experiment".) Young argued that his demonstration showed beyond doubt that light was a wave. What was more, the patterns varied according to the color of the diffracted light, which implied that colors were determined by wavelengths.

While Young's experiment revived debate about the nature of light, the wave theory of light was not yet fully accepted. Despite Hooke's early ideas, most scientists

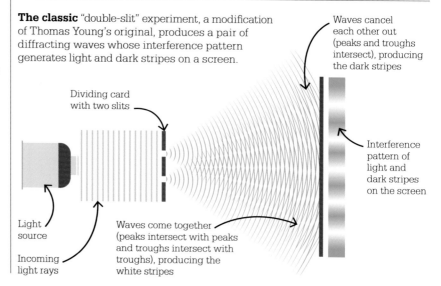

The classic "double-slit" experiment, a modification of Thomas Young's original, produces a pair of diffracting waves whose interference pattern generates light and dark stripes on a screen.

Dividing card with two slits

Waves cancel each other out (peaks and troughs intersect), producing the dark stripes

Interference pattern of light and dark stripes on the screen

Light source

Incoming light rays

Waves come together (peaks intersect with peaks and troughs intersect with troughs), producing the white stripes

Thomas Young

The eldest of 10 children, Young was born into a Quaker family in the English county of Somerset in 1773. A born polymath, he mastered several modern and ancient languages as a child, before going on to study medicine. After making a mark with his private scientific research, in 1801 he was appointed professor at the recently founded Royal Institution, but later resigned to avoid conflict with his medical practice.

Young continued to study and experiment and made valuable contributions in many fields. He made progress in the translation of Egyptian hieroglyphs, explained the working of the eye, investigated the properties of elastic materials, and developed a method for tuning musical instruments. His death in 1829 was widely mourned in the scientific community.

Key works

1804 "Experiments and Calculations Relative to Physical Optics"
1807 *A Course of Lectures on Natural Philosophy and the Mechanical Arts*

(including Young himself) at first assumed that if light was a wave, it must be longitudinal—like sound, where the disturbance of the medium moves back and forth parallel to the direction of propagation. This made some light phenomena, such as polarization, impossible to explain through waves.

A solution emerged in France in 1816, when André-Marie Ampère suggested to Augustin-Jean Fresnel that the waves might be transverse, which would explain the behavior of polarized light. Fresnel went on to produce a detailed wave theory of light that also explained diffraction effects.

Electromagnetic properties

Acceptance of light's wavelike nature coincided with developing studies of electricity and magnetism. By the 1860s, James Clerk Maxwell was able to describe light in a series of elegant equations as an electromagnetic disturbance moving at 186,282 miles (299,792 km) per second.

However, awkward questions still surrounded Huygens' so-called "luminiferous ether" through which light supposedly traveled. Most experts believed that Maxwell's equations described the speed at which light would enter the ether from a source. The failure of increasingly sophisticated experiments designed to detect this light-carrying medium raised fundamental questions and created a crisis that was only resolved by Einstein's theory of special relativity.

Waves and particles

Einstein was largely responsible for a final shift in our understanding of light. In 1905, he put forward an explanation for the photoelectric effect—a phenomenon in which currents flow from the surface of certain metals when they are exposed to certain kinds of light. Scientists had been puzzled by the fact that a feeble amount of blue or ultraviolet light would cause a current to flow from some metals while they remained inactive under even the most intense red light. Einstein suggested (building on the concept of light quanta previously used by Max Planck) that despite being fundamentally wavelike, light travels in small particle-like bursts, now called photons. The intensity of a light source depends on the number of photons it produces, but the energy of an individual photon depends on its wavelength or frequency—hence high-energy blue photons can provide electrons with enough energy to flow, while lower-energy red ones, even in large numbers, cannot. Since the early 20th century, the fact that light can behave like a wave or like a particle has been confirmed in numerous experiments. ∎

[W]e have good reason to conclude that light itself [...] is an electromagnetic disturbance in the form of waves propagating [...] according to the laws of electromagnetism.
James Clerk Maxwell

LIGHT IS NEVER KNOWN TO BEND INTO THE SHADOW

DIFFRACTION AND INTERFERENCE

IN CONTEXT

KEY FIGURE
Augustin-Jean Fresnel
(1788–1827)

BEFORE
1665 Robert Hooke compares the movement of light to the spreading of waves in water.

1666 Isaac Newton demonstrates that sunlight is composed of different colors.

AFTER
1821 Augustin-Jean Fresnel publishes his work on polarization, suggesting for the first time that light is a transverse (waterlike) wave rather than a longitudinal (soundlike) wave.

1860 In Germany, Gustav Kirchhoff and Robert Bunsen use diffraction gratings to link bright "emission lines" of light at specific wavelengths with different chemical elements.

D ifferent types of waves share similar behaviors, such as reflection (bouncing off a surface at the same angle as they approach the surface), refraction (changing direction at the boundaries between mediums), and diffraction—the way a wave spreads around obstacles or spreads out when it passes through an aperture. An example of diffraction is the way water waves spread into the shadow region beyond a barrier. The discovery that light also exhibits diffraction was key to proving its wavelike nature.

The diffraction of light was first systematically observed in the 1660s by Francesco Maria

See also: Reflection and refraction 168–169 ▪ Lumpy and wavelike light 176–179 ▪ Polarization 184–187 ▪ The Doppler effect and redshift 188–191 ▪ Electromagnetic waves 192–195 ▪ Particles and waves 212–215

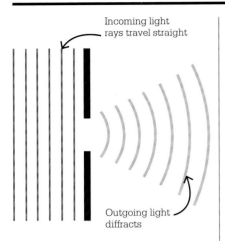

When linear light waves pass through a narrow aperture in a barrier, they diffract (spread out) into semicircular wave fronts.

Grimaldi, a Jesuit priest and physicist in Italy. Grimaldi built a dark room with a pinhole aperture through which a pencil-width beam of sunlight could enter and hit an angled screen. Where the beam struck the screen, it created an oval circle of light, which Grimaldi measured. He then held a thin rod in the path of the light beam and measured the size of the shadow it cast within the illuminated area. Grimaldi then compared his results to calculations based on the assumption that light travels in straight lines. He found that not only was the shadow larger than such calculations suggested it should be, but so was the illuminated oval.

The conclusion Grimaldi drew from these comparisons was that light was not composed of simple particles (corpuscles) that traveled in straight lines, but had wavelike properties, similar to water, that

allowed it to bend. It was Grimaldi who coined the term "diffraction" to describe this. His findings, published posthumously in 1665, were a thorn in the side of scientists trying to prove the corpuscular nature of light.

Competing theories
In the 1670s, Isaac Newton treated diffraction (which he called "inflexion") as a special type of refraction that occurred when light rays (which he believed were corpuscular) passed close to obstacles. At around the same time, Newton's rival Robert Hooke cast doubt on his theory by conducting successful demonstrations of Grimaldi's experiments, and Dutch scientist and inventor Christiaan Huygens put forward his own wave theory of light. Huygens argued that many optical phenomena could only be explained if light was treated as an advancing "wave front," along which any point was a source of secondary waves, whose interference and reinforcement determined the direction in which the wave front moved.

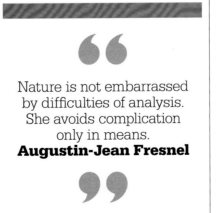

> Nature is not embarrassed by difficulties of analysis. She avoids complication only in means.
> **Augustin-Jean Fresnel**

Like light waves, wind-generated waves on a lake spread out in circular ripples when they pass through a narrow gap, illustrating the property of diffraction common to all waves.

Neither the wave theory of light nor Newton's corpuscular theory explained the phenomenon of colored fringing—something that Grimaldi had noted around both the edges of the illuminated circle and the shadow of the rod in his experiments. Newton's struggle with this question led him to propose some intriguing ideas. He maintained that light was strictly corpuscular, but that each fast-moving corpuscle could act as a source of periodic waves whose vibrations determined its color.

Young's experiments
Newton's improved theory of light, published in his book *Opticks* in 1704, did not entirely resolve the questions around diffraction, but the corpuscular model was widely accepted until 1803, when Thomas Young's demonstration of interference between diffracted light waves led to a resurgence of »

Augustin-Jean Fresnel

The second of four sons of a local architect, Fresnel was born at Broglie in Normandy, France, in 1788. In 1806, he enrolled to study civil engineering at the National School of Bridges and Roads, and went on to become a government engineer.

Briefly suspended from work for his political views during the last days of the Napoleonic Wars, Fresnel pursued his interest in optics. Encouraged by physicist François Arago, he wrote essays and a prize memoir on the subject, including a mathematical treatment of diffraction. Fresnel later explained polarization using a model of light as a transverse wave and developed a lens for focusing intense directional beams of light (mainly used in lighthouses). He died in 1827, at the age of 39.

Key works

1818 *Memoir on the Diffraction of Light*
1819 "On the Action of Rays of Polarized Light upon each other" (with François Arago)

interest in Huygens' ideas. Young proposed two modifications to the Huygens' model that could explain diffraction. One was the idea that the points on a wave front at the very edges of the open aperture produced wavelets that spread into the "shadow" region beyond the barrier. The other was that the observed pattern of diffraction came about through a wave passing near the edge of the aperture interfering with a wave bouncing off the sides of the barrier.

Young also proposed that there must be a distance where light was sufficiently far from the edge to be unaffected by diffraction. If this distance varied between the different colors of light, Young maintained, it could explain the colored fringes discovered by Grimaldi. For instance, if there

is a distance where red light is still diffracted, but blue light is not, then it is red fringing that would appear.

Crucial breakthrough

In 1818, the French Academy of Sciences offered a prize for a full explanation of the "problem of inflection" identified by Young. Civil engineer Augustin-Jean Fresnel had been working on precisely this question for several years, conducting a series of intricate homemade experiments and sharing his findings with academician François Arago. Some of his work unwittingly replicated that of Young, but there were new insights as well, and Arago encouraged Fresnel to submit an explanatory memoir for the prize competition. In this memoir, Fresnel gave complex mathematical

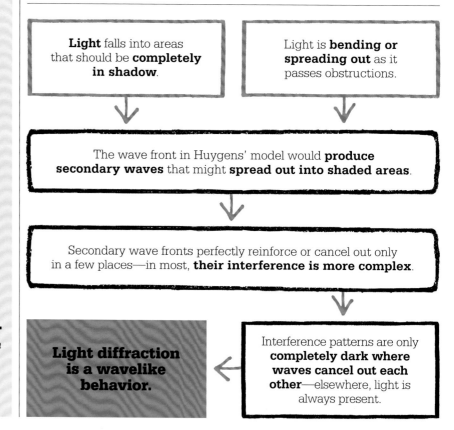

Light falls into areas that should be **completely in shadow**.

Light is **bending or spreading out** as it passes obstructions.

The wave front in Huygens' model would **produce secondary waves** that might **spread out into shaded areas**.

Secondary wave fronts perfectly reinforce or cancel out only in a few places—in most, **their interference is more complex**.

Interference patterns are only **completely dark where waves cancel out each other**—elsewhere, light is always present.

Light diffraction is a wavelike behavior.

The colors on a soap bubble are caused by the interruption of light waves as they reflect off the thin film of the bubble and interfere with one another.

equations to describe the position and strength of dark and light interference "fringes" and showed that they matched the results of real experiments.

Fresnel also demonstrated that the geometry of the fringes depended on the wavelength of the light that had produced them. For the first time, it was possible to measure interference fringes produced by a monochromatic (single-colored) light source and thereby calculate the wavelength of that specific color.

In the memoir, Fresnel summed up the difference between Huygens' wave theory and his own with remarkable simplicity. Huygens' error, he said, was to assume that light would only spread where the secondary waves exactly reinforced each other. In reality, however, complete darkness is only possible where the waves exactly cancel each other out.

Despite its elegance, Fresnel's prize memoir faced a skeptical reception from a judging panel who mostly supported the corpuscular theory of Newton. As one of the judges, Siméon-Denis Poisson, pointed out, Fresnel's equations predicted that the shadow cast by a circular obstacle illuminated by a point source of light, such as a pinhole, should have a bright spot at its very center—an idea Poisson believed to be absurd. The matter was resolved when Arago promptly constructed an experiment that produced just such a "Poisson spot." Despite this success, the wave theory was not broadly accepted until 1821, when Fresnel published his wave-based approach to explain polarization—how transverse light waves sometimes align in a particular direction.

Diffraction and dispersion

While Young and Fresnel were learning to understand the effects of diffraction and interference by directing light beams through two narrow slits, some instrument-makers were taking a different approach. As early as 1673, Scottish astronomer James Gregory had noted that sunlight passing through the fine gaps in a bird's feather spreads out into a rainbow-like spectrum similar to the behavior of white light passing through a prism.

In 1785, American inventor David Rittenhouse succeeded in replicating the effect of Gregory's bird feather artificially by winding fine hairs between two tightly threaded screws to create a mesh with a density of around 100 hairs per inch. This device was perfected in 1813 by German physicist and instrument-maker Joseph von Fraunhofer, who named it the "diffraction grating." Crucially, diffraction gratings offer a far more efficient means of splitting light than glass prisms because they absorb very little of the light striking them. They are therefore ideal for producing spectra from faint sources. Gratings and related instruments have gone on to be extremely useful in many branches of science.

Fraunhofer also built the first "ruling engine"—a machine that painstakingly scratched narrowly spaced transparent lines into a glass plate covered with an otherwise opaque surface layer. This device allowed him to spread sunlight into a far wider spectrum than could previously be achieved, revealing the presence of what became known as "Fraunhofer lines," which would later play a key role in understanding the chemistry of stars and atoms. ∎

> I think I have found the explanation and the law of colored fringes which one notices in the shadows of bodies illuminated by a luminous point.
> **Augustin-Jean Fresnel**

THE NORTH AND SOUTH SIDES OF THE RAY

POLARIZATION

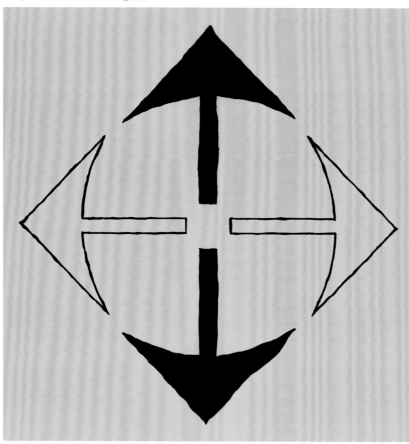

IN CONTEXT

KEY FIGURE
Étienne-Louis Malus
(1775–1812)

BEFORE
13th century References
to the use of "sunstones" in
Icelandic sagas suggest that
Viking sailors may have used
the properties of Iceland spar
crystals to detect polarization
for daytime navigation.

AFTER
1922 Georges Friedel
investigates the properties of
three kinds of "liquid crystal"
material and notes their ability
to alter the polarization plane
of light.

1929 American inventor
and scientist Edwin H. Land
invents "Polaroid," a plastic
whose polymer strands act as
a filter for polarized light, only
transmitting light in one plane.

Polarization is the alignment
of waves in a specific plane
or direction. The term is
usually applied to light, but any
transverse wave (oscillating at right
angles to the wave direction) can
be polarized. A number of natural
phenomena produce light that is
polarized—a portion of the sunlight
reflected off a flat and shiny surface
such as a lake, for example, is
polarized to match the angle of the
lake's surface—and light can also
be polarized artificially.

The investigation of different
effects caused by polarization
helped determine that light is
a wavelike phenomenon, and
also provided important evidence

See also: Force fields and Maxwell's equations 142–147 ▪ Focusing light 170–175 ▪ Lumpy and wavelike light 176–179 ▪ Diffraction and interference 180–183 ▪ Electromagnetic waves 192–195

> 66
>
> I believe that this phenomenon can serve lovers of nature and other interested persons for instruction or at least for pleasure.
> **Rasmus Bartholin**
> **on birefringence**
>
> 99

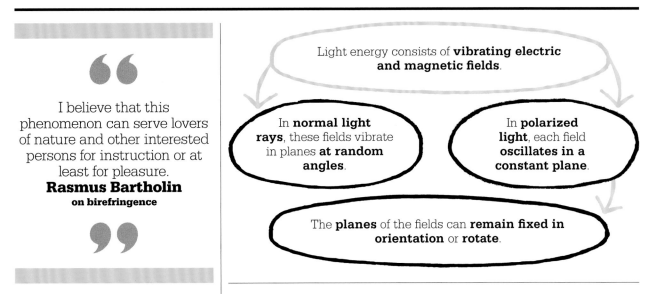

Light energy consists of **vibrating electric and magnetic fields**.

In **normal light rays**, these fields vibrate in planes **at random angles**.

In **polarized light**, each field **oscillates in a constant plane**.

The **planes** of the fields can **remain fixed in orientation** or **rotate**.

that it is electromagnetic in nature. The first account of identical light beams differing for seemingly inexplicable reasons was given by Danish scientist Rasmus Bartholin in 1669. He discovered that viewing an object through the crystals of a transparent mineral known as Iceland spar (a form of calcite) produced a double image of that object.

This phenomenon, known as birefringence, arises because the crystal has a refractive index (speed at which the light is transmitted) that changes according to the polarization of the light passing through it. Light reflected from most objects has no particular polarization, so on average, half the rays are split along each path, and a double image is created.

Early explanations

Bartholin published a detailed discussion of the birefringence effect in 1670, but he could not explain it in terms of any particular model of light. Christiaan Huygens,

an early champion of the wave theory, argued that the speed of light rays traveling through the crystal varied according to the direction in which they moved, and in 1690 used his "wave front" principle to model the image-doubling effect.

Huygens also carried out some new experiments, placing a second crystal of Iceland spar in front of the first and rotating it. When he did so, he found that the doubling effect disappeared at certain angles. He did not understand why, but he recognized that the two images produced by the first crystal were different from each other in some way.

Isaac Newton maintained that birefringence strengthened his case for light as a corpuscle—a particle with discrete "sides" that could be affected by its orientation. In Newton's view, the effect helped disprove the wave theory, as Huygens' model of light involved longitudinal waves (disturbances parallel to the wave direction) rather than transverse waves,

and Newton could not envisage how these could be sensitive to direction.

Sides and poles

In the early 19th century, French soldier and physicist Étienne-Louis Malus carried out his own version of Huygens' experiments. Malus was interested in applying »

Sunglasses with polarized lenses reduce the harsh glare of light reflecting off a snowy landscape by allowing only light waves that have been polarized in one direction to pass through.

mathematical rigor to the study of light. He developed a useful mathematical model for what was actually happening to light rays in three dimensions when they encountered materials with different reflective and refractive properties. (Previous models had simplified matters by considering only what happened to light rays moving in a single flat plane—two dimensions.)

In 1808, the French Institute of Science and Arts offered a prize for a full explanation of birefringence and encouraged Malus to participate. Like many in the French scientific establishment, Malus followed the corpuscular theory of light, and it was hoped that a corpuscular explanation for birefringence might help debunk the wave theory recently put forward in Britain by Thomas Young. Malus developed a theory in which light corpuscles had distinct sides and "poles" (axes of rotation). Birefringent materials, he claimed, refracted corpuscles along different paths depending on the direction of their poles. He coined the term "polarization" to describe this effect.

Have not the rays of light several sides, endued with several properties?
Isaac Newton

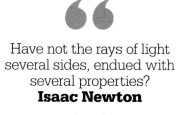

Malus discovered an important new aspect of the phenomenon while using a piece of Iceland spar to observe a reflection of the setting sun in a window of the Palais du Luxembourg in Paris. He noticed that the intensity of the sun's two images varied as he rotated the crystal, and one or other image disappeared completely with every 90-degree turn of the crystal. This, Malus realized, meant that the sunlight had already been polarized by reflection off the glass. He confirmed the effect by shining candlelight through a birefringent crystal and looking at how the resulting pair of light rays reflected in a bowl of water as he rotated the crystal—the reflections would disappear and reappear depending on the rotation of the crystal. He soon developed a law (Malus's law) describing how the intensity of a polarized image viewed through a crystal "filter" relates to the orientation of the crystal.

Further experiments revealed a similar effect with reflections from transparent materials. Malus noticed that interaction between the surface of the materials and unpolarized light (with a random mix of orientations) reflected light in one specific plane of polarization rather than others, while light polarized in the other plane passed into or through the new medium (such as water or glass) along a refracted path. Malus realized that the factor determining whether to reflect or refract must be the internal structure of the new medium, linked to its refractive index.

New insights

Malus's identification of a link between a material's structure and its effect on polarized light was important: one common application today is to study the internal

---

Étienne-Louis Malus

Born into a privileged Parisian family in 1775, Malus showed early mathematical promise. He served as a regular soldier before attending the École Polytechnique engineering school in Paris from 1794, after which he rose through the ranks of the army engineering corps and took part in Napoleon's expedition to Egypt (1798–1801). On returning to France, he worked on military engineering projects.

From 1806, a posting to Paris enabled Malus to mix with leading scientists. His talent for mathematically describing the behaviors of light, including

the "caustic" reflections created by curved reflecting surfaces, impressed his peers, and in 1810 his explanation of birefringence led to him being elected to the Académie des Sciences. A year later, he was awarded the Royal Society's Rumford Medal for the same work, despite Britain being at war with France. Malus died in 1812, aged 36.

Key works

1807 *Treatise on Optics*
1811 "Memoir on some new optical phenomena"

changes in materials under stress from the way they affect polarized light. He attempted to identify a relationship between a material's refractive index and the angle at which light reflecting from a surface would be perfectly "plane polarized" (aligned in a single plane). He found the correct angle for water but was thwarted by the poor quality of other materials he investigated; a general law was found a few years later in 1815 by Scottish physicist David Brewster.

The range of phenomena involving polarization grew. In 1811, French physicist François Arago found that passing polarized light through quartz crystals could rotate its axis of polarization (an effect now known as "optical activity"), and his contemporary Jean-Baptiste Biot reported that light from rainbows was highly polarized.

Biot went on to identify optical activity in liquids, formulating the concept of circular polarization (in which the polarization axis rotates as the light ray moves forward). He also discovered "dichroic" minerals, which are natural materials that allow light to pass through if it is polarized along one axis, but block

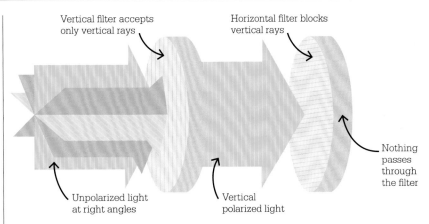

Vertical filter accepts only vertical rays

Horizontal filter blocks vertical rays

Unpolarized light at right angles

Vertical polarized light

Nothing passes through the filter

Unpolarized light waves oscillate at right angles to their direction of propagation (motion) in random directions. Plane-polarized light waves oscillate in a single direction, while the plane of circular-polarized waves rotates steadily as they propagate.

We find that light acquires properties which are relative only to the sides of the ray … I shall give the name of poles to these sides of the ray.
Étienne-Louis Malus

light polarized along the other axis. Using dichroic materials would become one of the ways in which polarized light could be easily produced. Others included using certain types of glass and specially shaped calcite prisms.

Polarized waves

In 1816, Arago and his protégé Augustin-Jean Fresnel made an unexpected and important discovery that undermined claims that polarization supported the corpuscular nature of light. They created a version of Thomas Young's "double-slit" experiment using two beams of light whose polarization could be altered, and found that patterns of dark and light caused by interference between the two beams were strongest when they had the same polarization. But the patterns faded as the difference in polarization increased, and they disappeared completely when the polarization planes were at right angles to each other.

This discovery was inexplicable through any corpuscular theory of light, and could not be explained by imagining light as a longitudinal wave. André-Marie Ampère suggested to Fresnel that a solution might be found by treating light as a transverse wave. If this were the case, it followed that the axis of polarization would be the plane in which the wave was oscillating (vibrating). Thomas Young reached the same conclusion when Arago told him about the experiment, but it was Fresnel who took inspiration to create a comprehensive wave theory of light, which would eventually displace all older models.

Polarization would also play a key role in the next major advance in the understanding of light. In 1845, British scientist Michael Faraday, seeking a way to prove the suspected link between light and electromagnetism, decided to see what would happen if he shone a beam of polarized light through a magnetic field. He discovered that he could rotate the plane of polarization by altering the strength of the field. This phenomenon, known as the Faraday effect, would ultimately inspire James Clerk Maxwell to develop his model of light as an electromagnetic wave. ∎

THE TRUMPETERS AND THE WAVE TRAIN

THE DOPPLER EFFECT AND REDSHIFT

IN CONTEXT

KEY FIGURE
Christian Doppler
(1803–1853)

BEFORE
1727 British astronomer James Bradley explains the aberration of starlight—a change in the angle at which light from distant stars approaches Earth, caused by Earth's motion around the sun.

AFTER
1940s "Doppler radar" is developed for aviation and weather forecasting after early forms of radar fail to account for rainfall and the Doppler shifts of moving targets.

1999 Based on observations of exploding stars, astronomers discover that some distant galaxies are further away than their Doppler shifts suggest, implying that the expansion of the universe is accelerating.

T oday, the Doppler effect is a familiar part of our everyday lives. We notice it as the shift in the pitch of sound waves when a vehicle with an emergency siren speeds toward us, passes us, and continues into the distance. The effect is named after the scientist who first proposed it as a theoretical prediction. When applied to light, it has turned out to be a powerful tool for learning about the universe.

Explaining star colors
The idea that waves (specifically, waves of light) might change their frequency depending on the relative motion of the source and the

See also: Lumpy and wavelike light 176–179 ▪ Discovering other galaxies 290–293 ▪ The static or expanding universe 294–295 ▪ Dark energy 306–307

> [The Doppler effect] will in the not too distant future offer astronomers a welcome means to determine the movements and distances of such stars.
> **Christian Doppler**

observer was first proposed by Austrian physicist Christian Doppler in 1842. Doppler was investigating the aberration of starlight—a slight shift in the apparent direction of light from distant stars, caused by Earth's motion around the sun—when he realized that motion through space would cause a shift in the frequency, and therefore color, of light.

The principle behind the theory is as follows. When the source and observer approach one another, "peaks" in the light waves from the source pass the observer at a higher frequency than they otherwise would, so the light's measured wavelength is shorter. In contrast, when the source and observer move apart, peaks in the wave reach the observer at lower frequency than they otherwise would. The color of light is dependent on its wavelength, so light from an approaching object appears bluer (due its shorter wavelengths) than it otherwise would, while light from a retreating object appears redder.

Doppler hoped this effect (which was soon named after him) would help explain the different colors of stars in the night sky when viewed from Earth. He elaborated his theory to take into account the motion of Earth through space, but did not realize the magnitude by which the speed of stellar motions is dwarfed by the speed of light, making the effect undetectable with the instruments of the time.

However, Doppler's discovery was confirmed in 1845, thanks to an ingenious experiment with sound. Making use of the recently opened Amsterdam–Utrecht railroad, Dutch scientist C.H.D. Buys Ballot asked a group of musicians to play a single »

Christian Doppler

Born in the Austrian city of Salzburg in 1803 to a wealthy family, Doppler studied mathematics at the Vienna Polytechnic Institute, then took physics and astronomy as a postgraduate. Frustrated at the bureaucracy involved in securing a permanent post, he came close to abandoning academia and emigrating to the US, but eventually found employment as a math teacher in Prague.

Doppler became a professor at the Prague Polytechnic in 1838, where he produced a wealth of mathematics and physics papers, including a celebrated study on the colors of stars and the phenomenon now known as the Doppler effect. By 1849, his reputation was such that he was given an important post at the University of Vienna. His health—delicate throughout much of his life—deteriorated, and he died of a chest infection in 1853.

Key work

1842 *Über das farbige Licht der Doppelsterne* (*Concerning the Colored Light of Double Stars*)

Sound waves spread out at the same speed in all directions when measured relative to the source. But if a siren approaches an observer, the wave frequency increases and the wavelength is shortened.

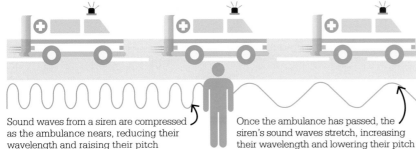

Sound waves from a siren are compressed as the ambulance nears, reducing their wavelength and raising their pitch

Once the ambulance has passed, the siren's sound waves stretch, increasing their wavelength and lowering their pitch

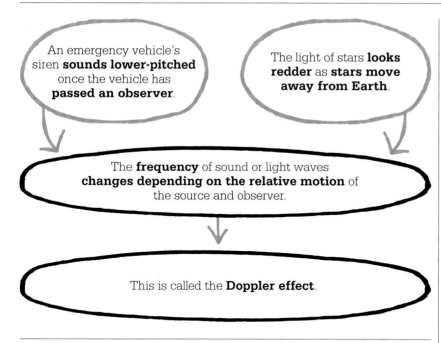

An emergency vehicle's siren **sounds lower-pitched** once the vehicle has **passed an observer**.

The light of stars **looks redder** as **stars move away from Earth**.

The **frequency** of sound or light waves **changes depending on the relative motion** of the source and observer.

This is called the **Doppler effect**.

continuous note while traveling on the train. As the carriage sped past him, Buys Ballot experienced the now-familiar pitch-shift of sound—the high pitch of the note as the train approached him grew lower as the train traveled away.

Measuring the shift

The effects of the Doppler shift on light were calculated by French physicist Armand Fizeau in 1848. Unaware of Doppler's own work, he provided a mathematical basis for the concept of red and blue light shifts. Doppler had suggested that his theory would bring about a new understanding of the motions, distances, and relationships of stars, but Fizeau suggested a practical way in which shifts in starlight might be detected. He correctly recognized that changes in a star's overall color would be both minute and hard to quantify. Instead, he proposed that astronomers could identify what we know as Doppler shifts through the positions of "Fraunhofer lines,"

narrow, dark lines—such as those already observed in the sun's spectrum, and presumed to exist in the spectra of other stars—that could act as defined reference points in a spectrum of light.

Putting this idea into practice, however, required significant advances in technology. At such vast distances, even the brightest stars are far more faint than the sun, and the spectra created when their light is split by diffraction—in order to measure individual spectral lines—are fainter still. British astronomers William and Margaret Huggins measured the spectrum of light from the sun in 1864, and their colleague William Allen Miller was able to use the measurements to identify elements in distant stars.

By this time, German scientists Gustav Kirchhoff and Robert Bunsen had shown that the sun's Fraunhofer lines were caused by specific elements absorbing light, which enabled astronomers to calculate the wavelengths where these absorption lines would

appear in the spectrum of a static star. In 1868, William Huggins successfully measured the Doppler shift of the spectral lines of Sirius, the brightest star in the sky.

Glimpsing deeper

In the late 19th century, astronomy's reliance on painstaking observations was transformed by improvements in photography, enabling scientists to measure the spectra and Doppler shifts of much fainter stars. Long-exposure photography gathered vastly more light than the human eye, and yielded images that could be stored, compared, and measured long after the original observation.

German astronomer Hermann Carl Vogel pioneered this marriage of photography and "spectroscopy" in the 1870s and 1880s, leading to important discoveries. In particular, he identified many apparently single stars with spectral lines that periodically "doubled," separating and then reuniting in a regular cycle. He showed that this line-splitting arose because the stars are binary pairs—visually inseparable stars whose close orbits around each other mean that as one star retreats from Earth

[The] color and intensity of […] light, and the pitch and strength of a sound will be altered by a motion of the source of the light or of the sound …
William Huggins

So-called "cosmological" redshift is not caused by galaxies moving apart in a conventional sense, but by the stretching of space itself as the universe has expanded over billions of years. This carries galaxies apart from each other and stretches the wavelengths of light as it moves between them.

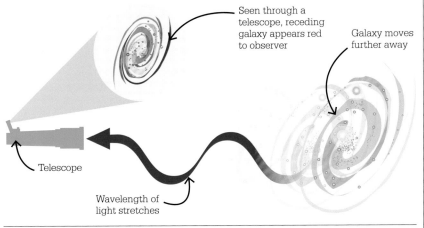

Seen through a telescope, receding galaxy appears red to observer

Galaxy moves further away

Telescope

Wavelength of light stretches

its light appears more red, while the other approaches Earth and its light appears more blue.

The expanding universe

As technology further improved, spectrographic techniques were applied to many other astronomical objects, including mysterious, fuzzy "nebulae." Some of these proved to be huge clouds of interstellar gas, emitting all their light at specific wavelengths, similar to those emitted by vapors in the laboratory. But others emitted light in a continuum (a broad swathe of light of all colors) with a few dark lines, suggesting they were made up of vast numbers of stars. These became known as "spiral nebulae" on account of their spiral shape.

From 1912, American astronomer Vesto Slipher began to analyze the spectra of distant spiral nebulae, finding that the majority of them had significant Doppler shifts to the red end of the spectrum. This showed that they were moving away from Earth at high speeds, regardless of which part of the sky they were seen in. Some astronomers

took this as evidence that spiral nebulae were vast, independent galaxies, located far beyond the gravitational grasp of the Milky Way. However, it was not until 1925 that American astronomer Edwin Hubble calculated the distance to spiral nebulae by measuring the brightness of stars within them.

Hubble began to measure galaxy Doppler shifts, and by 1929 he had discovered a pattern: the further a galaxy lies from Earth, the faster it moves away. This relationship, now known as Hubble's law but first predicted in 1922 by Russian physicist Alexander Friedmann, is best interpreted as an effect of the universe expanding as a whole.

This so-called "cosmological" redshift is not a result of the Doppler effect as Doppler would have understood it. Instead, expanding space drags galaxies apart, like raisins in a rising cake mix as it

Named after Edwin Hubble, the Hubble Space Telescope has been used to discover distant stars, measure their redshifts, and estimate the age of the universe at 13.8 billion years.

cooks. The rate of expansion is the same tiny amount for each light year of space, but it accumulates across the vastness of the universe until galaxies are dragged apart at close to the speed of light, and wavelengths of light stretch into the far red—and beyond visible light, into the infrared—end of the spectrum. The relationship between distance and redshift (denoted z) is so direct that astronomers use z, not light years, to indicate distance for the furthest objects in the universe.

Practical applications

Doppler's discovery has had many technological applications. Doppler radar (used in traffic-control "radar guns" and air-security applications) reveals the distance to—and relative speed of—an object that reflects a burst of radio waves. It can also track severe rainfall by calculating the speed of precipitation. GPS satellite navigation systems need to account for the Doppler shift of satellite signals in order to accurately calculate a receiver unit's position in relation to the satellites' orbits. They can also use this information to make precise measurements of the receiver unit's own motion. ■

THESE MYSTERIOUS WAVES WE CANNOT SEE

ELECTROMAGNETIC WAVES

IN CONTEXT

KEY FIGURES
Heinrich Hertz (1857–1894),
Wilhelm Röntgen
(1845–1923)

BEFORE
1672 Isaac Newton splits
white light into a spectrum
using a prism and then
recombines it.

1803 Thomas Young suggests
that the colors of visible
light are caused by rays of
different wavelengths.

AFTER
c. 1894 Italian engineer
Guglielmo Marconi achieves
the first long-distance
communication using
radio waves.

1906 American inventor
Reginald Fessenden uses
an "amplitude modulation"
system to make the first
radio broadcast.

I n 1865, James Clerk Maxwell
interpreted light as a moving
electromagnetic wave with
transverse electric and magnetic
fields—waves locked in step at
right angles to each other (shown
opposite). Maxwell's theory caused
scientists to ask further questions.
How far did the "electromagnetic
spectrum" extend beyond the range
visible to the human eye? And
what properties might distinguish
waves with much longer or shorter
wavelengths than visible light?

Early discoveries
German-born astronomer William
Herschel found the first evidence for
the existence of radiation beyond

See also: Force fields and Maxwell's equations 142–147 ▪ Lumpy and wavelike light 176–179 ▪ Diffraction and interference 180–183 ▪ Polarization 184–187 ▪ Seeing beyond light 202–203 ▪ Nuclear rays 238–239

Heinrich Hertz

Born in Hamburg, Germany, in 1857, Heinrich Hertz went on to study sciences and engineering in Dresden, Munich, and Berlin, under eminent physicists such as Gustav Kirchhoff and Hermann von Helmholtz. He obtained his doctorate from the University of Berlin in 1880.

Hertz became a full professor of physics at the University of Karlsruhe in 1885. Here, he conducted groundbreaking experiments to generate radio waves. He also contributed to the discovery of the photoelectric effect (the emission of electrons when light hits a material) and carried out important work investigating the way that forces are transferred between solid bodies in contact. Hertz's growing reputation led to an appointment as director of the Physics Institute at Bonn in 1889. Three years later, Hertz was diagnosed with a rare disease of the blood vessels and died in 1894.

Key works

1893 *Electric Waves*
1899 *Principles of Mechanics*

---

the range of visible light in 1800. While measuring temperatures associated with the different (visible) colors in sunlight, Herschel allowed the spectrum of light he was projecting onto a thermometer to drift beyond the visible red light. He was surprised to find the temperature reading had shot up—a sign that much of the heat in radiation from the sun is carried by invisible rays.

In 1801, German pharmacist Johann Ritter reported evidence for what he called "chemical rays." Ritter's experiment involved studying the behavior of silver chloride, a light-sensitive chemical, which was relatively inactive when exposed to red light, but would darken in blue light. Ritter showed that exposing the chemical to radiation beyond violet in the visible spectrum (now known as "ultraviolet" radiation) produced an even faster darkening reaction.

Maxwell published his model of light as an electromagnetic wave that is self-reinforcing (supports itself continuously). Maxwell's theories linked wavelength and color with electromagnetic waves, which meant that his model could also be applied to infrared and ultraviolet rays (with wavelengths shorter or longer than those within the visible spectrum), allowing ultraviolet and infrared rays to be treated as natural extensions of the visible spectrum.

Finding proof

Maxwell's ideas remained theoretical, as at the time there were no technologies suitable to prove them. However, he was still able to predict phenomena that would be associated with his model of light—such as the existence of waves with radically different wavelengths. Most scientists concluded that the best way to prove Maxwell's model would be to look for evidence of these predicted phenomena.

Heinrich Hertz, in 1886, experimented with an electrical circuit that consisted of two separate spiral-wound conducting wires placed near each other. Both ends of each wire terminated in a metal ball, and when a current was applied to one of the wires, a spark would leap between the metal ball terminals of the other. The effect was an example of electromagnetic induction, with the spiral-wound wires acting as "induction coils." »

Electromagnetic waves are made up of two matching waves at right angles—one is an oscillating electric field, and the other is an oscillating magnetic field.

Fields oscillate at right angles to each other

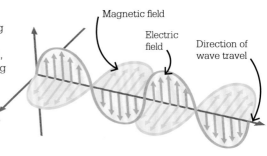

Magnetic field

Electric field

Direction of wave travel

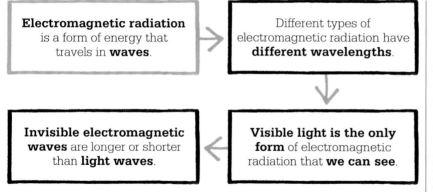

Electromagnetic radiation is a form of energy that travels in **waves**.

Different types of electromagnetic radiation have **different wavelengths**.

Invisible electromagnetic waves are longer or shorter than **light waves**.

Visible light is the only form of electromagnetic radiation that **we can see**.

Current flowing in one coiled wire generated a magnetic field that caused current to flow in the other—but as Hertz studied this experiment more closely, he formulated an idea for a circuit that could test Maxwell's theories.

Hertz's circuit, completed in 1888, had a pair of long wires running toward each other with a tiny "spark gap" between their ends. The other end of each wire was attached to its own 12-inch (30-cm) zinc sphere. Running current through an "induction coil" nearby induced sparks across the spark gap, creating a high voltage difference between the two ends of each wire and an electric current

that oscillated rapidly back and forth. With careful adjustments of the currents and voltages, Hertz was able to "tune" his circuit to an oscillation frequency of around 50 million cycles per second.

According to Maxwell's theory, this oscillating current would produce electromagnetic waves with wavelengths of a few feet, which could be detected at a distance. The final element of Hertz's experiment was a "receiver" (which would receive the wave signals)—this was a separate rectangle of copper wire with its own spark gap, mounted at some distance from the main circuit. Hertz found that when current

flowed in the induction coil of the circuit, it triggered sparks in the spark gap of that main circuit, and also in the receiver's spark gap. The receiver was well beyond the range of any possible induction effects, so something else had to be causing current to oscillate within it—the electromagnetic waves.

Hertz carried out a variety of further tests to prove that he had indeed produced electromagnetic waves similar to light—such as showing that the waves traveled at the speed of light. He published his results widely.

Tuning into radio

Other physicists and inventors soon began investigating these "Hertzian waves" (later called radio waves) and found countless applications. As the technology improved so did the range and quality of signals, and the ability to broadcast many different streams of radio waves from a single mast. Wireless telegraphy (simple signals transmitted as Morse code) was followed by voice communication and eventually television. Hertz's radio waves still play a vital role in modern-day technology.

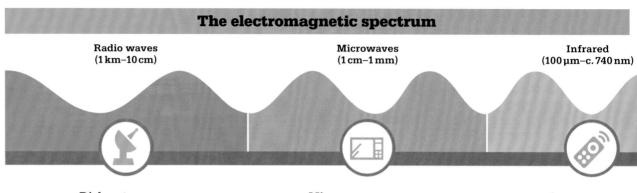

The electromagnetic spectrum

Radio waves (1 km–10 cm)

Microwaves (1 cm–1 mm)

Infrared (100 µm–c. 740 nm)

Dish antennas can capture radio waves to help astronomers detect stars.

Microwave ovens heat up food by using microwaves to make the water molecules within the food vibrate.

A remote control sends signals to a television using infrared waves.

By the time radio communications were becoming a reality, another different type of electromagnetic radiation had been discovered. In 1895, German engineer Wilhelm Röntgen was experimenting with the properties of cathode rays. These are streams of electrons observed in vacuum tubes, which are emitted from a cathode (the electrode connected to the negative terminal of a voltage supply) and then released inside a discharge tube (a glass vessel with a very high voltage between its ends). To avoid any possible effects of light on the tube, Röntgen wrapped it in cardboard. A nearby fluorescent

I have seen
my death!
Anna Röntgen
**on seeing the first X-ray image
revealing the bones of her hand**

The first X-ray image ever taken was made by the silhouette of Anna Bertha Röntgen's hand on a photographic plate, in 1895. Her wedding ring is visible on her fourth finger.

detector screen was glowing during his experiments, and this was caused by unknown rays being emitted from inside the tube, which were passing straight through the cardboard casing.

X-ray imaging

Tests revealed that Röntgen's new rays (which he called "X" rays, indicating their unknown nature) could also affect photographic film, meaning that he could permanently record their behavior. He carried out experiments to test what types of material the rays would pass through and discovered that they were blocked by metal. Röntgen asked his wife, Anna, to place her hand on a photographic plate while he directed the X-rays at it; he found that bones blocked the rays but soft tissue did not.

Scientists and inventors raced to develop new applications for X-ray imaging. In the early 1900s, the damaging effects of X-ray

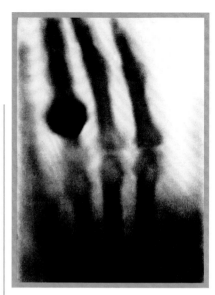

overexposure on living tissue were noticed, and steps were gradually introduced to limit exposure. This also included a return to X-ray photographs that required only a brief burst of rays. The true nature of X-rays was debated until 1912 when Max von Laue used crystals to successfully diffract X-rays (diffraction occurs when any wave encounters an obstacle). This proved that X-rays were waves and a high-energy form of electromagnetic radiation. ∎

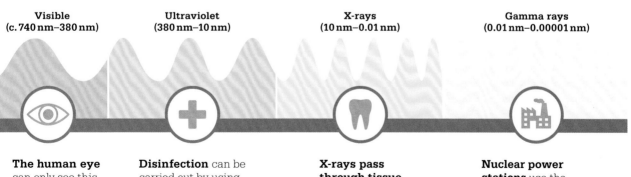

Visible
(c. 740 nm–380 nm)

Ultraviolet
(380 nm–10 nm)

X-rays
(10 nm–0.01 nm)

Gamma rays
(0.01 nm–0.00001 nm)

The human eye can only see this short range of the electromagnetic spectrum.

Disinfection can be carried out by using some wavelengths of UV light to kill bacteria.

X-rays pass through tissue to reveal the teeth or bones underneath.

Nuclear power stations use the energy of gamma radiation to generate electricity.

THE LANGUAGE OF SPECTRA IS A TRUE MUSIC OF THE SPHERES

LIGHT FROM THE ATOM

IN CONTEXT

KEY FIGURE
Niels Bohr (1885–1962)

BEFORE
1565 In Spain, Nicolás Monardes describes the fluorescent properties of a kidney wood infusion.

1669 Hennig Brand, a German chemist, discovers phosphorus, which glows in the dark after being illuminated.

AFTER
1926 Austrian physicist Erwin Schrödinger shows that electron orbits resemble fuzzy clouds rather than circular paths.

1953 In the US, Charles Townes uses stimulated emission from electron movements to create a microwave amplifier, demonstrating the principle that foreshadows the laser.

The ability of materials to generate light rather than simply reflect it from luminous sources such as the sun was initially regarded with mild curiosity. Eventually, however, it proved essential to understanding the atomic structure of matter, and in the 20th century it led to valuable new technologies.

Discovering fluorescence
Substances with a natural ability to glow in certain conditions were recorded from at least the 16th century, but scientists did not attempt to investigate this phenomenon until the early 1800s. In 1819, English clergyman and

See also: Electric potential 128–129 ▪ Electromagnetic waves 192–195 ▪ Energy quanta 208–211 ▪ Particles and waves 212–215 ▪ Matrices and waves 218–219 ▪ The nucleus 240–241 ▪ Subatomic particles 242–243

Some minerals are fluorescent. Like this fluorite, they absorb certain types of light, such as ultraviolet, and then release it at a different wavelength, changing its perceived color.

mineralogist Edward Clarke described the properties of the mineral fluorspar (now better known as fluorite), which glows in certain circumstances.

In 1852, Irish physicist George Gabriel Stokes showed that the mineral's glow was caused by exposure to ultraviolet light, while the emitted light was limited to a single blue color and wavelength, appearing as a line when it was split apart into a spectrum. He argued that fluorspar transformed shorter-wavelength ultraviolet light directly into visible light by somehow stretching its wavelength, and coined the term "fluorescence" to describe this particular behavior.

Meanwhile, early electrical experimenters had discovered another means of creating luminous matter. When they attempted to pass electric current between metal electrodes at either end of a glass tube that had as much air removed as possible, the thin gas that remained between the electrodes began to glow when the voltage difference between the electrodes was sufficiently high. Working with

Michael Faraday, Stokes explained the phenomenon in terms of gas atoms becoming energized, allowing current to flow through them, then releasing light as they shed their energy.

A few years later, German glassblower Heinrich Geissler invented a new means of creating much better vacuums in glass vessels. By adding specific gases to a vacuum, he found he could cause a lamp to glow in various colors. This discovery lies behind the invention of fluorescent lights, which became widespread in the 20th century.

Elementary emissions

The reason why some atoms produce light remained a mystery until the late 1850s, when fellow Germans chemist Robert Bunsen and physicist Gustav Kirchhoff joined forces to investigate the phenomenon. They focused on the colors generated when different elements were heated to incandescent temperatures in the searing flame of Bunsen's recently invented laboratory gas burner. They found that the light produced was not a blended continuum of different wavelengths and colors—like sunlight or the light from stars—but was a mix of a few bright lines emitted with very specific wavelengths and colors (emission lines).

The precise pattern of emissions was different for each element and was therefore a unique

chemical "fingerprint". In 1860 and 1861, Bunsen and Kirchhoff identified two new elements— cesium and rubidium—from their emission lines alone.

Balmer's breakthrough

Despite the success of Kirchhoff and Bunsen's method, they had yet to explain why the emission lines were produced at specific wavelengths. However, they realized it had something to do with the properties of atoms in individual elements. This was a new concept. At the time, atoms were generally thought of as solid, indivisible particles of a particular element, so it was difficult to imagine an internal process that generated light or altered the wavelengths of the emission lines.

An important breakthrough came in 1885, when Swiss mathematician Johann Jakob Balmer identified a pattern in the series of emission lines created by hydrogen. The wavelengths, which until then had seemed like a random collection of lines, could be predicted by a mathematical formula involving »

> A chemist who is
> not a physicist
> is nothing at all.
> **Robert Bunsen**

We have an intimate knowledge of the constituents of … individual atoms.
Niels Bohr

two series of whole numbers. Emission lines linked with higher values of either or both of these numbers were only generated in higher-temperature environments.

Balmer's formula, called the Balmer series, immediately proved its worth when the high-energy absorption lines associated with hydrogen were identified at the predicted wavelengths in the spectrum of the sun and other stars. In 1888, Swedish physicist Johannes Rydberg developed a more generalized version of the formula, subsequently called the Rydberg formula, which could be used (with a few tweaks) to predict emission lines from many different elements.

Clues from the atom
The 1890s and the beginning of the 20th century saw major advances in the understanding of atoms. Evidence emerged that atoms were not the solid, uniform lumps of matter previously assumed. First, in 1897, J.J. Thomson discovered the electron (the first subatomic particle), and then in 1909, building on the work of Hans Geiger and Ernest Marsden, Ernest Rutherford discovered the nucleus, in which most of an atom's mass is concentrated. Rutherford imagined

the atom with a tiny central nucleus surrounded by electrons scattered at random throughout the rest of the volume.

Discussions with Rutherford inspired a young Danish scientist, Niels Bohr, to work on his own atomic model. Bohr's breakthrough lay in combining Rutherford's nuclear model with Max Planck's radical suggestion, in 1900, that, in particular circumstances, radiation was emitted in bite-size chunks called quanta. Planck had suggested this theory as a mathematical means of explaining the characteristic light output of stars and other incandescent objects. In 1905, Albert Einstein had gone a step further than Planck, suggesting that these small bursts of electromagnetic radiation were not merely a consequence of certain types of radiation emission, but were fundamental to the nature of light.

Bohr did not, at this point in time, accept Einstein's theory that light always traveled in bursts called photons, but he did wonder whether something about atomic structure, and in particular the arrangement of electrons, could

sometimes produce light in small bursts of specific wavelength and energy.

Bohr's model
In 1913, Bohr found a means of linking atomic structure to the Rydberg formula for the first time. In an influential trilogy of papers, he suggested that the movement of electrons in an atom was constrained so that they could only have certain values of angular momentum (the momentum due to their orbit around the nucleus). In practice this meant that electrons could only orbit at certain fixed distances from the nucleus. Because the strength of the electromagnetic attraction between the positively charged nucleus and the negatively charged electrons would also vary, depending on the electron's orbit, each electron could be said to have a certain energy, with lower-energy orbits close to the nucleus and higher-energy ones further out.

Each orbit could hold a maximum number of electrons, and orbits close to the nucleus "filled up" first. Empty spaces

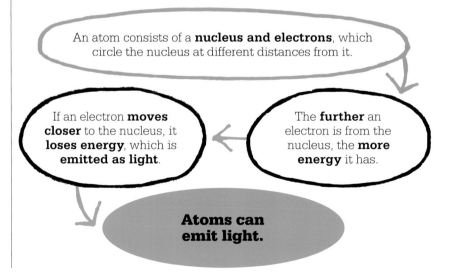

An atom consists of a **nucleus and electrons**, which circle the nucleus at different distances from it.

The **further** an electron is from the nucleus, the **more energy** it has.

If an electron **moves closer** to the nucleus, it **loses energy**, which is **emitted as light**.

Atoms can emit light.

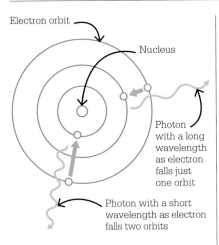

Electron orbit

Nucleus

Photon with a long wavelength as electron falls just one orbit

Photon with a short wavelength as electron falls two orbits

The wavelength of light that an atom emits depends on how much energy an electron loses in falling from one orbit to another. The further it falls, the shorter the wavelength of the light and the higher its energy.

(and entire empty orbits) further out were available for electrons that were closer in to "jump" into if they received a boost of energy—from a ray of incoming light, for example. If a gap remained in the closer orbit, the transition would usually be brief, and the "excited" electron would almost immediately fall back into the lower energy state. In this case, it would emit a small burst of light whose frequency, wavelength, and color were determined by the equation $\Delta E = h\nu$ (called Planck's equation, after Max Planck), where ΔE is the energy difference in the transition, ν (the Greek letter nu) is the frequency of the emitted light, and h is the Planck constant, relating the frequency and energy of electromagnetic waves.

Bohr convincingly applied this new model to the simplest atom, hydrogen, and showed how it could produce the familiar lines of the Balmer series. However, the most convincing evidence that confirmed he was on the right

track came with an explanation for a series of lines in the spectrum of super-hot stars, known as the Pickering series. Bohr correctly explained that these lines were associated with electrons jumping between orbits in ionized helium (He^+)—atoms of helium that had already been stripped of one of their two electrons.

Roots of the laser

The following decades showed that Bohr's model was an oversimplification of the strange quantum processes going on inside real atoms, but it nevertheless marked a huge step forward in scientific knowledge. The Bohr model not only explained the link between atomic structure and spectral lines for the first time, but also paved the way for new technologies that would take advantage of this emission, such as the laser—an intense beam of photons created by triggering a cascade of emission events within an energized material. The behavior upon which the laser is based was predicted by Einstein as early as 1917, and proven to exist in 1928 by Rudolf W. Ladenburg, but a working laser beam was not achieved in practice until 1960. ■

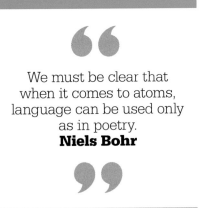

> We must be clear that when it comes to atoms, language can be used only as in poetry.
> **Niels Bohr**

Niels Bohr

Born in Copenhagen, Denmark, in 1885, Bohr studied physics at the city's university. His PhD thesis, completed in 1911, was a groundbreaking investigation of the distribution of electrons in metals. A visit to British laboratories in the same year inspired Bohr to formulate the model of atomic structure, for which he was awarded the Nobel Prize in Physics in 1922. By this time he was director of the new Danish Institute for Theoretical Physics.

In 1940, Denmark was occupied by the Nazis. Three years later, Bohr, whose mother was Jewish, fled to the US, where he contributed to the Manhattan Project to build an atomic bomb. Returning to Denmark in 1945, he helped establish the International Atomic Energy Agency. Bohr died at his home in Copenhagen in 1962.

Key works

1913 "On the Constitution of Atoms and Molecules"
1924 "The Quantum Theory of Radiation"
1939 "The Mechanism of Nuclear Fission"

SEEING WITH SOUND
PIEZOELECTRICITY AND ULTRASOUND

Crystals are **piezoelectric** if heating them and/or distorting their structure can **create an electric current**.

High-frequency electric currents can cause piezoelectric crystals to produce high-pitched **ultrasound**.

Ultrasound **echoes** bouncing off an object are affected by its shape, composition, and distance.

When the crystals compress, they produce **electric signals** that can be **transformed into images**.

These echoes cause the piezoelectric crystals to **compress**.

Using echolocation—the detection of hidden objects by perceiving the sound waves reflected from those objects as echoes—was first proposed in 1912 by British physicist Lewis Fry Richardson. Shortly after the RMS *Titanic* disaster of that year, Richardson applied for a patent on a method for warning ships about icebergs and other large submerged or semi-submerged objects. Richardson noted that a ship emitting sound waves with higher frequencies and shorter wavelengths—later called ultrasound—would be able to detect submerged objects with greater accuracy than normal sound waves allowed. He

See also: Electric potential 128–129 ▪ Music 164–167 ▪ The Doppler effect and redshift 188–191 ▪ Quantum applications 226–231

>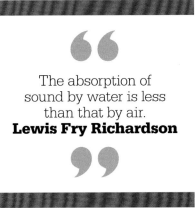
>
> ❝ The absorption of sound by water is less than that by air. ❞
> **Lewis Fry Richardson**

envisaged a mechanical means of producing such waves and showed that they would carry further in water than in air.

Curie's crystals

Richardson's invention never became a reality. The RMS *Titanic* disaster faded from the spotlight and the need became less urgent. Furthermore, a more practical means of creating and detecting ultrasound had already been discovered through the work of French brothers Jacques and Pierre Curie.

In around 1880, while investigating the way certain crystals can create an electric current when heated, they found that using pressure to deform the crystal structure also created an electric potential difference—called piezoelectricity. A year later, they confirmed a prediction made by French physicist Gabriel Lippmann that passing an electric current through a crystal would cause a reverse effect—physical deformation of the crystal.

The practical applications for piezoelectricity were not immediately realized, but Paul

Langevin, a former pupil of Pierre Curie, continued to investigate. With the outbreak of World War I and the appearance of a new form of warfare, the German U-boat (submarine), Langevin realized that piezoelectricity could be used to produce and detect ultrasound.

New technology

Pulses of high-frequency sound waves could be driven out into the water using powerful oscillating currents running through stacks of crystals sandwiched between metal sheets, in a device called a transducer. The reverse principle could convert the compression caused by returning echoes into an electric signal. Langevin's application of the Curie brothers' discovery forms the foundation of sonar and other echolocation systems that are still used today.

At first, the interpretation of echolocation data entailed turning the returning pulse back into sound through a speaker, but in the early 20th century electronic displays were developed. These eventually evolved into the sonar display systems used in navigation, defense, and medical ultrasound. ▪

In piezoelectric crystals such as amethyst, the structure of the cells is not symmetrical. When pressure is applied, the structure deforms and the atoms move, creating a small voltage.

Pierre Curie

The son of a doctor, Curie was born in Paris in 1859 and educated by his father. After studying mathematics at the University of Paris, he worked as a laboratory instructor in the university's science faculty. The experiments he conducted in the lab with his brother Jacques led to the discovery of piezoelectricity and his invention of an "electrometer" to measure the weak electrical currents involved.

For his doctorate, Curie researched the relationship between temperature and magnetism. Through this, he began working with Polish physicist Maria Skłodowska, whom he married in 1895. They spent the rest of their lives researching radioactivity, leading to a Nobel prize in 1903. Pierre died in a traffic accident in 1906.

Key work

1880 "Développement, par pression, de l'électricité polaire dans les cristaux hémièdres à faces inclinée" ("Development of electricity through pressure on the inclined faces of hemihedral crystals")

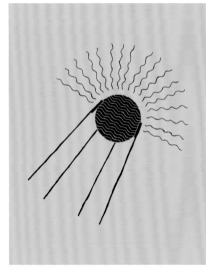

A LARGE FLUCTUATING ECHO

SEEING BEYOND LIGHT

IN CONTEXT

KEY FIGURE
Jocelyn Bell Burnell (1943–)

BEFORE
1800 In the UK, William Herschel accidentally discovers the existence of infrared radiation.

1887 Heinrich Hertz successfully generates radio waves for the first time.

AFTER
1967 US military Vela satellites, designed to detect nuclear tests, record the first gamma-ray bursts from violent events in the distant universe.

1983 NASA, the UK, and the Netherlands launch the first infrared space telescope, IRAS (Infrared Astronomical Satellite).

2019 Using aperture synthesis to construct a collaboration of telescopes around the world, astronomers observe radiation around a supermassive black hole in a distant galaxy.

Physicists' discovery of electromagnetic radiations beyond visible light in the 19th and early 20th centuries introduced new ways of observing nature and the universe. They had to overcome many challenges in their quest, particularly since

---

Invisible radio waves from space penetrate Earth's atmosphere.

⬇

Radio telescopes can gather radio waves to identify their **approximate location**.

⬇

This helps astronomers locate **distant stars and galaxies**.

Earth's atmosphere blocks or swamps many of these radiations. In the case of radio waves from planets, stars, and other celestial phenomena (which can penetrate the atmosphere), the main problem was their huge wavelengths, which made it difficult to locate their source.

The first steps in radio astronomy were made in 1931, when American physicist Karl Jansky erected a large sensitive antenna on a turntable. By measuring how signals changed through the day as Earth rotated, revealing different parts of the sky above the horizon, he was able to prove that the center of the Milky Way was a strong source of radio waves.

Mapping radio waves
In the 1950s, the first giant "dish" telescope was built in the UK by Bernard Lovell at Jodrell Bank in Cheshire. With a diameter of 250 ft (76.2 m), the telescope was able to produce blurry images of the radio sky, pinning down the location and shape of individual radio sources for the first time.

Significant improvements in the resolving power of radio telescopes were achieved in the 1960s, with

See also: Electromagnetic waves 192–195 ▪ Black holes and wormholes 286–289 ▪ Dark matter 302–305 ▪ Dark energy 306–307 ▪ Gravitational waves 312–315

Radio maps of the sky show the source of waves arriving from deep space. The most intense radio emissions (red) come from the central regions of our Milky Way galaxy.

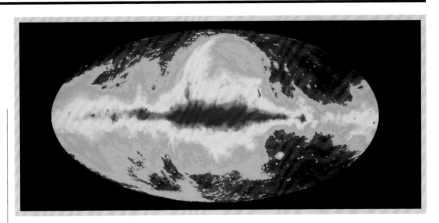

the work of Martin Ryle, Antony Hewish, and graduate student Jocelyn Bell Burnell at the University of Cambridge. Ryle invented a technique called aperture synthesis, in which a grid, or "array," of individual radio antennae function as a single giant telescope. The principle behind it involved measuring the varying amplitude (strength) of incoming radio waves received by the different antennae at the same moment. This method, and some basic assumptions about the shape of the radio waves, allowed the direction of their source to be calculated with much greater precision.

Using Ryle's discovery, Hewish and Bell Burnell set out to build a radio telescope with several thousand antennae, ultimately covering 4 acres (1.6 ha) of ground. Hewish hoped to use this telescope

to discover interplanetary scintillation (IPS)—predicted variations in radio signals from distant sources as they interact with the sun's magnetic field and solar wind.

Bell Burnell's project involved monitoring the IPS telescope's measurements and looking for the occasional variations that were predicted. Among her findings, however, was a much shorter and more predictable radio signal, lasting 1/25th of a second and repeating every 1.3 seconds, keeping pace with the movement of the stars.

This ultimately proved to be a pulsar—the first known example of a rapidly rotating neutron star, an object whose existence had first been proposed in the 1930s.

Bell Burnell's discovery of pulsars was the first of many breakthroughs for radio astronomy. As technology improved, aperture synthesis would be used to make increasingly detailed images of the radio sky, helping to reveal the structure of the Milky Way, distant colliding galaxies, and the material surrounding supermassive black holes. ▪

Jocelyn Bell Burnell

Born in Belfast, Northern Ireland, in 1943, Jocelyn Bell became interested in astronomy after visiting the Armagh Planetarium as a young girl. She went on to graduate in physics from the University of Glasgow in 1965, and moved to the University of Cambridge to study for a PhD under Antony Hewish. It was here that she discovered the first pulsar.

Despite being listed as the second author on the paper that announced the discovery, Bell Burnell was overlooked when her colleagues were awarded the Nobel Prize in 1974. She has since

gone on to a highly successful career in astronomy and been awarded several other prizes, including the 2018 Special Breakthrough Prize in Fundamental Physics, for both her scientific research and her work promoting the role of women and minorities in science and technology.

Key work

1968 "Observation of a Rapidly Pulsating Radio Source" (*Nature* paper, with Antony Hewish and others)

THE QUAN

WORLD

OUR UNCERTAIN

UNIVERSE

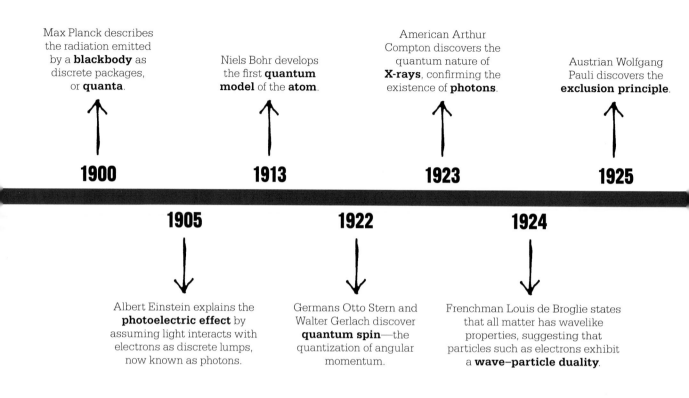

Max Planck describes the radiation emitted by a **blackbody** as discrete packages, or **quanta**.

Niels Bohr develops the first **quantum model** of the **atom**.

American Arthur Compton discovers the quantum nature of **X-rays**, confirming the existence of **photons**.

Austrian Wolfgang Pauli discovers the **exclusion principle**.

1900

1913

1923

1925

1905

1922

1924

Albert Einstein explains the **photoelectric effect** by assuming light interacts with electrons as discrete lumps, now known as photons.

Germans Otto Stern and Walter Gerlach discover **quantum spin**—the quantization of angular momentum.

Frenchman Louis de Broglie states that all matter has wavelike properties, suggesting that particles such as electrons exhibit a **wave–particle duality**.

Our world is deterministic, following laws that set out definitively how a system will evolve. Usually through trial (such as playing sports and predicting a ball's trajectory) and error (getting hit by a few balls), we innately learn these deterministic laws for specific everyday situations.

Physicists design experiments to uncover these laws and so enable us to predict how our world or things within it will change over time. These experiments have led to the deterministic physics we have talked about in the book up until this point. However, it was unnerving at the start of the 20th century to discover that at its heart this is not how nature behaves. The deterministic world we see each day is just a blurred picture, an average, of a far more unsettling world that exists on the smallest scales. Our door to this new realm was unlocked by finding out that light behaves in unfamiliar ways.

Waves or particles?
Since the 16th century, a debate had raged as to the nature of light. One camp vowed that light was made up of tiny particles, an idea championed by English physicist Isaac Newton, but others viewed light as a wave phenomenon. In 1803, British physicist Thomas Young's double-slit experiment seemed to provide definitive evidence for wavelike light as it was seen to exhibit interference, a behavior that could not be explained by particles. In the early years of the 20th century, German physicists Max Planck and Albert Einstein revisited Newton's ideas and suggested that light must be made from discrete lumps. Einstein had to invoke this seemingly strange conclusion to describe the observed phenomenon of the photoelectric effect.

There was now evidence for both sides of the particle–wave debate. Something very fishy was happening—welcome to the world of quanta, objects that behave as both particles and waves, depending upon the situation. Most quanta are elementary, or fundamental, subatomic particles, not composed of other particles. When looking at them, they appear particle-like, but between observations they behave as if they are waves. The wavelike behavior is not like that of water. If two waves combine to make a larger wave, this does not increase the energy of a

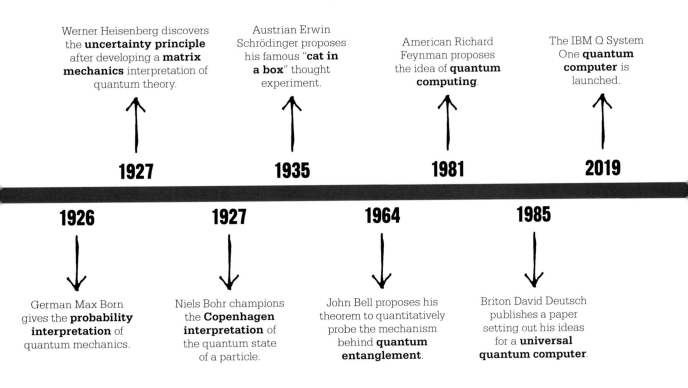

Werner Heisenberg discovers the **uncertainty principle** after developing a **matrix mechanics** interpretation of quantum theory.

Austrian Erwin Schrödinger proposes his famous "**cat in a box**" thought experiment.

American Richard Feynman proposes the idea of **quantum computing**.

The IBM Q System One **quantum computer** is launched.

1927

1935

1981

2019

1926

1927

1964

1985

German Max Born gives the **probability interpretation** of quantum mechanics.

Niels Bohr champions the **Copenhagen interpretation** of the quantum state of a particle.

John Bell proposes his theorem to quantitatively probe the mechanism behind **quantum entanglement**.

Briton David Deutsch publishes a paper setting out his ideas for a **universal quantum computer**.

---

quantum as it would a water wave. Instead, it boosts the probability of a wave being seen in that particular location. When we view a quantum, it cannot be in all locations at the same time; instead, it decides upon a single location dependent upon the probabilities outlined by its wave.

This was a new, probabilistic way of behaving, one where we could know all about a quantum at one point in time but never be able to predict definitively where it would be later. These quanta do not behave like sports balls—the flight of which is predictable—there is always a probability that wherever we stand we might get hit.

The mechanism for how quantum objects transition between wavelike behavior and particle-like measurements has been fiercely debated. In 1927, Danish physicist

Niels Bohr championed the Copenhagen interpretation of quantum physics. This series of ideas maintains that the wave function represents probabilities for final outcomes and it collapses into just one possible outcome when measured by an observer. It has been followed up since with a collection of more complicated interpretations. And still the debate rages on.

Nothing is certain

That is not where the strangeness ends. You can never truly know everything about a quantum. German physicist Werner Heisenberg's uncertainty principle explains that it is impossible to know certain pairs of properties, such as momentum and position, with exacting precision. The more

accurately you measure one property, the less accurately you will know the other.

Even more bizarre is the phenomenon of quantum entanglement, which allows one particle to affect another in an entirely different place. When two particles are entangled, even if separated by a great distance, they are effectively a single system. In 1964, Northern Irish physicist John Stewart Bell presented evidence that quantum entanglement really existed, and French physicist Alain Aspect demonstrated this "action at a distance" in 1981.

Today, we are learning how to utilize these strange quantum behaviors to do some fantastic things. New technology using quantum principles is set to change the world of the near future. ∎

THE ENERGY OF LIGHT IS DISTRIBUTED DISCONTINUOUSLY IN SPACE

ENERGY QUANTA

IN CONTEXT

KEY FIGURE
Max Planck (1858–1947)

BEFORE
1839 French physicist
Edmond Becquerel makes
the first observation of the
photoelectric effect.

1899 British physicist
J.J. Thomson confirms that
ultraviolet light can generate
electrons from a metal plate.

AFTER
1923 American physicist
Arthur Compton succeeds
in scattering X-rays off
electrons, demonstrating
that they act like particles.

1929 American chemist
Gilbert Lewis coins the name
"photons" for light quanta.

1954 American scientists
working at Bell Laboratories
invent the first practical
solar cell.

On October 19, 1900, Max
Planck gave a lecture to
the German Physical
Society (Deutsche Physikalische
Gesellschaft) in Berlin. Although it
would be a few years before the full
implications of his pronouncements
became apparent, they marked the
beginning of a new age for physics:
the era of the quantum.

Planck submitted a solution to
a problem that had been vexing
physicists up until then. The
problem involved blackbodies—
objects that are perfect absorbers
and emitters of all frequencies
of electromagnetic radiation. A
blackbody is so called because all
the radiation striking its surface is

See also: Thermal radiation 112–117 ▪ Lumpy and wavelike light 176–179 ▪ Diffraction and interference 180–183
▪ Electromagnetic waves 192–195 ▪ Light from the atom 196–199 ▪ Particles and waves 212–215

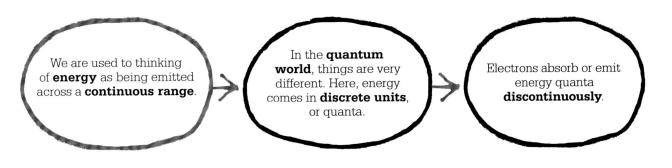

We are used to thinking of **energy** as being emitted across a **continuous range**.

In the **quantum world**, things are very different. Here, energy comes in **discrete units**, or quanta.

Electrons absorb or emit energy quanta **discontinuously**.

absorbed; no radiation is reflected and the energy it emits depends solely on its temperature. Perfect blackbodies don't exist in nature.

A theory formulated by British physicists James Jeans and John Strutt, Lord Rayleigh, accurately explained the behavior of blackbodies at low frequencies, but predicted that, as there was effectively no upper limit to the higher frequencies that could be generated, the amount of energy radiated from a blackbody should continue to increase infinitely. This was dubbed the "ultraviolet catastrophe" as it concerned short-wavelength radiation beyond the ultraviolet. Everyday observation shows the prediction to be wrong. If it was correct, bakers would be exposed to lethal doses of radiation every time they opened their ovens. But in the late 19th century, no one could explain why it was wrong.

Planck made the radical assumption that the vibrating atoms in a blackbody emit energy in discrete packets, which he called quanta. The size of these quanta is proportional to the frequency of vibration. Although there is in theory an infinite number of higher frequencies, it takes increasingly large amounts of energy to release quanta at those levels. For example,

a quantum of violet light has twice the frequency, and therefore twice the energy, of a quantum of red light. This proportionality explains why a blackbody does not give off energy equally across the electromagnetic spectrum.

Planck's constant

Planck denoted the constant of proportionality as **h**—now known as Planck's constant—and related the energy of a quantum to its frequency by the simple formula $E = h\nu$, where **E** equals energy and **ν** equals frequency. The energy of a quantum can be calculated by multiplying its frequency by Planck's constant, which is $6.62607015 \times 10^{-34}$ J s (joule-seconds).

Planck's solution worked—the results of experiments were in accord with the predictions made by his theory. Yet Planck was not entirely happy and resisted the idea for years that his quanta had any basis in reality, viewing them instead as more of a mathematical "fix" for a difficult problem. He could give no good reason why quanta should be true and by his own admission had introduced them as "an act of desperation," but it led to the quantum revolution that transformed physics.

The photoelectric effect

When Albert Einstein heard about Planck's theory, he commented, "It was as if the ground had been »

The photoelectric effect is the emission of electrons by some metals when they are hit by a beam of light. The higher the frequency of the light, the higher-energy photons it has and the higher the energy of the electrons emitted.

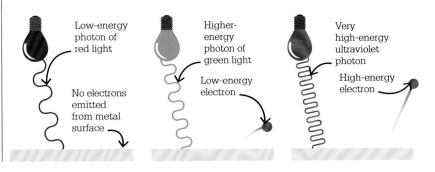

Low-energy photon of red light

No electrons emitted from metal surface

Higher-energy photon of green light

Low-energy electron

Very high-energy ultraviolet photon

High-energy electron

Electric charges can build up on the exterior of spacecraft such as the SpaceX Dragon unless systems are put in place to drain them.

The photoelectric effect in space

When a spacecraft is exposed to prolonged sunlight on one side, high-energy ultraviolet photons, or light quanta, striking its metallic surface cause a steady stream of electrons to be ejected. The loss of electrons causes the spacecraft to develop a positive charge on the sunlit side, while the other, shaded, side has a relative negative charge.

Without conductors to prevent charge from building up, the difference in charge across the surface will cause an electric current to flow from one side of the spacecraft to the other. In the early 1970s, before measures were taken to counter this phenomenon, it disrupted the delicate circuitry of several Earth-orbiting satellites, even causing the complete loss of a military satellite in 1973. The rate of electron loss depends on the surface material of the spacecraft, the angle at which the sun's rays strike it, and the amount of solar activity, including sunspots.

---

pulled out from under us." In 1904, Einstein wrote to a friend that he had discovered "in a most simple way the relation between the size of elementary quanta ... and the wavelengths of radiation." This relationship was the answer to a curious aspect of radiation that had previously defied explanation. In 1887, German physicist Heinrich Hertz had discovered that certain types of metal would emit electrons when a beam of light was directed at them. This "photoelectric effect" is similar to the phenomenon harnessed for use in fiber-optic communications (although optical fibers are made of semiconductor materials rather than metals).

A problem with electrons

Physicists at first assumed that the electric field (a region of space where electric charge is present) part of the electromagnetic wave provides the energy that electrons need to break free. If that were the case, then the brighter the light, the more high-energy the emitted electrons should be. That was found not to be the case, however. The energy of the released electrons depends not on the intensity of the light, but on its frequency. Shifting the beam to higher frequencies, from blue to violet and beyond, produces higher-energy electrons; a low-frequency red light beam, even if it is blindingly bright, produces no electrons. It is as if fast-moving ripples can readily move the sand on a beach but a slow-moving wave, no matter how big, leaves it untouched. In addition, if the electrons are going to jump at all, they jump right away: no buildup of energy is involved. This made no

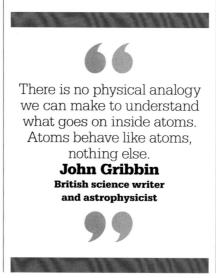

> " There is no physical analogy we can make to understand what goes on inside atoms. Atoms behave like atoms, nothing else.
> **John Gribbin**
> **British science writer and astrophysicist**

sense until the revelations of Planck and Einstein. In March 1905, Einstein published a paper, which took Planck's quanta and married them to the photoelectric effect, in the monthly *Annals of Physics* journal. This paper would eventually win him the Nobel Prize in 1921. Einstein was particularly interested in the differences between particle theories and wave theories. He compared the formulae that describe the way the particles in a gas behave as it is compressed or allowed to expand with those describing the similar changes as radiation spreads through space. He found that both obey the same rules, and the mathematics underpinning both phenomena is the same. This gave Einstein a way to calculate the energy of a light quantum of a particular frequency; his results agreed with Planck's.

From here, Einstein went on to show how the photoelectric effect could be explained by the existence of light quanta. As Planck had established, the energy of a quantum was determined by its frequency. If a single quantum transfers its energy to an electron, then the higher the energy of the

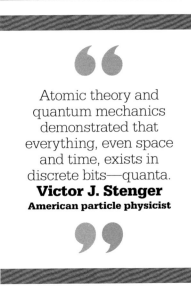

> Atomic theory and quantum mechanics demonstrated that everything, even space and time, exists in discrete bits—quanta.
> **Victor J. Stenger**
> American particle physicist

quantum, the higher the energy of the emitted electron. High-energy blue photons, as light quanta were later named, have the heft to punch out electrons; red photons simply didn't. Increasing the intensity of the light produces greater numbers of electrons, but it does not create more energetic ones.

Further experiments

Whereas Planck had viewed the quantum as little more than a mathematical device, Einstein was now suggesting that it was an actual physical reality. That didn't go down well with many other physicists who were reluctant to give up the idea that light was a wave, and not a stream of particles. In 1913, even Planck commented about Einstein, "That sometimes … he may have gone overboard in his speculations should not be held against him."

Skeptical American physicist Robert Millikan performed experiments on the photoelectric effect that were aimed at proving that Einstein's assertion was wrong, but they actually produced results entirely in line with the latter's predictions. Yet Millikan still spoke of Einstein's "bold, not to say reckless, hypothesis." It wasn't until the experiments conducted by American physicist Arthur Compton in 1923 that the quantum theory finally began to gain acceptance. Compton observed the scattering of X-rays from electrons and provided plausible evidence that in scattering experiments, light behaves as a stream of particles and cannot be explained purely as a wave phenomenon. In his paper, published in the *Physical Review* journal, he explained that "this remarkable agreement between our formulas and the experiments can leave but little doubt that the scattering of X-rays is a quantum phenomenon."

Einstein's explanation of the photoelectric effect could be verified by experiment—light, it seemed, acted as if it were a stream of particles. However, light also acted like a wave in familiar and well-understood phenomena such as reflection, refraction, diffraction, and interference. So, for physicists, the question still remained: what was light? Was it a wave or was it a particle? Could it possibly be both? ∎

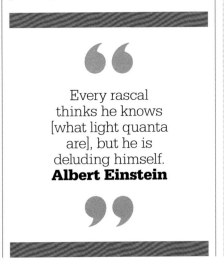

> Every rascal thinks he knows [what light quanta are], but he is deluding himself.
> **Albert Einstein**

Max Planck

Born in Kiel, Germany, in 1858, Max Planck studied physics at the University of Munich, graduating at 17 and gaining his doctorate four years later. Developing a keen interest in thermodynamics, in 1900 he produced what is now known as Planck's radiation formula, introducing the idea of quanta of energy. This marked the beginning of quantum theory, one of the cornerstones of 20th-century physics, although its far-reaching consequences weren't understood for several years.

In 1918, Planck received the Nobel Prize for Physics for his achievement. After Adolf Hitler came to power in 1933, Planck pleaded in vain with the dictator to abandon his racial policies. He died in Göttingen, Germany, in 1947.

Key works

1900 "On an Improvement of Wien's Equation for the Spectrum"
1903 *Treatise on Thermodynamics*
1920 *The Origin and Development of the Quantum Theory*

THEY DO NOT BEHAVE LIKE ANYTHING THAT YOU HAVE EVER SEEN

PARTICLES AND WAVES

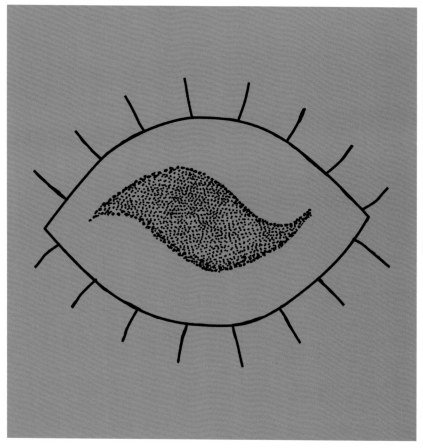

IN CONTEXT

KEY FIGURE
Louis de Broglie (1892–1987)

BEFORE
1670 Isaac Newton develops his corpuscle (particle) theory of light.

1803 Thomas Young performs his double-slit experiment, demonstrating that light behaves like a wave.

1897 British physicist J.J. Thomson announces that electricity is made up of a stream of charged particles, now called "electrons."

AFTER
1926 Austrian physicist Erwin Schrödinger publishes his wave equation.

1927 Danish physicist Niels Bohr develops the Copenhagen interpretation, stating that a particle exists in all possible states until it is observed.

The nature of light lies at the heart of quantum physics. People have tried for centuries to explain what it is. The ancient Greek thinker Aristotle thought of light as a wave traveling through an invisible ether that filled space. Others thought it was a stream of particles that were too small and fast moving to be perceived individually. In 55 BCE, the Roman philosopher Lucretius wrote: "The light and heat of the sun; these are composed of minute atoms which, when they are shoved off, lose no time in shooting right across the interspace of air." However, the particle theory did not find much favor and so, for the

Physics isn't just about writing equations on a board and sitting in front of a computer. Science is about exploring new worlds.
Suchitra Sebastian
Indian physicist

next 2,000 years or more, it was generally accepted that light traveled in waves.

Isaac Newton was fascinated by light and carried out many experiments. He demonstrated, for example, that white light could be split into a spectrum of colors by passing it through a prism. He observed that light travels in straight lines and that shadows have sharp edges. It seemed obvious to him that light was a stream of particles, and not a wave.

The double-slit experiment
The formidably talented British scientist Thomas Young theorized that if the wavelength of light was sufficiently short, then it would appear to travel in straight lines as if it were a stream of particles. In 1803, he put his theory to the test.

First, he made a small hole in a window blind to provide a point source of illumination. Next, he took a piece of board and made two pinholes in it, placed close together. He positioned his board

so that the light coming through the hole in the window blind would pass through the pinholes and onto a screen. If Newton was right, and light was a stream of particles, then two points of light would be visible on the screen where the particles traveled through the pinholes. But that is not what Young saw.

Rather than two discrete points of light, he saw a series of curved, colored bands separated by dark lines, exactly as would be expected if light were a wave. Young himself had investigated interference patterns in waves two years earlier. He had described how the peak of one wave meeting the peak of another add together to make a higher

peak, while two troughs make a deeper trough. If a trough and a peak coincide, then they cancel each other out. Unfortunately, Young's findings were not well received as they disagreed with the great Isaac Newton's view that light was carried by a stream of particles.

Particles of light
In the 1860s, Scottish scientist James Clerk Maxwell declared that light was an electromagnetic wave. An electromagnetic wave is made up of two adjoined waves traveling in the same direction but at right angles to each other. One of these waves is an oscillating magnetic field, the other an oscillating electric field. The two »

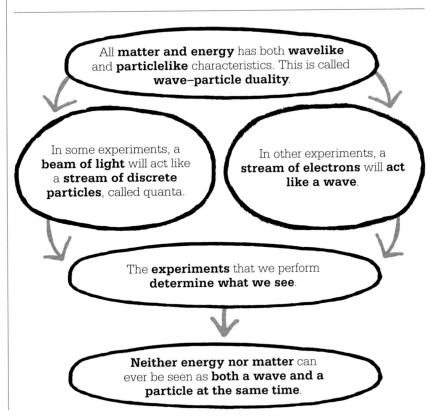

All **matter and energy** has both **wavelike** and **particlelike** characteristics. This is called **wave–particle duality**.

In some experiments, a **beam of light** will act like a **stream of discrete particles**, called quanta.

In other experiments, a **stream of electrons** will **act like a wave**.

The **experiments** that we perform **determine what we see**.

Neither energy nor matter can ever be seen as **both a wave and a particle at the same time**.

> We have two contradictory pictures of reality; separately neither of them fully explains the phenomena of light, but together they do.
> **Albert Einstein**

fields keep in step with each other as the wave travels along. When, in 1900, Max Planck solved the blackbody radiation problem by assuming that electromagnetic energy was emitted in quanta, he didn't actually believe they were real. The evidence that light was a wave was too overwhelming for Planck to accept that it was actually composed of particles.

In 1905, Albert Einstein showed how the photoelectric effect—which could not be explained by the wave theory of light—could be explained by

assuming that light was indeed composed of photons, or discrete energy quanta. As far as Einstein was concerned, light quanta were a physical reality, but he struggled without success for the rest of his life to resolve the apparent paradox that light was also demonstrably wavelike. In an experiment carried out in 1922, American physicist Arthur Compton succeeded in scattering X-rays off electrons. The small change in the frequency of the X-rays that resulted became known as the Compton effect and showed that both the X-rays and the electrons were acting like particles when they collided.

Symmetry of nature
In 1924, Louis de Broglie put forward a theory in his doctoral thesis that all matter and energy—not just light—has the characteristics of both particles and waves. De Broglie believed intuitively in the symmetry of nature and Einstein's quantum theory of light. He asked, if a wave can behave like a particle then why can't a particle such as an electron also behave like a wave?

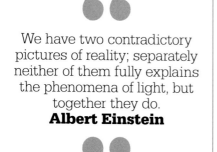

Wavelength of particle   Velocity at which particle is moving

$$\lambda = h/mv$$

Planck's constant   Mass of particle

De Broglie's 1924 equation is used to calculate the wavelength of a particle by dividing Planck's constant by the particle's momentum—its mass multiplied by its velocity.

De Broglie considered Einstein's famous $E = mc^2$, which relates mass to energy, and the fact that Einstein and Planck had related energy to the frequency of waves. By combining the two, de Broglie suggested that mass should have a wavelike form as well and came up with the concept of a matter wave: any moving object has an associated wave. The kinetic energy of the particle is proportional to its frequency and the speed of the particle is inversely proportional to its wavelength—with faster particles having shorter wavelengths.

Einstein supported de Broglie's idea as it seemed a natural continuation of his own theories. De Broglie's claim that electrons could behave like waves was experimentally verified in 1927 when British physicist George Thomson and American physicist Clinton Davisson both demonstrated that

This image shows the X-ray diffraction pattern for platinum. X-rays are wavelike forms of electromagnetic energy carried by particles which transmit light called "photons." The diffraction experiment here shows X-rays behaving like rays, but a different experiment might show them behaving like particles.

a narrow electron beam directed through a thin crystal of nickel formed a diffraction pattern as it passed through the crystal lattice.

How can it be like that?

Thomas Young had demonstrated that light was a wave by showing how it formed interference patterns. In the early 1960s, American physicist Richard Feynman described a thought experiment in which he imagined what would happen if one photon or electron at a time was sent toward twin slits that could be opened or closed. The expected result would be that the photons would travel as particles, arrive as particles, and be detected on the screen as individual dots. Rather than interference patterns, there should be two bright areas when both slits were open, or just one if one of the slits was closed. However, Feynman predicted an alternative outcome, that the pattern on the screen builds up, particle by particle, into interference patterns when both slits are open—but not if one of the slits is closed.

Even if subsequent photons are fired off after the earlier ones have hit the screen, they still somehow

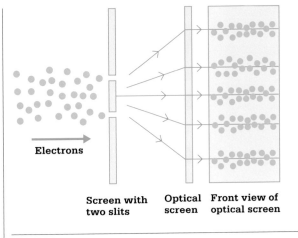

Electrons

Screen with two slits | **Optical screen** | **Front view of optical screen**

When particles, such as electrons or atoms, are passed through a dual-slit apparatus, interference patterns of light and dark bands are produced, just as happens with waves. This shows that particles have wavelike properties and exhibit wavelike behavior.

"know" where to go to build up the interference pattern. It is as if each particle travels as a wave, passes through both slits simultaneously, and creates interference with itself. But how does a single particle traveling through the left-hand slit know whether the right-hand slit is open or closed?

Feynman advised against even attempting to answer these questions. In 1964 he wrote: "Do not keep saying to yourself, if you can possibly avoid it 'But how can it be like that?' because you will go down the drain into a blind alley from which nobody has yet escaped.

Nobody knows how it can be like that." Feynman's predicted findings have since been confirmed by other scientists.

What is apparent is that both the wave theory and the particle theory of light are correct. Light acts as a wave when it is traveling through space, but as a particle when it is being measured. There is no single model that can describe light in all its aspects. It is easy enough to say that light has "wave–particle duality" and leave it at that, but what that statement actually means is something that no one can answer satisfactorily. ∎

Louis de Broglie

Born in 1892 in Dieppe, France, Louis de Broglie obtained a degree in history in 1910 and a science degree in 1913. He was conscripted during World War I, and when the war ended he resumed his physics studies. In 1924, at the Faculty of Sciences at Paris University, de Broglie delivered his doctoral thesis, "Recherches sur la théorie des quanta" ("Researches on the quantum theory"). This thesis, which was first received with astonishment, but later confirmed by the discovery of electron diffraction in 1927, served as the basis for developing the theory of

wave mechanics. De Broglie taught theoretical physics at the Institut Henri Poincaré in Paris until he retired. In 1929, he was awarded the Nobel Prize for Physics for his discovery of the wave nature of electrons. He died in 1987.

Key works

1924 "Recherches sur la théorie des quanta" ("Researches on the quantum theory"), *Annals of Physics*
1926 *Ondes et Mouvements* (*Waves and Motions*)

A NEW IDEA OF REALITY
QUANTUM NUMBERS

IN CONTEXT

KEY FIGURE
Wolfgang Pauli (1900–1958)

BEFORE
1672 Isaac Newton splits white light into a spectrum.

1802 William Hyde Wollaston sees dark lines in the solar spectrum.

1913 Niels Bohr puts forward his shell model of the atom.

AFTER
1927 Niels Bohr proposes the Copenhagen interpretation, stating that a particle exists in all possible states until it is observed.

1928 Indian astronomer Subrahmanyan Chandrasekhar calculates that a large enough star could collapse to form a black hole at the end of its life.

1932 British physicist James Chadwick discovers the neutron.

In 1802, British chemist and physicist William Hyde Wollaston noticed that the spectrum of sunlight was overlaid by a number of fine dark lines. German lens-maker Joseph von Fraunhofer first examined these lines in detail in 1814, listing more than 500. In the 1850s, German physicist Gustav Kirchhoff and German chemist Robert Bunsen found that each element produces its own unique set of lines, but did not know what caused them.

Quantum leaps

In 1905, Albert Einstein had explained the photoelectric effect, by which light can cause electrons to be emitted from some metals, saying that light could behave like a stream of packets of energy, called quanta. In 1913, Danish physicist Niels Bohr proposed a model of the atom that accounted for both quanta and the spectra of elements. In Bohr's atom, electrons traveled around the core nucleus in fixed, or quantized orbits. Light quanta (later called photons) striking the atom could be absorbed by electrons, which then moved to higher orbits (further from the nucleus). A sufficiently energetic photon may eject an electron from its orbit altogether. Conversely, when an electron gives up its "extra" energy as a photon of light, it falls back down to its original energy level, closer to the

Fraunhofer lines are fine dark lines that overlay the spectrum of visible light. Each element produces its own unique set of Fraunhofer lines (designated by one or more letters), determined by the quantum numbers of its electrons. The most prominent of these lines are shown here.

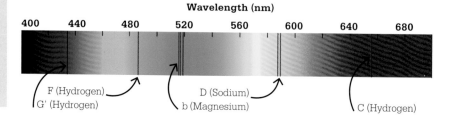

Wavelength (nm)

400 440 480 520 560 600 640 680

F (Hydrogen)
G' (Hydrogen)
D (Sodium)
b (Magnesium)
C (Hydrogen)

See also: Magnetic monopoles 159 ▪ Electromagnetic waves 192–195 ▪ Light from the atom 196–199 ▪ Energy quanta 208–211 ▪ Subatomic particles 242–243

The laws of quantum physics **prevent identical particles** from **occupying the same space at the same time**.

→

Each electron around an atom has a unique code made up of **four quantum numbers**.

↓

Two electrons with the same quantum numbers would have to **occupy different energy levels** in an atom.

←

The quantum numbers define **characteristics of the electron**—energy, spin, angular momentum, and magnetism.

Wolfgang Pauli

The son of a chemist, Wolfgang Pauli was born in 1900 in Vienna, Austria. When he was a schoolboy, he is said to have smuggled Albert Einstein's papers on special relativity into school to read at his desk. As a student at the University of Munich in Germany, Pauli published his first paper on relativity in 1921. This was praised by Einstein himself.

After graduating, Pauli assisted physicist Max Born at the University of Göttingen, Germany. The threat of Nazi persecution led him to move to Princeton, New Jersey, in 1940 and he later became an American citizen. He was awarded the Nobel Prize in 1945 for his discovery of the exclusion principle. Pauli moved to Zurich, Switzerland, in 1946 and worked as a professor at the city's Eidgenössische Technische Hochschule until his death in 1958.

Key works

1926 *Quantum Theory*
1933 "Principles of Wave Mechanics"

atom's nucleus. These steps up and down are called "quantum leaps." Their size is unique to each atom.

Atoms give off light at specific wavelengths, and so each element produces a characteristic set of spectral lines. Bohr proposed that these lines were related to the energy levels of the electron orbits, and that they were produced by an electron absorbing or emitting a photon at the frequency that corresponds to the spectral line.

The energy level that an electron can have in an atom is denoted by the primary, or principal, quantum number, n, in which $n = 1$ equals the lowest possible orbit, $n = 2$ the next highest, and so on. Using this scheme, Bohr was able to describe the energy level in the simplest atom, hydrogen, in which a single electron orbits a single proton. Later models that incorporated the wavelike properties of electrons were able to describe larger atoms.

Pauli exclusion principle

In 1925, Wolfgang Pauli was trying to explain the structure of atoms. What decided an electron's energy level and the number of electrons that each energy level, or shell, could hold? The reason he came up with was that each electron had a unique code, described by its four quantum numbers—energy, spin, angular momentum, and magnetism. The exclusion principle stated that no two electrons in an atom could share the same four quantum numbers. No two identical particles could occupy the same state at the same time. Two electrons could occupy the same shell but only if they had opposite spins, for example. ▪

66

Physics is puzzle solving, but of puzzles created by nature, not by the mind of man.
Maria Goeppert Mayer
German–American physicist

99

ALL IS WAVES
MATRICES AND WAVES

IN CONTEXT

KEY FIGURE
Erwin Schrödinger
(1887–1961)

BEFORE
1897 J.J. Thomson discovers
the electron and suggests that
it is the carrier of electricity.

1924 Louis de Broglie
proposes that particles
have wavelike properties.

AFTER
1927 Werner Heisenberg
publishes his uncertainty
principle.

1927 Niels Bohr endorses the
Copenhagen interpretation,
claiming that observation
determines the quantum
state of a particle.

1935 Erwin Schrödinger sets
out the scenario of a cat that is
simultaneously alive and dead
to illustrate a problem with the
Copenhagen interpretation.

By the 1920s, scientists were
starting to challenge the
model of the atom that
Danish physicist Niels Bohr had
proposed in 1913. Experiments had
begun to show not only that light
could behave like a stream of
particles, but also that electrons
could act like waves.

German physicist Werner
Heisenberg tried to develop a
system of quantum mechanics
that relied only on what could be
observed. It was impossible to see
an electron orbiting an atom directly,
but it was possible to observe the
light absorbed and emitted by atoms
when an electron jumps from one
orbit to another. He used these
observations to produce tables of
numbers to represent the positions
and momentum of electrons and
worked out rules to calculate the
values of these properties.

Matrix mechanics
In 1925, Heisenberg shared his
calculations with German–Jewish
physicist Max Born, who realized
that this type of table was known
as a matrix. Together with Born's
student, Pascual Jordan, Born and
Heisenberg worked out a new
theory of matrix mechanics that
could be used to link the energies
of electrons to the lines that had
been observed in the visible
light spectrum.

One of the interesting things
about matrix mechanics is that
the order in which calculations are
made is important. Calculating the
momentum and then the position of
a particle will give a different result
from calculating the position first
and then the momentum. It was
this difference that would lead
Heisenberg to his uncertainty
principle, which stated that in
quantum mechanics the velocity

> Why do all experiments that
> involve, say, the position of a
> particle make the particle
> suddenly be somewhere
> rather than everywhere?
> No one knows.
> **Christophe Galfard**
> **French physicist**

See also: Electromagnetic waves 192–195 ▪ Light from the atom 196–199 ▪ Energy quanta 208–211 ▪ Heisenberg's uncertainty principle 220–221 ▪ Antimatter 246

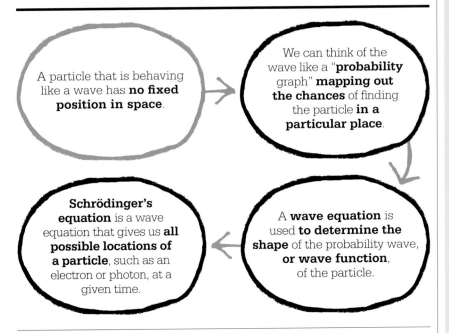

A particle that is behaving like a wave has **no fixed position in space**.

We can think of the wave like a "**probability** graph" **mapping out the chances** of finding the particle **in a particular place**.

A **wave equation** is used **to determine the shape** of the probability wave, **or wave function**, of the particle.

Schrödinger's equation is a wave equation that gives us **all possible locations of a particle**, such as an electron or photon, at a given time.

Erwin Schrödinger

Erwin Schrödinger was born in 1887 in Vienna, Austria, and studied theoretical physics. After serving in World War I, he held posts at universities in Zurich, Switzerland, and Berlin, Germany. In 1933, as the Nazis came to power in Germany, he moved to Oxford in the UK.

That same year, he shared the Nobel Prize with British theoretical physicist Paul Dirac for "the discovery of new productive forms of atomic theory." In 1939, he became director of theoretical physics at the Institute of Advanced Studies in Dublin, Ireland. He retired to Vienna in 1956, and died in 1961. Fascinated by philosophy, Schrödinger remains best known for his 1935 thought experiment, Schrödinger's cat, which examined the idea that a quantum system could exist in two different states simultaneously.

Key works

1926 "An undulatory theory of the mechanics of atoms and molecules," *Physical Review*
1935 "The Present Situation in Quantum Mechanics"

of an object and its position cannot both be measured exactly at the same time.

In 1926, Austrian physicist Erwin Schrödinger devised an equation that determines how probability waves, or wave functions (mathematical descriptions of a quantum system) are shaped and how they evolve. The Schrödinger equation is as important to the subatomic world of quantum mechanics as Newton's laws of motion are for events on a large scale. Schrödinger tested his equation on the hydrogen atom and found that it predicted its properties with great accuracy.

In 1928, British physicist Paul Dirac married Schrödinger's equation to Einstein's special relativity, which had demonstrated the link between mass and energy, encapsulated in the famous $E = mc^2$ equation. Dirac's equation was consistent with both special relativity and quantum mechanics in its description of electrons and other particles. He proposed that electrons should be viewed as arising from an electron field, just as photons arose from the electromagnetic field.

A combination of Heisenberg's matrices and Schrödinger's and Dirac's equations laid the basis for two fundamentals of quantum mechanics, the uncertainty principle, and the Copenhagen interpretation. ▪

Can nature possibly be so absurd as it seemed to us in these atomic experiments?
Werner Heisenberg

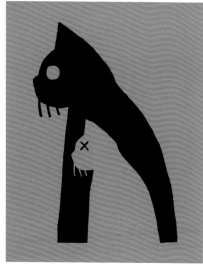

THE CAT IS BOTH ALIVE AND DEAD

HEISENBERG'S UNCERTAINTY PRINCIPLE

IN CONTEXT

KEY FIGURE
Werner Heisenberg
(1901–1976)

BEFORE
1905 Einstein proposes that light is made up of discrete packets called photons.

1913 Niels Bohr sets out a model of the atom with electrons around a nucleus.

1924 French physicist Louis de Broglie suggests that particles of matter can also be considered as waves.

AFTER
1935 Einstein, American–Russian physicist Boris Podolsky, and American-born Israeli physicist Nathan Rosen challenge the Copenhagen interpretation.

1957 American physicist Hugh Everett puts forward his "many worlds theory" to explain the Copenhagen interpretation.

According to **quantum theory**, until a particle is observed and measured, it **simultaneously exists** in all the possible locations and states that it could be in. This is called **superposition**.

All the possible states a particle might be in are described by its **wave function**.

Erwin Schrödinger compared this to a cat that is **simultaneously alive and dead**.

The particle's properties have no definite value until we **measure** them—when we do so, the **wave function collapses**.

The collapse of the wave function **fixes the properties of the particle**.

I n classical physics, it was generally accepted that the accuracy of any measurement was only limited by the precision of the instruments used. In 1927, Werner Heisenberg showed that this just wasn't so.

Heisenberg asked himself what it actually meant to define the position of a particle. We can only know where something is by interacting with it. To determine the position of an electron, we bounce a photon off it. The accuracy of the measurement is determined by the wavelength of the photon; the higher the frequency of the photon, the more accurate the position of the electron.

Max Planck had shown that the energy of a photon is related to its frequency by the formula $E = h\nu$,

Werner Heisenberg

Werner Heisenberg was born in 1901 in Würzburg, Germany. He went to work with Niels Bohr at Copenhagen University in 1924. Heisenberg's name will always be associated with his uncertainty principle, published in 1927, and he was awarded the Nobel Prize for Physics in 1932 for the creation of quantum mechanics.

In 1941, during World War II, Heisenberg was appointed director of the Kaiser Wilhelm Institute for Physics (later renamed the Max Planck Institute) in Munich, and was put in charge of Nazi Germany's atomic bomb project. Whether this failed due to a lack of resources or because Heisenberg wanted it to fail is unclear. He was eventually taken prisoner by American troops and sent to the UK. After the war, he served as the director of the Max Planck Institute until he resigned in 1970. He died in 1976.

Key works

1925 "On Quantum Mechanics"
1927 "On the Perceptual Content of Quantum Theoretical Kinematics and Mechanics"

where E equals energy, v equals frequency, and h is Planck's constant. The higher the frequency of the photon, the more energy it carries and the more it will knock the electron off course. We know where the electron is at that moment, but we cannot know where it is going. If it were possible to measure the electron's momentum with absolute precision, its location would become completely uncertain, and vice versa.

Heisenberg showed that the uncertainty in the momentum multiplied by the uncertainty in the position can never be smaller than a fraction of Planck's constant. The uncertainty principle is a fundamental property of the universe that puts a limit on what we can know simultaneously.

Copenhagen interpretation

What came to be known as the "Copenhagen interpretation" of quantum physics was championed by Niels Bohr. It accepted that, as Heisenberg had shown, there are some things we simply cannot know about the universe. The properties of a quantum particle have no definite value until a measurement is made. It is impossible to design an experiment that would allow us to see an electron as wave and particle at the same time, for example. The wave and particle nature of matter are two sides of the same coin, Bohr suggested. The Copenhagen interpretation opened a sharp divide between classical physics and quantum physics over whether or not physical systems have definite properties prior to being measured.

Schrödinger's cat

According to the Copenhagen interpretation, any quantum state can be seen as the sum of two or more distinct states, known as superposition, until it is observed, at which point it becomes either one or the other.

Erwin Schrödinger asked: when is the switch made from superposition into one definite reality? He described the scenario of a cat in a box, along with a poison that will be released when a quantum event occurs. According to the Copenhagen interpretation, the cat is in the superposition state of being both alive and dead until someone looks in the box, which Schrödinger thought was ridiculous. Bohr retorted that there was no reason why the rules of classical physics should also apply to the quantum realm—this was just the way things were. ▪

> The atoms or elementary particles themselves are not real; they form a world of potentialities or possibilities rather than one of things or facts.
> **Werner Heisenberg**

SPOOKY ACTION AT A DISTANCE
QUANTUM ENTANGLEMENT

IN CONTEXT

KEY FIGURE
John Stewart Bell
(1928–1990)

BEFORE
1905 Albert Einstein publishes his theory of special relativity, which is based in part on the idea that nothing can travel faster than the speed of light.

1926 Erwin Schrödinger publishes his wave equation.

1927 Niels Bohr champions the Copenhagen interpretation of the way in which quantum systems interact with the large-scale world.

AFTER
1981 Richard Feynman proposes superposition and the entanglement of particles as basis for a quantum computer.

1995 Austrian quantum physicist Anton Zeilinger demonstrates wave/particle switching in an experiment using entangled photons.

One of the main tenets of quantum mechanics is the idea of uncertainty—we cannot measure all the features of a system simultaneously, no matter how perfect the experiment. The Copenhagen interpretation of quantum physics championed by Niels Bohr effectively says that the very act of measurement selects the characteristics that are observed.

Another peculiar property of quantum mechanics is called "entanglement." If two electrons, for example, are ejected from a quantum system, then conservation of momentum laws tell us that the momentum of one particle is equal and opposite to that of the other. According to the Copenhagen interpretation, neither particle will have a definite state until it is measured, but measuring the momentum of one will determine the state and momentum of the other, regardless of the distance between the particles.

This is known as "non-local behavior," though Albert Einstein called it "spooky action at a distance." In 1935, Einstein attacked entanglement, claiming that there

When any **two subatomic particles**, such as electrons, **interact with each other**, their states become interdependent—they are **entangled**.

The particles **remain connected** even when **physically separated** by an enormous distance (for example, in different galaxies).

As a result, manipulating one particle **instantaneously alters** its partner.

Measuring the properties of one particle **gives information** about the properties of the other.

Particle A and particle B have interacted with
each other and become entangled. They will remain
entangled even if they are sent in different directions.

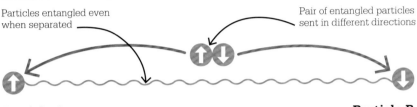

Particles entangled even
when separated

Pair of entangled particles
sent in different directions

Particle A   **Particle B**

John Stewart Bell

John Stewart Bell was born
in 1928 in Belfast, Northern
Ireland. After graduating from
Queen's University Belfast,
he went on to gain a PhD in
nuclear physics and quantum
field theory at the University
of Birmingham. He then
worked at the Atomic Research
Establishment in Harwell,
UK, and later at the European
Organization for Nuclear
Research (CERN) in Geneva,
Switzerland. There, he worked
on theoretical particle science
and accelerator design. After
a year spent at Stanford,
Wisconsin-Madison, and
Brandeis universities in
the US, Bell published his
breakthrough paper in 1964,
which proposed a way to
distinguish between quantum
theory and Einstein's notion
of local reality. He was elected
a member of the American
Academy of Arts and Sciences
in 1987. Bell's early death in
1990 at the age of 62 meant
he did not live to see his ideas
tested experimentally.

Key work

1964 "On the Einstein-
Podolsky-Rosen Paradox,"
Physics

are "hidden variables" at work that
make it unnecessary. He argued
that for one particle to affect the
other, a faster-than-light signal
between them (which was forbidden
by Einstein's theory of special
relativity) would be required.

Bell's theorem
In 1964, Northern Irish physicist
John Stewart Bell proposed an
experiment that could test whether
or not entangled particles actually
did communicate with each other
faster than light. He imagined the
case of a pair of entangled electrons,
one with spin up and the other
with spin down. According to
quantum theory, the two electrons
are in a superposition of states until
they are measured—either one
of them could be spin up or spin
down. However, as soon as one is
measured, we know with certainty
that the other has to be its opposite.
Bell derived formulas, called the Bell
inequalities, which determine how
often the spin of particle A should

correlate with that of particle B if
normal probability (as opposed to
quantum entanglement) is at work.
The statistical distribution of his
results proved mathematically that
Einstein's "hidden variables"
idea isn't true, and that there is
instantaneous connection between
entangled particles. Physicist Fritjof
Capra maintains that Bell's theorem
shows that the universe is
"fundamentally interconnected."

Experiments such as the one
carried out by French physicist
Alain Aspect in the early 1980s
(which used entangled photon
pairs generated by laser)
demonstrate convincingly that
"action at a distance" is real—the
quantum realm is not bound by
rules of locality. When two particles
are entangled, they are effectively
a single system that has a single
quantum function. ▪

This conceptual computer artwork
shows a pair of particles that have
become entangled: manipulating one
will result in manipulating the other,
regardless of the distance between
them. Entanglement has applications
in the new technologies of quantum
computing and quantum cryptography.

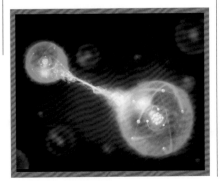

THE JEWEL OF PHYSICS
QUANTUM FIELD THEORY

IN CONTEXT

KEY FIGURE
Richard Feynman
(1918–1988)

BEFORE
1873 James Clerk Maxwell
publishes his equations
describing the properties
of the electromagnetic field.

1905 Einstein proposes that
light, as well as acting like
a wave, can be imagined
as a stream of particles
called quanta.

AFTER
1968 Theoretical physicists
Sheldon Glashow, Abdus
Salam, and Steven Weinberg
present their theory of the
electroweak force, uniting
the electromagnetic and
weak nuclear forces.

1968 Experiments at the
Stanford Linear Accelerator
Laboratory, CA, discover
evidence of quarks, the
building blocks of
subatomic particles.

A **field** maps out the
strength of a force across
space and time.

↓

Quantum field theory says
that forces are transmitted by
force-carrier particles.

↓

The force carriers of the
electromagnetic force
are **photons**.

↓

The ways in which **photon
exchange interactions** take
place can be visualized using
a **Feynman diagram**.

One of quantum mechanics'
biggest shortcomings is
that it fails to take
Einstein's relativity theories into
account. One of the first to try
reconciling these cornerstones
of modern physics was British
physicist Paul Dirac. Published
in 1928, the Dirac equation
viewed electrons as excitations of
an electron field, in the same way
that photons could be seen as
excitations of the electromagnetic
field. The equation became one
of the foundations of quantum
field theory.

The idea of fields carrying
forces across a distance is well
established in physics. A field can
be thought of as anything that has
values that vary across space and
time. For example, the pattern
made by iron filings scattered
around a bar magnet maps out the
lines of force in a magnetic field.

In the 1920s, quantum field
theory proposed a different
approach, which suggested that
forces were carried by means
of quantum particles, such as
photons (particles of light that
are the carrier particles of
electromagnetism). Other particles
that were discovered subsequently,

See also: Force fields and Maxwell's equations 142–147 ▪ Quantum applications 226–231 ▪ The particle zoo and quarks 256–257 ▪ Force carriers 258–259 ▪ Quantum electrodynamics 260 ▪ The Higgs boson 262–263

>
> What we observe is not nature itself, but nature exposed to our method of questioning.
> **Werner Heisenberg**

such as the quark, the gluon, and the Higgs boson (an elementary particle which gives particles their mass) are believed to have their own associated fields.

QED

Quantum electrodynamics, or QED, is the quantum field theory that deals with electromagnetic force. The theory of QED was fully (and independently) developed by Richard Feynman and Julian Schwinger in the US and Shin'ichirō

Tomonaga in Japan. It proposes that charged particles, such as electrons, interact with each other by emitting and absorbing photons. This can be visualized by means of Feynman diagrams, developed by Richard Feynman (see right).

QED is one of the most astonishingly accurate theories ever formulated. Its prediction for the strength of the magnetic field associated with an electron is so close to the value produced, that if the distance from London to Timbuktu was measured to the same precision, it would be accurate to within a hair's breadth.

The Standard Model

QED was a stepping stone toward building quantum field theories for the other fundamental forces of nature. The Standard Model combines two theories of particle physics into a single framework, which describes three of the four known fundamental forces (the electromagnetic, weak, and strong interactions), but not gravity. The strongest of these is the strong

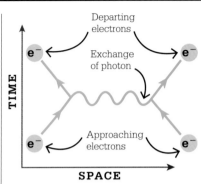

This **Feynman diagram** represents electromagnetic repulsion. Two electrons (e^-) approach, exchange a photon, then move apart.

interaction, or strong nuclear force, which binds together the protons and neutrons in the nucleus of an atom. The electroweak theory proposes that the electromagnetic and weak interactions can be considered two facets of a single, "electroweak," interaction. The Standard Model also classifies all known elementary particles. Reconciling the gravitational force with the Standard Model remains one of physics' biggest challenges. ▪

Richard Feynman

Born in 1918, Richard Feynman grew up in New York City. Fascinated by math from an early age, he won a scholarship to the Massachusetts Institute of Technology and achieved a perfect score in his PhD application to Princeton. In 1942, he joined the Manhattan Project to develop the first atomic bomb.

After the war, Feynman worked at Cornell University, where he developed the QED theory, for which he was jointly awarded the Nobel Prize in 1965. He moved to the California Institute of Technology in 1960.

His autobiography, *Surely You're Joking, Mr. Feynman!*, is one of the best-selling books written by a scientist. One of his last major achievements was uncovering the cause of NASA's 1986 Challenger space shuttle disaster. He died in 1988.

Key works

1967 *The Character of Physical Law*
1985 *QED: The Strange Theory of Light and Matter*
1985 *Surely You're Joking, Mr. Feynman!*

COLLABORATION

BETWEEN PARALLEL

UNIVERSES

QUANTUM APPLICATIONS

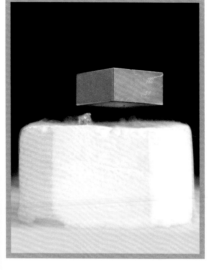

The grand realm of quantum physics seems a long way from the common-sense everyday world, but it has given rise to a surprising number of technological advances that play a vital role in our lives. Computers and semiconductors, communications networks and the internet, GPS and MRI scanners—all depend on the quantum world.

Superconductors
In 1911, Dutch physicist Heike Kamerlingh Onnes made a remarkable discovery when he was experimenting with mercury at very low temperatures. When the temperature of the mercury reached −268.95°C, its electrical resistance disappeared. This meant that, theoretically, an electrical current could flow through a loop of super-cooled mercury forever.

American physicists John Bardeen, Leon Cooper, and John Schrieffer came up with an explanation for this strange phenomenon in 1957. At very low temperatures, electrons form so-called "Cooper pairs." Whereas a single electron has to obey the Pauli exclusion principle, which forbids two electrons from sharing the same quantum state, the Cooper pairs form a "condensate." This means that the pairs act as if they were a single body, with no resistance from the conducting material, rather than a collection of electrons flowing through the conductor. Bardeen, Cooper, and Schrieffer won the 1972 Nobel Prize in Physics for this discovery.

In 1962, Welsh physicist Brian Josephson predicted that Cooper pairs should be able to tunnel through an insulating barrier between two superconductors. If a voltage is applied across one

The Meissner effect is the expulsion of a magnetic field from a superconducting material, causing a magnet to levitate above it.

of these junctions, known as "Josephson junctions," the current vibrates at a very high frequency. Frequencies can be measured with much greater precision than voltages, and Josephson junctions have been used in devices such as SQUIDS (superconducting quantum interference devices) to examine the tiny magnetic fields produced by the human brain. They also have potential for use in ultrafast computers.

Without superconductivity, the powerful magnetic forces harnessed in magnetic resonance imaging (MRI) scanners would not be possible. Superconductors also act to expel magnetic fields. This is the Meissner effect, a phenomenon that has allowed the building of magnetically-levitated trains.

Superfluids
Kamerlingh Onnes was also the first person to liquefy helium. In 1938, Russian physicist Pyotr Kapitsa and British physicists John Allen and Don Misener discovered that below

> Technology at the forefront of human endeavor is hard. But that is what makes it worth it.
> **Michelle Yvonne Simmons**
> Professor of quantum physics

See also: Fluids 76–79 ▪ Electric current and resistance 130–133 ▪ Electronics 152–155 ▪ Particles and waves 212–215
▪ Heisenberg's uncertainty principle 220–221 ▪ Quantum entanglement 222–223 ▪ Gravitational waves 312–315

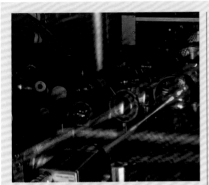

A blue laser beam excites strontium atoms in the 3-D quantum gas atomic clock developed by physicists at JILA in Boulder, Colorado.

Quantum clocks

Reliable timekeeping is crucial in synchronizing the activities that our technological world depends on. The most accurate clocks in the world today are atomic clocks, which use atoms "jumping" back and forth between energy states as their "pendulum."

The first accurate atomic clock was built in 1955 by physicist Louis Essen at the UK's National Physical Laboratory. The late 1990s saw major advances in atomic clock technology, including the cooling of atoms to near absolute zero temperatures by slowing them with lasers.

Today, the most accurate atomic clocks measure time using the transitions between spin states in the nucleus of an atom. Until recently, they used atoms of the cesium-133 isotope. These clocks can achieve an accuracy equivalent to gaining or losing a second over 300 million years and are crucial to GPS navigation and telecommunications. The most recent atomic clock uses strontium atoms and is believed to be even more accurate.

−270.98°C, liquid helium completely lost its viscosity, seemingly flowing without any apparent friction and possessing a thermal conductivity that far outstripped that of the best metal conductors. When helium is in its superfluid state, it does not behave like it does at higher temperatures. It flows over the rim of a container and leaks through the smallest of holes. When set rotating, it does not stop spinning. The Infrared Astronomical Satellite (IRAS) launched in 1983 was cooled by superfluid helium.

A fluid becomes a superfluid when its atoms begin to occupy the same quantum states; in essence, they lose their individual identities and become a single entity. Superfluids are the only quantum phenomenon that can be observed by the naked eye.

Tunneling and transistors

Some of the devices we use daily, such as the touchscreen technology of mobile phones, would not work if not for the strange phenomenon of quantum tunneling. First explored in the 1920s by German physicist Friedrich Hund and others, this phenomenon allows particles to pass through barriers that would traditionally be impassable. This oddity arises from considering electrons, for example, as waves of probability rather than particles existing at a particular point. In transistors, for example, quantum tunneling allows electrons to pass across a junction between semiconductors. This can cause problems, as chips become smaller and the insulating layers between components become too thin to stop the electrons at all—effectively making it impossible to turn the device off.

Quantum imaging

An electron microscope depends on the wave/particle duality of electrons. It works in a similar »

In classical physics, an object—for example, a rolling ball—cannot move across a barrier without gaining enough energy to get over it. A quantum particle, however, has wavelike properties, and we can never know exactly how much energy it has. It may have enough energy to pass through a barrier.

Rolling ball

Quantum particle passes through barrier

Classical physics

Quantum tunneling

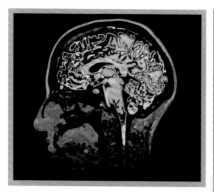

Magnetic resonance imaging (MRI) scanners produce detailed images, such as this image of the human brain, using a magnetic field and radio waves.

way to a light microscope, but instead of using lenses to focus light, it employs magnets to focus beams of electrons. As the stream of electrons passes through the magnetic "lens," they behave like particles as the magnet bends their path, focusing them on the sample to be examined. Now they act like waves, diffracting around the object and on to a fluorescent screen where an image is formed. The wavelength of an electron is much smaller than that of light, allowing the image resolution of objects millions of times smaller than can be seen with a light microscope.

Magnetic resonance imaging (MRI) was developed in the 1970s by researchers such as Paul Lauterbur in the US and Peter Mansfield in the UK. Inside an MRI scanner, the patient is surrounded by a magnetic field many thousands of times more powerful than that of Earth, produced by a superconducting electromagnet. This magnetic field affects the spin of the protons that make up the hydrogen atoms in the water molecules that constitute so much of a human body, magnetizing them in a particular way. Radio waves are then used to alter the spin of the protons,

changing the way they are magnetized. When the radio waves are switched off, the protons return to the previous spin state, emitting a signal which is recorded electronically and turned into images of the body tissues they form a part of.

Quantum dots

Quantum dots are nanoparticles, typically consisting of just a few dozen atoms, made of semiconducting materials. They were first created by Russian physicist Alexey Ekimov and American physicist Louis Brus, who were working independently in the early 1980s. The electrons within semiconductors are locked inside the crystal lattice that makes up the material, but can be freed if they are excited by photons. When the electrons are free, the electrical resistance of the semiconductor falls rapidly, allowing current to flow more easily.

Quantum dot technology can be used in television and computer screens to produce finely detailed images. Quantum dots can be precisely controlled to do all kinds of useful things. The electrons in a quantum dot each occupy a different quantum state, so the dot has discrete energy levels just like

an individual atom. The quantum dot can absorb and emit energy as electrons move between levels. The frequency of the light emitted depends on the spacing between the levels, which is determined by the size of the dot—larger dots glow at the red end of the spectrum and smaller dots at the blue. The luminosity of the dots can be tuned precisely by varying their size.

Quantum computing

Quantum dot technology is likely to be used in constructing quantum computers. Computers depend on binary bits of information, corresponding to the on (1) and off (0) positions of their electronic switches. Spin is a quantum property that seems to turn up quite often in quantum technology. It is the spin of the electron that gives some materials their magnetic properties. Using lasers, it is possible to get electrons into a superposition state where they have both up and down spin at the same time. These superposition electrons can theoretically be used as qubits (quantum bits) which effectively can be "on," "off," and something in between, all at the same time. Other particles, such as polarized photons, can also be used as qubits. It was Richard Feynman

In display technology, quantum dots change their size and shape to emit light of specific colors when they are stimulated by blue light from LEDs.

Blue light from LED

Core absorbs blue light and emits red light

Red light from quantum dot

Surface molecules stabilize quantum dot and increase its efficiency

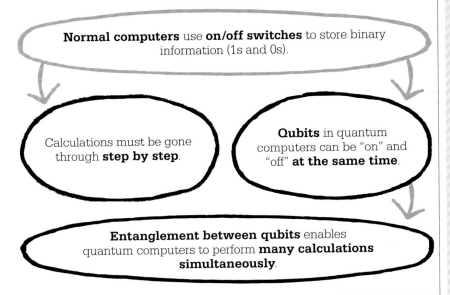

Normal computers use **on/off switches** to store binary information (1s and 0s).

Calculations must be gone through **step by step**.

Qubits in quantum computers can be "on" and "off" **at the same time**.

Entanglement between qubits enables quantum computers to perform **many calculations simultaneously**.

David Deutsch

Born in Haifa, Israel, in 1953, David Deutsch is one of the pioneers of quantum computing. He studied physics in the UK, at Cambridge and Oxford. He then spent several years working in the US at the University of Texas at Austin, before returning to Oxford University. Deutsch is a founding member of the Centre for Quantum Computation at Oxford University.

In 1985, Deutsch wrote a groundbreaking paper, "Quantum Theory, the Church–Turing Principle and the Universal Quantum Computer," which set out his ideas for a universal quantum computer. His discovery of the first quantum algorithms, his theory of quantum logic gates, and his ideas on quantum computational networks are among the most important advances in the field.

Key works

1985 "Quantum Theory, the Church–Turing Principle and the Universal Quantum Computer," *Proceedings of the Royal Society*
1997 *The Fabric of Reality*
2011 *The Beginning of Infinity*

who first suggested, in 1981, that enormous computing power would be unleashed if the superposition state could be exploited. Potentially, qubits can be used to encode and process vastly more information than the simple binary computer bit.

In 1985, British physicist David Deutsch began to set out ideas on how such a quantum computer could actually work. The field of computer science is in large part built on the idea of the "universal computer," first suggested by British mathematician Alan Turing in the 1930s. Deutsch pointed out that Turing's concept was limited by its reliance on classical physics and would therefore represent only a subset of possible computers. Deutsch proposed a universal computer based on quantum physics and began rewriting Turing's work in quantum terms.

The power of qubits is such that just 10 are sufficient to allow the simultaneous processing of 1,023 numbers; with 40 qubits, the possible number of parallel computations will exceed 1 trillion. Before quantum computers

become a reality, the problem of decoherence will have to be solved. The smallest disturbance will collapse, or decohere, the superposition state. Quantum computing could avoid this by making use of the phenomenon of quantum entanglement, which is what Einstein called "spooky action at a distance." This allows one particle to affect another somewhere else and could allow the value of the qubits to be determined indirectly. ∎

> It's as if you were trying to do a complex jigsaw puzzle in the dark with your hands tied behind your back.
> **Brian Clegg**
> **British science writer, on quantum computing**

NUCLEAR PARTICLE

INSIDE THE ATOM

AND PHYSICS

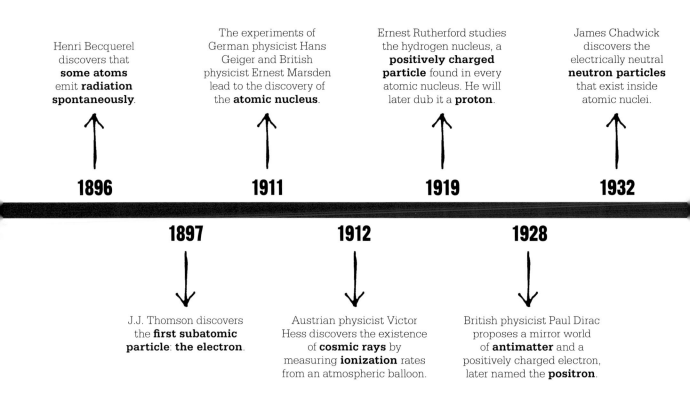

Henri Becquerel discovers that **some atoms** emit **radiation spontaneously**.

1896

The experiments of German physicist Hans Geiger and British physicist Ernest Marsden lead to the discovery of the **atomic nucleus**.

1911

Ernest Rutherford studies the hydrogen nucleus, a **positively charged particle** found in every atomic nucleus. He will later dub it a **proton**.

1919

James Chadwick discovers the electrically neutral **neutron particles** that exist inside atomic nuclei.

1932

1897

J.J. Thomson discovers the **first subatomic particle: the electron**.

1912

Austrian physicist Victor Hess discovers the existence of **cosmic rays** by measuring **ionization** rates from an atmospheric balloon.

1928

British physicist Paul Dirac proposes a mirror world of **antimatter** and a positively charged electron, later named the **positron**.

The idea of atoms as tiny particles of matter dates back to the ancient world, their seemingly indivisible nature enshrined in their Greek name *atomos* ("uncuttable"). British physicist John Dalton, who proposed his atomic theory in 1803, remained convinced of their indestructible nature—as did most 19th-century scientists. However, in the late 1890s, some researchers began to challenge this view.

In 1896, French physicist Henri Becquerel discovered by chance, when experimenting with X-rays, that the uranium salt coating his photographic plate emitted radiation spontaneously. A year later, British physicist J.J. Thomson deduced that the rays he produced in a cathode ray experiment were made up of negatively charged particles more than a thousand times lighter than a hydrogen atom; these subatomic particles were later named electrons.

Exploring the nucleus

Becquerel's student Marie Curie proposed that such rays came from within the atom, rather than being the result of chemical reactions—an indication that atoms contained smaller particles. In 1899, New Zealand-born physicist Ernest Rutherford confirmed that there are different types of radiation. He named two—alpha rays, which he later recognized were positively charged helium atoms, and beta rays, negatively charged electrons. In 1900, French scientist Paul Villard discovered a high-energy light, which Rutherford called gamma rays to complete the trio of subatomic particles he named by the first three letters of the Greek alphabet.

Rutherford and other physicists used alpha particles as tiny projectiles, firing them into atoms to search for smaller structures. Most passed through the atoms, but a small fraction bounced almost entirely backward in the direction they were fired from. The only possible explanation appeared to be that a densely packed region of positive charge within the atom was repelling them. Danish physicist Niels Bohr worked with Rutherford to produce, in 1913, a new model of an atom with electrically positive nucleus surrounded by light electrons, orbiting like planets.

Further research led physicists to suggest that other particles must exist to make up the mass

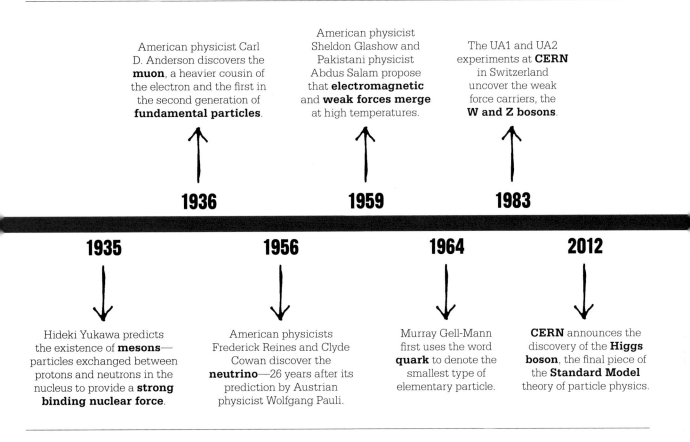

American physicist Carl D. Anderson discovers the **muon**, a heavier cousin of the electron and the first in the second generation of **fundamental particles**.

American physicist Sheldon Glashow and Pakistani physicist Abdus Salam propose that **electromagnetic** and **weak forces merge** at high temperatures.

The UA1 and UA2 experiments at **CERN** in Switzerland uncover the weak force carriers, the **W and Z bosons**.

1936

1959

1983

1935

1956

1964

2012

Hideki Yukawa predicts the existence of **mesons**—particles exchanged between protons and neutrons in the nucleus to provide a **strong binding nuclear force**.

American physicists Frederick Reines and Clyde Cowan discover the **neutrino**—26 years after its prediction by Austrian physicist Wolfgang Pauli.

Murray Gell-Mann first uses the word **quark** to denote the smallest type of elementary particle.

CERN announces the discovery of the **Higgs boson**, the final piece of the **Standard Model** theory of particle physics.

of the nucleus: Rutherford's 1919 discovery of protons provided the positive electric charge, while electrically neutral neutrons were identified by British physicist James Chadwick in 1932.

More particles revealed

Among the next conundrums to solve was why positively charged protons within a nucleus did not push the nucleus apart. Japanese physicist Hideki Yukawa supplied an answer in 1935, proposing that an ultra short-range force (the strong nuclear force), carried by a particle called a meson, held them together.

When two nuclear bombs were dropped on Japan to end World War II in 1945, the lethal force of nuclear power became patently clear. Yet its use in peacetime to produce domestic energy also became a

reality as the first commercial nuclear power reactors opened in the US and UK in the 1950s.

Meanwhile, particle research continued, and ever more powerful particle accelerators uncovered a host of new particles, including kaons and baryons, which decayed more slowly than expected. Various scientists, including American physicist Murray Gell-Mann, dubbed this quality of long decay "strangeness" and used it to classify the subatomic particles that exhibit it into familiar groups, according to their properties. Gell-Mann later coined the name quark for the constituents that determined such properties, and established different "flavors" of quarks—initially up, down, and strange quarks. Charm, top, and bottom quarks came later.

The neutrino, first proposed in 1930 to explain missing energy from beta radiation, was detected in 1956. Heavier versions of the electron and quarks were detected; from their interactions, physicists began to piece together a picture of how these particles exchanged forces and changed from one type to another. The go-betweens that made everything happen—the force-carrying bosons—were also discovered, and the Higgs boson completed the picture in 2012.

However, the Standard Model—the theory explaining the four basic forces, force carriers, and the fundamental particles of matter—has known limitations. Modern particle physics is starting to push these limits in pursuit of dark matter, dark energy, and a clue as to the origin of matter itself. ∎

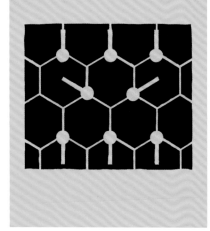

MATTER IS NOT INFINITELY DIVISIBLE

ATOMIC THEORY

IN CONTEXT

KEY FIGURE
John Dalton (1766–1844)

BEFORE
c. 400 BCE The ancient Greek philosophers Leucippus and Democritus theorize that everything is composed of "uncuttable" atoms.

1794 French chemist Joseph Proust finds that elements always join in the same proportions to form compounds.

AFTER
1811 Italian chemist Amedeo Avogadro proposes gases are composed of molecules of two or more atoms, drawing a distinction between atoms and molecules.

1897 British physicist J.J. Thomson discovers the electron.

1905 Albert Einstein uses mathematics to provide evidence for Dalton's theory.

The concept of atoms dates back to the ancient world. The Greek philosophers Democritus and Leucippus, for instance, proposed that eternal "atoms" (from *atomos*, meaning uncuttable) make up all matter. These ideas were revived in Europe in the 17th and 18th centuries as scientists experimented with combining elements to create other materials. They shifted away from historic models of four or five elements (usually earth, air, fire, and water) and categorized elements as oxygen, hydrogen, carbon, and others, but they had yet to discover what made each unique.

Dalton's theory

An explanation arrived with atomic theory, which was developed in the early 19th century by British scientist John Dalton. Dalton proposed that if the same pair of

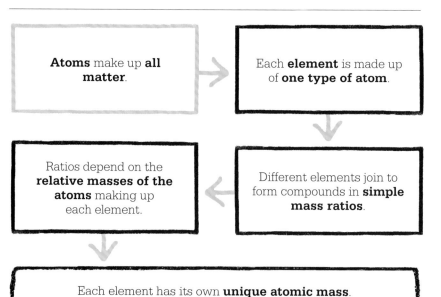

Atoms make up **all matter**.

Each **element** is made up of **one type of atom**.

Different elements join to form compounds in **simple mass ratios**.

Ratios depend on the **relative masses of the atoms** making up each element.

Each element has its own **unique atomic mass**.

See also: Models of matter 68–71 ▪ Light from the atom 196–199 ▪ Particles and waves 212–215 ▪ The nucleus 240–241
▪ Subatomic particles 242–243 ▪ Antimatter 246 ▪ Nuclear bombs and power 248–251

elements could be combined in different ways to form different compounds, the ratio of these elements' masses could be represented with whole numbers. He noticed, for instance, that the mass of oxygen in pure water was nearly eight times the mass of hydrogen, so whatever made up oxygen must weigh more than whatever made up hydrogen. (It was later shown that one oxygen atom weighs 16 times that of a hydrogen atom—since a water molecule contains one oxygen atom and two hydrogen atoms, this fits with Dalton's discovery.)

Dalton concluded that each element was composed of its own unique particles: atoms. These could be joined to or broken apart from other atoms, but they could not be divided, created, or destroyed. "No new creation or destruction of matter is within the reach of chemical agency," he wrote. "We might as well attempt to introduce a new planet into the solar system, or to annihilate one already in existence, as to create

John Dalton used wooden balls in public demonstrations of his theory of atoms. He imagined atoms as hard, consistent spheres.

or destroy a particle of hydrogen." However, matter could be changed by separating particles that are combined and joining them to other particles to form new compounds.

Dalton's model made accurate, verifiable predictions and marked the first time that scientific experimentation had been used to demonstrate atomic theory.

Mathematical evidence
Evidence to support Dalton's theory followed in 1905 when a young Albert Einstein published a paper explaining the erratic jiggling of pollen particles in water—the so-called Brownian motion described in 1827 by Scottish botanist Robert Brown.

Einstein used mathematics to describe how pollen was bombarded by individual water molecules. Although the overall random movement of these many

water molecules couldn't be seen, Einstein proposed that occasionally a small group would mostly move in the same direction and that this was enough to "push" a grain of pollen. Einstein's mathematical description of Brownian motion made it possible to calculate the size of an atom or molecule based on how fast the pollen grain moved.

While later scientific advances revealed that there was more inside the atom than Dalton could ever have imagined, his atomic theory set the foundation for chemistry and many areas of physics. ▪

John Dalton

John Dalton was born in 1766 in England's Lake District to a poor Quaker family. He began working to support himself at the age of 10 and taught himself science and mathematics. He took up teaching at Manchester's New College in 1793 and went on to propose a theory (later disproved) on the cause of color blindness, a condition he shared with his brother. In 1800, he became secretary of the Manchester Literary and Philosophical Society and wrote influential essays describing his experiments with gases. He was also an avid

meteorologist. He became a high-profile scientist, giving lectures at the Royal Institution in London. After his death in 1844, he was granted a public funeral with full honors.

Key works

1806 *Experimental Enquiry into the Proportion of the Several Gases or Elastic Fluids, Constituting the Atmosphere*
1808 and 1810 *A New System of Chemical Philosophy, 1 and 2*

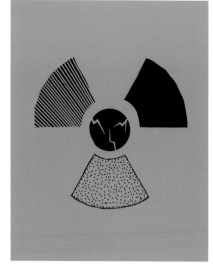

A VERITABLE TRANSFORMATION OF MATTER

NUCLEAR RAYS

Until the end of the 19th century, scientists believed that matter emits radiation (such as visible and ultraviolet light) only when stimulated, such as by heating. This changed in 1896 when French physicist Henri Becquerel conducted an experiment into a newly discovered type of radiation—X-rays. He expected to find that uranium salts emit radiation as a result of absorbing sunlight. However, overcast weather in Paris forced him to put the experiment on hold, and he left a wrapped photographic plate coated with a uranium salt (potassium uranyl-sulfate) in a drawer. Despite remaining in darkness, strong outlines of the sample developed on the plate. Becquerel concluded that the uranium salt was emitting radiation on its own.

Radioactive insights

Becquerel's doctoral student Marie Curie threw herself into studying this phenomenon (which she later termed radioactivity) with her husband Pierre. In 1898, they extracted two new radioactive

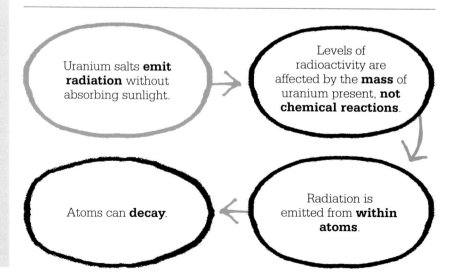

Uranium salts **emit radiation** without absorbing sunlight.

Levels of radioactivity are affected by the **mass** of uranium present, **not chemical reactions**.

Radiation is emitted from **within atoms**.

Atoms can **decay**.

See also: Particles and waves 212–215 ▪ The nucleus 240–241 ▪ Subatomic particles 242–243 ▪ Nuclear bombs and power 248–251

There are many different types of radiation, each with unique properties. Ernest Rutherford identified alpha and beta radiation; Paul Villard discovered gamma radiation.

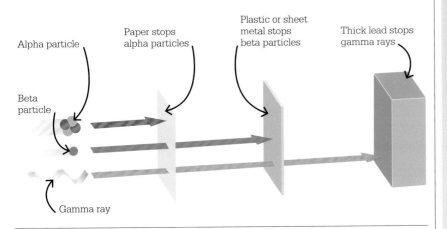

Alpha particle

Beta particle

Gamma ray

Paper stops alpha particles

Plastic or sheet metal stops beta particles

Thick lead stops gamma rays

Marie Curie

Marie Curie (née Skłodowska) was born in Warsaw, Poland, in 1867, to a struggling family of teachers. She traveled to France in 1891 and enrolled at the University of Paris, where she met her future collaborator and husband, Pierre Curie. The Curies shared the Nobel Prize in Physics with Henri Becquerel in 1903. In 1911, Marie was awarded the Nobel Prize in Chemistry.

Later in life, Curie led the Radium Institute in Paris and developed and organized mobile X-ray units that were used to treat over a million soldiers in World War I. She sat on the League of Nations' committee on academic cooperation with Albert Einstein. She died in 1934 from complications likely to have been caused by a lifetime's exposure to radioactivity.

Key works

1898 *On a new radioactive substance contained in pitchblende*
1898 *Rays emitted by compounds of uranium and of thorium*
1903 *Research on radioactive substances*

elements, polonium and radium, from uranium ore. Marie noticed that the level of electrical activity in the air surrounding uranium ore was associated only with the mass of radioactive substance present. She postulated that radiation was not caused by chemical reactions but came from within atoms—a bold theory when science still held that atoms could not be divided.

The radiation produced by uranium was found to be a result of the "decay" of individual atoms. There is no way to predict when an individual atom decays. Instead, physicists measure the time it takes for half of the atoms in a sample to decay—this is the half-life of that element and can be anything from an instant to billions of years. This concept of half-life was proposed by New Zealand–born physicist Ernest Rutherford.

In 1899, Rutherford confirmed suspicions raised by Becquerel and Curie that there are different categories of radiation. He named and described two: alpha and beta.

Alpha emissions are positively charged helium atoms and are unable to penetrate more than several inches of air; beta emissions are streams of negatively charged electrons that can be blocked by an aluminum sheet. Gamma radiation (discovered by French chemist Paul Villard in 1900 as a high-frequency ray) is electrically neutral. It can be blocked with several inches of lead.

Changing elements

Rutherford and his collaborator Frederick Soddy found that alpha and beta radiation were linked with subatomic changes: elements transmute (change from one element to another) via alpha decay. Thorium, for example, changed to radium. They published their law of radioactive change in 1903.

This series of rapid discoveries overturned the age-old concept of the indivisible atom, leading scientists to probe inside it and establish new fields of physics and world-changing technologies. ▪

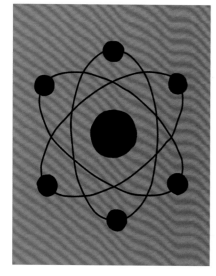

THE CONSTITUTION OF MATTER

THE NUCLEUS

IN CONTEXT

KEY FIGURE
Ernest Rutherford
(1871–1937)

BEFORE
1803 John Dalton proposes the atomic theory, stating that all matter is composed of atoms.

1903 William Crookes, a British chemist and physicist, invents the spinthariscope for detecting ionizing radiation.

1904 J.J. Thomson proposes the "plum pudding" model of the atom.

AFTER
1913 Danish physicist Niels Bohr develops a new model of the atom, in which electrons travel in orbits around the central nucleus and can move between orbits.

1919 Ernest Rutherford reports that the hydrogen nucleus (proton) is present in other nuclei.

The discoveries of the electron and of radiation emanating from within atoms at the end of the 19th century raised the need for a more sophisticated atomic model than had previously been described.

In 1904, British physicist J.J. Thomson, who had discovered the electron in 1897, proposed the "plum pudding" model of the atom. In this model, negatively charged electrons are scattered throughout a much larger, positively charged atom, like dried fruit in a Christmas dessert. Four years later, New Zealand–born physicist Ernest

When we have found how the nucleus of atoms is built up, we shall have found the greatest secret of all—except life.
Ernest Rutherford

Rutherford began to dismantle the plum pudding model through a series of tests conducted at his Manchester laboratory with Ernest Marsden and Hans Geiger. It became known as the gold foil experiment.

Flashing particles
Rutherford and his colleagues observed the behavior of alpha radiation targeted at a thin sheet of gold foil, just 1,000 or so atoms thick. They fired a narrow beam of alpha particles at the foil from a radioactive source that was enclosed by a lead shield. The gold foil was surrounded by a zinc sulfide–coated screen that emitted a small flash of light (scintillation) when struck by the alpha particles. Using a microscope, the physicists could watch as alpha particles struck the screen. (The set-up was similar to a spinthariscope, which had been developed by William Crookes in 1903 for detecting radiation).

Geiger and Marsden noticed that most of the alpha particles shot straight through the gold foil, implying that—in contradiction to Thomson's model—most of an atom was empty space. A small fraction

See also: Models of matter 68–71 ▪ Atomic theory 236–237 ▪ Nuclear rays 238–239 ▪ Subatomic particles 242–243 ▪ Nuclear bombs and power 248–251

of the alpha particles was deflected at large angles with some even bouncing back toward the source. Approximately one in every 8,000 alpha particles were deflected by an average angle of 90 degrees. Rutherford was stunned by what he and his colleagues observed—it seemed these particles were being electrically repelled by a small, heavy, positive charge inside the atoms.

A new model of the atom

The results of the gold foil experiment formed the basis for Rutherford's new model of the atom, which he published in 1911. According to this model, the vast majority of an atom's mass is concentrated in a positively charged core called the nucleus. This nucleus is orbited by electrons that are electrically bound to it.

Rutherford's model had some similarities to the Saturnian model of the atom proposed by Japanese physicist Hantaro Nagaoka in 1904. Nagaoka had described electrons revolving around a positively charged center, rather like Saturn's icy rings.

Science is an iterative process, however, and in 1913, Rutherford played a major role in replacing his own model of the atom with a new one that incorporated quantum mechanics in its description of the behavior of electrons. This was the Bohr model, which was developed with Niels Bohr and had electrons orbiting in different "shells." Regardless, Rutherford's discovery of the atomic nucleus is widely considered one of the most significant discoveries in physics, laying the foundation for nuclear and particle physics. ∎

Ernest Rutherford

Ernest Rutherford was born in Brightwater, New Zealand. In 1895, he won a grant that allowed him to travel to the University of Cambridge, UK, and work with J.J. Thomson at its Cavendish Laboratory. While there, he found a way to transmit and receive radio waves over long distances, independently of Italian inventor Guglielmo Marconi.

Rutherford became a professor at McGill University in Canada in 1898 but returned to the UK in 1907—this time to Manchester, where he conducted his most famous research. He was awarded the Nobel Prize in Physics in 1908. Later in life, he became director of the Cavendish Laboratory and President of the Royal Society. Following his death, he was awarded the honor of a burial at Westminster Abbey.

Key works

1903 "Radioactive Change"
1911 "The Scattering of Alpha and Beta Particles by Matter and the Structure of the Atom"
1920 "Nuclear Constitution of Atoms"

The gold foil experiment

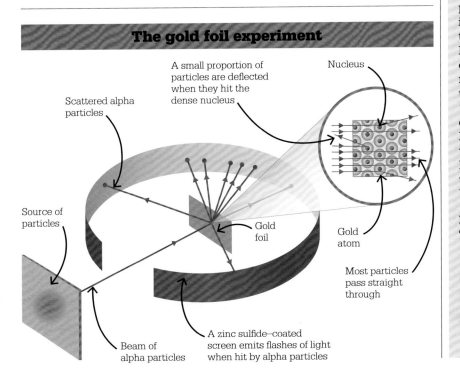

A small proportion of particles are deflected when they hit the dense nucleus

Nucleus

Scattered alpha particles

Source of particles

Gold foil

Gold atom

Most particles pass straight through

Beam of alpha particles

A zinc sulfide–coated screen emits flashes of light when hit by alpha particles

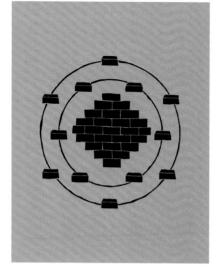

THE BRICKS OF WHICH ATOMS ARE BUILT UP

SUBATOMIC PARTICLES

For millennia, atoms were thought to be unbreakable units. A succession of discoveries by three generations of Cambridge-based physicists dismantled this idea, revealing smaller particles within the atom.

The first subatomic particle was discovered in 1897 by British physicist J.J. Thomson while he was experimenting with cathode rays. These rays are produced by the negative electrode (cathode) of an electrically charged vacuum tube and are attracted to a positive electrode (anode). The cathode rays caused the glass at the far end of the tube to glow, and Thomson deduced that the rays were composed

> 66
>
> … the radiation consists of neutrons: particles of mass 1, and charge 0.
> **James Chadwick**
>
> 99

of negatively charged particles more than 1,000 times lighter than a hydrogen atom. He concluded that these particles were a universal component of atoms, naming them "corpuscles" (later called electrons).

Delving into the atom

Thomson incorporated electrons into his 1904 "plum pudding" model of the atom. However, in 1911 a new model was proposed by Thomson's student, Ernest Rutherford. This had a dense, positively charged nucleus orbited by electrons. Further refinements made by Niels Bohr, along with Rutherford, produced the Bohr model in 1913.

In 1919, Rutherford discovered that when nitrogen and other elements were struck with alpha radiation, hydrogen nuclei were emitted. He concluded that the hydrogen nucleus—the lightest nucleus—is a constituent of all other nuclei and named it the proton.

Physicists struggled to explain the properties of atoms with only protons and electrons, because these particles accounted for only half the measured mass of atoms. In 1920, Rutherford suggested that a neutral particle composed

See also: Models of matter 68–71 ▪ Electric charge 124–127 ▪ Light from the atom 196–199 ▪ Particles and waves 212–215 ▪ The nucleus 240–241 ▪ Nuclear bombs and power 248–251 ▪ The particle zoo and quarks 256–257

Models of the atom

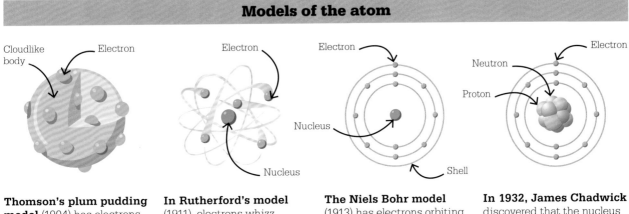

Thomson's plum pudding model (1904) has electrons dotted randomly in a positively charged atom.

In Rutherford's model (1911), electrons whizz around a dense, positively charged nucleus.

The Niels Bohr model (1913) has electrons orbiting the nucleus in inner and outer shells.

In 1932, James Chadwick discovered that the nucleus of an atom is made up of protons and neutrons.

of a bound-together proton and electron—which he called a neutron—may exist inside nuclei. While this offered a simple explanation for how electrons radiated from nuclei, it violated principles of quantum mechanics: there was simply not enough energy to trap electrons inside nuclei.

In Paris in 1932, Irène Joliot-Curie and her husband Frédéric experimented with a newly

discovered type of neutral radiation (thought to be a form of gamma radiation), which emerged when alpha radiation struck light elements such as beryllium. The Joliot-Curies found this radiation carried enough energy to eject high-energy protons from compounds rich in hydrogen. But neither Rutherford nor James Chadwick—Rutherford's former student—believed that these were gamma rays. Chadwick repeated

the experiments with more accurate measurements. He showed that this radiation had about the same mass as protons and concluded that this new radiation was composed of neutrons—neutral particles contained in the nucleus. This discovery completed the atomic model in a way that made sense. It also set in motion the development of new medical technologies and the dawn of the nuclear age. ▪

James Chadwick

When Cheshire-born James Chadwick won a scholarship to the University of Manchester, he was accidentally enrolled in physics instead of mathematics. He studied under Ernest Rutherford and wrote his first paper on how to measure gamma radiation. In 1913, Chadwick traveled to Berlin to study under Hans Geiger but was imprisoned in the Ruhleben detention camp during World War I.

After Chadwick's release, he joined Rutherford at the University of Cambridge in 1919. In 1935, he was awarded the Nobel Prize in

Physics for the discovery of the neutron. During World War II, Chadwick headed the British team working on the Manhattan Project to develop nuclear weapons for the Allies. He later served as British scientific adviser to the UN Atomic Energy Commission.

Key works

1932 "Possible Existence of a Neutron"
1932 "The existence of a neutron"

LITTLE WISPS OF CLOUD
PARTICLES IN THE CLOUD CHAMBER

IN CONTEXT

KEY FIGURE
Charles T.R. Wilson
(1869–1959)

BEFORE
1894 Charles T.R. Wilson creates clouds in chambers while studying meteorological phenomena at the Ben Nevis observatory, Scotland.

1910 Wilson realizes a cloud chamber can be used to study subatomic particles emitted by radioactive sources.

AFTER
1912 Victor Hess proposes that high-energy ionizing radiation enters the atmosphere from space as "cosmic rays."

1936 American physicist Alexander Langsdorf modifies the cloud chamber by adding dry ice.

1952 The bubble chamber supersedes the cloud chamber as a basic tool in particle physics.

Subatomic particles are ghostly objects, usually made visible only through their interactions. The invention of the cloud chamber, however, allowed physicists to witness the movements of these particles for the first time and determine their properties.

Interested in designing a tool to study cloud formation, Scottish physicist and meteorologist Charles T.R. Wilson experimented with

Wilson's cloud chamber, on display at the Cavendish Laboratory museum, Cambridge, produced tracks inside "as fine as little hairs."

expanding moist air in a sealed chamber, making it supersaturated. He realized that when ions (charged atoms) collide with water molecules, they knock away electrons to create a path of ions around which mist forms, and this leaves a visible trail in the chamber. By 1910, Wilson had perfected his cloud chamber and he demonstrated it to scientists in 1911. Combined with magnets and electric fields, the apparatus enabled physicists to calculate properties such as mass and electrical charge from the cloudy trails left by particles. By 1923, he added stereoscopic photography to

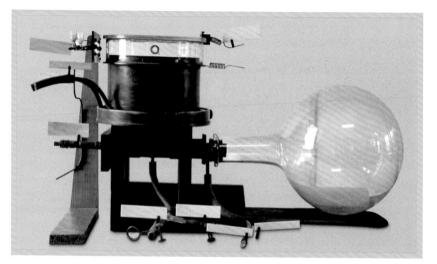

See also: Electric charge 124–127 ▪ Nuclear rays 238–239 ▪ Antimatter 246 ▪ The particle zoo and quarks 256–257 ▪ Force carriers 258–259

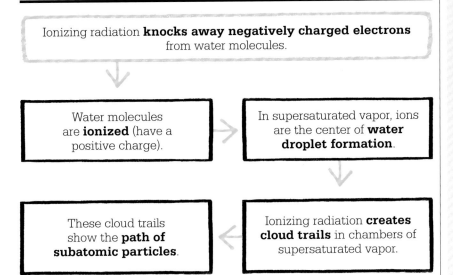

Ionizing radiation **knocks away negatively charged electrons** from water molecules.

↓

Water molecules are **ionized** (have a positive charge). → In supersaturated vapor, ions are the center of **water droplet formation**.

↓

These cloud trails show the **path of subatomic particles**. ← Ionizing radiation **creates cloud trails** in chambers of supersaturated vapor.

Charles T.R. Wilson

Born into a family of farmers in Midlothian, Scotland, Charles T.R. Wilson moved to Manchester after his father's death. He had planned to study medicine but won a scholarship to the University of Cambridge and switched to natural sciences. He began to study clouds and worked for some time at the Ben Nevis meteorological observatory, which inspired him to develop the cloud chamber. In 1927, he shared the Nobel Prize in Physics with Arthur Compton for its invention.

Despite Wilson's huge contribution to particle physics, he remained more interested in meteorology. He invented a method to protect British barrage balloons from lightning strikes during World War II and proposed a theory of thunderstorm electricity.

Key works

1901 *On the Ionization of Atmospheric Air*
1911 *On a Method of Making Visible the Paths of Ionizing Particles Through a Gas*
1956 *A Theory of Thundercloud Electricity*

---

record them. While Wilson observed radiation from radioactive sources, cloud chambers can also be used to detect cosmic rays (radiation from beyond the solar system and Milky Way).

The concept of cosmic rays was confirmed in 1911–1912 when physicist Victor Hess measured ionization rates from an atmospheric balloon. During risky ascents to 17,380 ft (5,300 m), day and night, he found increased levels of ionization and concluded: "radiation of very high penetrating power enters from above into our atmosphere." These cosmic rays (mostly composed of protons and alpha particles) collide with the nuclei of atoms and create cascades of secondary particles when they impact Earth's atmosphere.

The anti-electron
While investigating cosmic rays in 1932, American physicist Carl D. Anderson identified something that seemed to be an exact reflection of an electron—of equal mass with an equal but opposite charge. He eventually concluded the trails belonged to an anti-electron (the positron). Four years later, Anderson discovered another particle—the muon. Its trails showed it was more than 200 times heavier than an electron but had the same charge. This hinted at the possibility of multiple "generations" of particles that are linked by similar properties.

Discovery of the weak force
In 1950, physicists from Melbourne University found a neutral particle among cosmic rays, which decayed into a proton and other products. They named it the lambda baryon (Λ°). There are several baryons—composite particles each affected by the strong force that acts within the nucleus. Physicists predicted that the lambda baryon should decay in 10^{-23} seconds, but it survived much longer. This finding led them to conclude that another fundamental force, which worked at a short range, was involved. It was named the weak interaction, or weak force. ▪

OPPOSITES CAN EXPLODE

ANTIMATTER

IN CONTEXT

KEY FIGURE
Paul Dirac (1902–1984)

BEFORE
1898 German-born British physicist Arthur Schuster speculates on the existence of antimatter.

1928 Paul Dirac proposes that electrons could have both a positive and negative charge.

1931 Dirac publishes a paper that predicts the positively charged "anti-electron."

AFTER
1933 Dirac proposes the antiproton—the antimatter equivalent to the proton.

1955 Research at the University of California, Berkeley confirms the antiproton exists.

1965 CERN physicists describe the production of bound antimatter in the form of antideuterium.

In the 1920s, British physicist Paul Dirac proposed a mirror world of antimatter. In a 1928 paper, he demonstrated it is equally valid for electrons to have positive or negative energy states. An electron with negative energy would behave the opposite way to an ordinary electron. For instance, it would be repelled by a proton rather than attracted to it, so it has positive charge.

Dirac ruled out the possibility that this particle was a proton, given that a proton's mass is far greater than that of an electron. Instead, he proposed a new particle with the same mass as an electron but a positive charge. He called it an "anti-electron." On meeting, an electron and anti-electron would annihilate each other, producing a mass of energy.

Carl D. Anderson is credited with confirming the existence of this anti-electron (which he renamed the positron) in 1932. Anderson allowed cosmic rays to pass through a cloud chamber under the influence of a magnetic field, causing them to bend in different directions based on their charge. He discovered a particle with a curving trail like that of an electron but pointing in the opposite direction.

Anderson's discovery was followed by other antimatter particles and atoms. It is now known that all particles have an equivalent antimatter particle, but the question of why ordinary matter is dominant in the universe remains unresolved. ■

> I think that the discovery of antimatter was perhaps the biggest jump of all the big jumps in physics in our century.
> **Werner Heisenberg**

See also: Subatomic particles 242–243 ■ Particles in the cloud chamber 244–245 ■ The particle zoo and quarks 256–257 ■ Matter–antimatter asymmetry 264

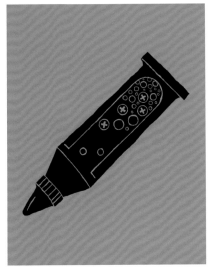

IN SEARCH OF ATOMIC GLUE

THE STRONG FORCE

IN CONTEXT

KEY FIGURE
Hideki Yukawa (1907–1981)

BEFORE
1935 Yukawa predicts
the existence of a new force
within the atomic nucleus.

1936 Carl D. Anderson
discovers the muon, thought
for a while to be the carrier of
this new force.

AFTER
1948 A team of physicists
at University of California,
Berkeley, produce pions
artificially by firing alpha
particles at carbon atoms.

1964 American physicist
Murray Gell-Mann predicts
the existence of quarks,
which interact via the
strong force.

1979 The gluon is discovered
at the PETRA (Positron-
Electron Tandem Ring
Accelerator) particle
accelerator in Germany.

The discovery of subatomic
particles at the beginning
of the 20th century raised
as many questions as it answered.
One such question was how are
positively charged protons bound
together in a nucleus despite their
natural electric repulsion?

In 1935, Japanese physicist
Hideki Yukawa provided an answer
when he predicted an ultra short-
range force that acts within the
atomic nucleus to bind its
components (protons and neutrons).
He said this force is carried by a
particle he called a meson. In fact
there are many mesons. The first
to be discovered was the pion (or
pi-meson), which physicists from
the UK and Brazil found in 1947 by
studying cosmic rays raining down
on the Andes. Their experiments
confirmed that the pion was
involved in the strong interactions
Yukawa had described.

By far the strongest of
the four fundamental forces (the
others being electromagnetic,
gravitational, and the weak force),
this newly discovered strong force

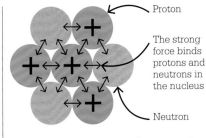

Neutrons and protons (nucleons)
are bound together inside the nucleus
by the strong force carried by mesons.
Within protons and neutrons, smaller
particles (quarks) are bound by gluons.

is in fact responsible for the vast
energy unleashed in nuclear
weapons and nuclear reactors
as atoms are split. Mesons are
the carriers of this force between
nucleons. The strong force between
quarks (which combine in threes to
make up a proton) is carried by
gluons (elementary particles, or
bosons), named for their ability
to "glue together" quarks with
different "colors" (a property
unrelated to normal colors) to
form "colorless" particles such
as protons and pions. ∎

See also: The nucleus 240–241 ▪ Subatomic particles 242–243 ▪ The particle
zoo and quarks 256–257 ▪ Force carriers 258–259

DREADFUL AMOUNTS OF ENERGY

NUCLEAR BOMBS AND POWER

IN CONTEXT

KEY FIGURES
Enrico Fermi (1901–1954),
Lise Meitner (1878–1968)

BEFORE
1898 Marie Curie discovers
how radioactivity emanates
from materials like uranium.

1911 Ernest Rutherford
proposes that the atom has
a dense nucleus at its core.

1919 Rutherford shows that
one element can be changed
to another by bombarding it
with alpha particles.

1932 James Chadwick
discovers the neutron.

AFTER
1945 The first atomic bomb
is tested in New Mexico.
A-bombs are dropped on
Hiroshima and Nagasaki.

1951 The first nuclear reactor
for electricity generation opens.

1986 The Chernobyl disaster
highlights nuclear power risks.

At the turn of the 20th
century, physicists
were unknowingly laying
the groundwork for scientists to
understand and eventually seize
the immense power trapped
inside the atom. By 1911, Ernest
Rutherford had proposed a model
of the atom that placed a dense
nucleus at its core. In Paris, Marie
Curie and her collaborators,
including husband Pierre, had
discovered and described how
radioactivity emanated from within
the atom of natural materials such
as uranium.

See also: Generating electricity 148–151 ▪ Atomic theory 236–237 ▪ Nuclear rays 238–229 ▪ The nucleus 240–241
▪ Subatomic particles 242–243

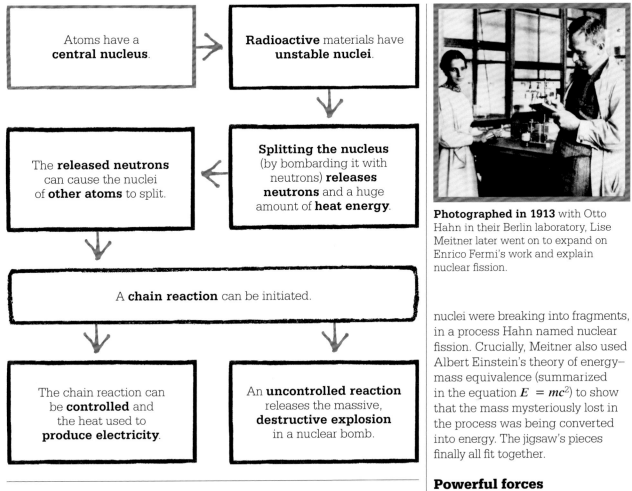

Atoms have a **central nucleus**.	**Radioactive** materials have **unstable nuclei**.

Splitting the nucleus (by bombarding it with neutrons) **releases neutrons** and a huge amount of **heat energy**.

The **released neutrons** can cause the nuclei of **other atoms** to split.

A **chain reaction** can be initiated.

The chain reaction can be **controlled** and the heat used to **produce electricity**.

An **uncontrolled reaction** releases the massive, **destructive explosion** in a nuclear bomb.

Photographed in 1913 with Otto Hahn in their Berlin laboratory, Lise Meitner later went on to expand on Enrico Fermi's work and explain nuclear fission.

nuclei were breaking into fragments, in a process Hahn named nuclear fission. Crucially, Meitner also used Albert Einstein's theory of energy–mass equivalence (summarized in the equation $E = mc^2$) to show that the mass mysteriously lost in the process was being converted into energy. The jigsaw's pieces finally all fit together.

Powerful forces

The forces within a nucleus are powerful and precariously balanced. The nuclear force binds the nucleus together while positively charged protons inside the nucleus repel each other with approximately 230N of force. Holding together the nucleus requires a huge amount of "binding energy," which is released when the nucleus breaks apart. The loss of equivalent mass is measurable: for the uranium fission reaction under investigation, around one-fifth the mass of a proton seemed to vanish in a burst of heat. »

In 1934, Italian physicist Enrico Fermi bombarded uranium with neutrons (subatomic particles discovered just two years earlier). Fermi's experiment seemed to transform uranium into different elements—not the new, heavier ones he had expected, but isotopes (variants with a different neutron number) of lighter elements. Fermi had divided the apparently indivisible atom, although it took years for the scientific community to grasp the magnitude of what he had done. German chemist Ida Noddack suggested that the new elements were fragments from the original uranium nuclei, but her proposal was dismissed in the scramble for scientists to understand Fermi's experiment.

In 1938, German chemists Otto Hahn and Fritz Strassmann expanded on Fermi's work. They found that bombarding uranium with neutrons produced barium, apparently losing 100 protons and neutrons in the process. Hahn conveyed the baffling findings to his former colleague Lise Meitner, who had fled Nazi Germany for Sweden. Meitner proposed that the uranium

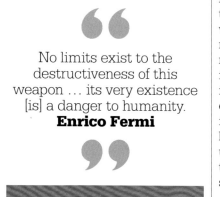

> No limits exist to the destructiveness of this weapon ... its very existence [is] a danger to humanity.
> **Enrico Fermi**

Meitner and other physicists recognized that neutrons "boiling off" from nuclear fission opened up the possibility of chain reactions in which the free nuclei caused successive fission reactions, releasing more energy and neutrons with each reaction. A chain reaction could either release a steady flow of energy or mount a vast explosion. As nations lurched toward World War II, and fearful of how this power might be used in the wrong hands, researchers immediately began to study how to sustain fission.

The Manhattan Project

US president Franklin D. Roosevelt wanted the US and its allies to be the first to harness atomic power as conflict consumed much of the world. Although the US did not enter World War II until Japan's attack on Pearl Harbor in 1941, the secretive Manhattan Project, which employed many of the greatest scientists and mathematicians of the 20th century, was established in 1939 to develop nuclear weapons under the scientific direction of J. Robert Oppenheimer.

Having fled Fascist Italy in 1938, Fermi moved to the US and restarted work on applications of his discovery, as part of the Manhattan Project. He and his colleagues found that slow-moving (thermal) neutrons were more likely to be absorbed by nuclei and cause fission. U-235 (a natural isotope of uranium) was identified as an ideal fuel because it releases three thermal neutrons every time it fragments. U-235 is a rare isotope that makes up less than 1 percent of natural uranium, so natural uranium has to be painstakingly enriched to sustain a chain reaction.

The critical factor

The Manhattan Project pursued multiple methods for enriching uranium and two possible nuclear reactor designs: one based at Columbia University that used heavy water (water containing a hydrogen isotope) to slow down neutrons; and a second led by Fermi at the University of Chicago that used graphite. The scientists were aiming for criticality—when the rate at which neutrons are produced from fission is equal to the rate at which neutrons are lost through absorption and leaking. Producing too many neutrons would mean the reactions spiraling out of control, while producing too few would cause the reactions to die out. Criticality requires a careful balance of fuel mass, fuel density, temperature, and other variables.

Fermi's nuclear reactor went live in 1942. The Chicago Pile-1 was built on a university squash court, using almost 5 tons of unenriched uranium, 44 tons of uranium oxide, and 363 tons of graphite bricks. It was crude, unshielded, and low-power, but it marked the first time scientists had sustained a fission chain reaction.

While nuclear reactors must sustain criticality for civilian use (as in nuclear power stations), nuclear weapons must surpass criticality to release deadly quantities of binding energy in a flash. Scientists working under Oppenheimer at Los Alamos were responsible for designing such weapons. One design used implosion, in which explosives around a fissile core ignited and so produced shock waves. It compressed the core to a smaller, denser volume that passed criticality. An alternative design—a "gun-type" weapon—blasted two

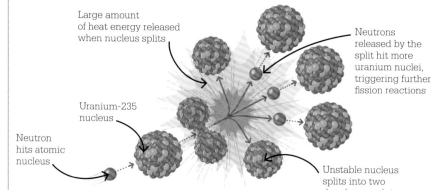

Large amount of heat energy released when nucleus splits

Neutrons released by the split hit more uranium nuclei, triggering further fission reactions

Uranium-235 nucleus

Neutron hits atomic nucleus

Unstable nucleus splits into two daughter nuclei

Uranium-235 (U-235) is a purified isotope of uranium. This isotope is naturally unstable and emits neutrons and heat. Nuclear fission of U-235 starts when a neutron hits a U-235 nucleus.

The nucleus splits to produce smaller, lighter "daughter" nuclei plus a few more neutrons, each of which can go on to produce fission in more uranium nuclei.

The world's first atomic bomb (code-named Trinity) was detonated at a test site in Alamogordo, New Mexico, on July 16, 1945, producing a massive fireball and mushroom cloud of debris.

smaller pieces of fissile material together at high speed to create a large mass that exceeded criticality.

In July 1945, in the New Mexico desert, Manhattan Project scientists detonated a nuclear weapon for the first time. A massive fireball was followed by a radioactive cloud of debris and water vapor. Most observers were silent. Oppenheimer later admitted that the words of the god Vishnu from Hindu scripture leapt to mind: "Now I am become Death, the destroyer of worlds." The only two nuclear weapons ever used in armed conflict to date were dropped on Japan weeks later: the gun-type "Little Boy" on Hiroshima on August 6 and the implosion-type "Fat Man" (which used Pu-239, an isotope of plutonium, as the fissile material) on Nagasaki on August 9.

Nuclear power for energy

The end of the war signaled a partial shift to nuclear fission for peaceful purposes, and the Atomic Energy Commission was created in the US in 1946 to oversee the development of civilian nuclear applications. The first nuclear reactor for electricity generation opened in 1951 in Idaho, and civil nuclear power stations multiplied in the next two decades. Nuclear reactors use controlled chain reactions—sped up or slowed down using control rods that capture free neutrons—to release energy gradually, boiling water into steam to turn electrical generators. The amount of energy that can be extracted from nuclear fuel is millions of times that found in similar amounts of fuels like coal, making it an efficient carbon-neutral energy source.

However, in 1986, explosions in a reactor core at Chernobyl in Ukraine, then part of the Soviet Union, released radioactive material into the atmosphere, eventually killing thousands of people across Europe. This and other nuclear disasters, together with concerns about how to store long-lived and highly radioactive waste, have undermined nuclear power's environmental credentials.

From fission to fusion?

Some physicists hope that nuclear fusion could be the sustainable energy source of the future. This is the binding of two nuclei to form a larger nucleus, releasing excess energy in the form of photons. Scientists have struggled for decades to induce fusion—the powerful repulsive force of protons can only be overcome amid extreme heat and density. The most promising method uses a doughnut-shaped device called a tokamak. This generates a powerful magnetic field to hold plasma, matter so hot that electrons are stripped from their atoms, rendering it conductive and easy to manipulate with magnetic fields. ∎

Enrico Fermi

Born in Rome in 1901, Enrico Fermi was best known for developing early nuclear applications, but was also admired as a theoretical physicist. Fermi studied at the University of Pisa and then left Italy to collaborate with physicists such as Max Born. Returning to lecture at the University of Florence in 1924, he helped to develop Fermi-Dirac statistics. In 1926, he became a professor at the University of Rome. Here, he proposed the theory of the weak interaction and demonstrated nuclear fission.

In 1938, the year Fermi won the Nobel Prize in Physics, he fled Fascist Italy to escape anti-Jewish laws that restricted the rights of his wife and colleagues. Despite his close involvement with the Manhattan Project, he later became a vocal critic of nuclear weapons development. He died in Chicago in 1954.

Key works

1934 "Artificial Radioactivity Produced by Neutron Bombardment"
1939 "Simple Capture of Neutrons by Uranium"

A WINDOW ON CREATION

PARTICLE ACCELERATORS

IN CONTEXT

KEY FIGURE
John Cockcroft (1897–1967)

BEFORE
1919 Ernest Rutherford
artificially induces nuclear
fission (splitting the nucleus
of an atom into two nuclei).

1929 Ukrainian–American
physicist George Gamow lays
out his theory of quantum
tunneling for alpha particles
emitted in alpha decay.

AFTER
1952 The Cosmotron, the first
proton synchrotron, begins
operating at Brookhaven
National Laboratory in the US.

2009 The Large Hadron
Collider (LHC) at CERN in
Switzerland becomes fully
operational and breaks the
record as the highest-energy
particle accelerator.

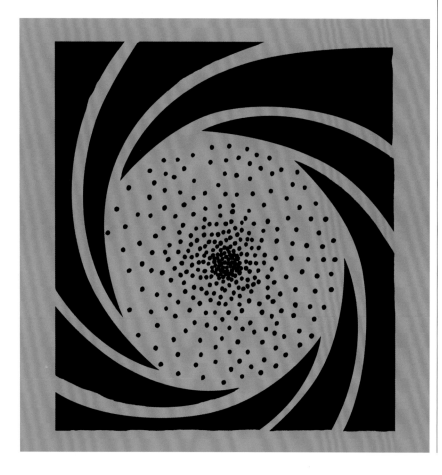

I n 1919, Ernest Rutherford's
investigation into the
disintegration of nitrogen
atoms proved that it is possible
to break apart particles bound
by the nuclear force—one of the
strongest forces in the universe.
Soon, physicists were wondering
whether they could explore deeper
inside the atom by smashing it to
pieces and examining the remains.

In the late 1920s, former British
soldier and engineer John Cockcroft
was one of the young physicists
assisting Rutherford in such
research at Cambridge University's
Cavendish Laboratory. Cockcroft
was intrigued by the work of
George Gamow, who in 1928

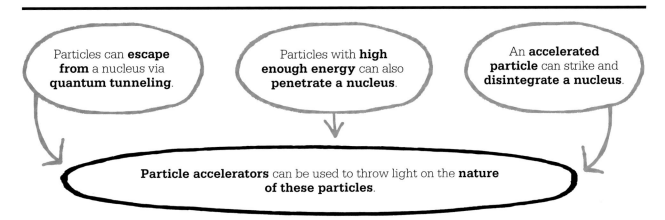

Particles can **escape from** a nucleus via **quantum tunneling**.

Particles with **high enough energy** can also **penetrate a nucleus**.

An **accelerated particle** can strike and **disintegrate a nucleus**.

Particle accelerators can be used to throw light on the **nature of these particles**.

described the phenomenon of quantum tunneling. This is the idea that subatomic particles, such as alpha particles, can escape from the nucleus, despite the strong nuclear force restraining them, because they have wavelike attributes, enabling some to go beyond the nuclear force barrier, escaping its attractive power.

Reversing the principle
Gamow visited the Cambridge laboratory at Cockcroft's invitation, and the men discussed whether Gamow's theory could be applied in reverse; was it possible to accelerate a proton with enough energy to penetrate and burst the nucleus of an element? Cockcroft told Rutherford he believed they could penetrate a boron nucleus using protons accelerated with 300 kilovolts, and that a nucleus of lithium would potentially require less energy. Boron and lithium have light nuclei, so the energy barriers to be overcome are lower than those of heavier elements.

Rutherford gave the go-ahead and, in 1930, joined by Irish physicist Ernest Walton, Cockcroft experimented with accelerating

beams of protons from a canal ray tube (essentially a back-to-front cathode ray tube). When they failed to detect the gamma rays noted by French scientists engaged in similar research, they realized that their proton energy was too low.

Ever-increasing energy
The quest for more powerful particle accelerators began. In 1932, Cockcroft and Walton built a new apparatus that was able to accelerate a beam of protons to higher energies using lower voltages. This Cockcroft–Walton accelerator begins by accelerating charged particles through an initial diode (a semiconductor device) to charge a capacitor (a component that stores electrical energy) to peak voltage. The voltage is then reversed, boosting the particles through the next diode and effectively doubling their energy. Through a series of capacitors and diodes, charge is stacked up to multiple times what would normally be possible, by applying the maximum voltage.

Using this pioneering device, Cockcroft and Walton bombarded lithium and beryllium nuclei with

high-speed protons and monitored the interactions on a detector, a zinc fluoride fluorescent screen. They expected to see the gamma rays that French scientists Irene Joliot-Curie and her husband Frédéric had reported. Instead, they inadvertently produced neutrons (as British physicist James Chadwick would later prove). Cockcroft and Walton then performed the first artificial disintegration of an atomic nucleus on a nucleus of lithium, reducing it to alpha particles. »

> Particles were coming out of the lithium, hitting the screen, and producing scintillations. They looked like stars suddenly appearing and disappearing.
> **Ernest Walton**
> **on splitting the atom**

Particle accelerators use electric and magnetic fields to produce a beam of high-energy subatomic particles, such as protons, which are crashed together or fired at a metal target.

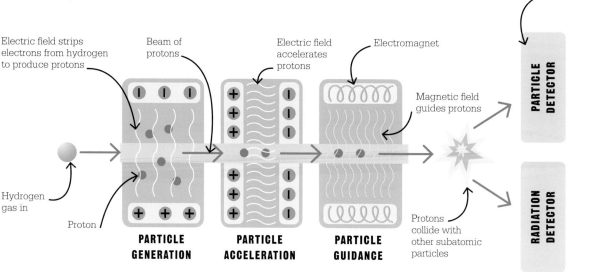

Detectors capture radiation and particles from the collision

Electric field strips electrons from hydrogen to produce protons

Beam of protons

Electric field accelerates protons

Electromagnet

Magnetic field guides protons

Hydrogen gas in

Proton

PARTICLE GENERATION

PARTICLE ACCELERATION

PARTICLE GUIDANCE

Protons collide with other subatomic particles

PARTICLE DETECTOR

RADIATION DETECTOR

With this historic first, Cockcroft and Walton showed the value of using particle accelerators (nicknamed "atom smashers") to probe the atom and discover new particles, offering a more controlled alternative to observing cosmic rays (high-energy particles moving through space).

High-powered accelerators

Cockcroft–Walton machines and all early particle accelerators were electrostatic devices, using a static electric field to accelerate particles. These are still widely used today in academic, medical, and industrial low-energy particle studies, and in everyday electronic applications, such as microwave ovens. Their energy limit, however, precludes their use for research in modern particle physics. At a certain energy point, raising voltages to push particles any further causes the insulators used in construction of the particle accelerators to experience electrical breakdown and start conducting.

Most accelerators in particle physics research use oscillating electromagnetic fields to accelerate charged particles. In electrodynamic accelerators, a particle is accelerated toward a plate, and as it passes through it, the charge on the plate switches, repelling the particle toward the next plate. This process is repeated with faster and faster oscillations to push the particles to speeds comparable to that of light. These oscillating fields are typically produced via one of two mechanisms—either magnetic induction or radio frequency (RF) waves. Magnetic induction uses a magnetic field to induce movement of charged particles to create a circulating electric field. An RF cavity is a hollow metallic chamber in which resonating radio waves create an electromagnetic field to boost charged particles as they pass through.

Modern particle accelerators come in three types: linear accelerators (linacs), cyclotrons,

and synchrotrons. Linacs, such as the Stanford Linear Accelerator in California, accelerate particles in a straight line toward a target at one end. Cyclotrons are composed of two hollow D-shaped plates and a magnet, which bends particles into a circular path as they spiral outward toward a target. Synchrotrons accelerate charged particles continuously in a circle, until they reach the required energies, using many magnets to guide the particles.

Dizzying particle speeds

In 1930, Cockcroft and Walton had recognized that the more they could accelerate particles, the deeper into matter they could see. Physicists can now use synchrotron accelerators that boost particles to dizzying speeds, which nudge toward the speed of light. Such speeds create relativistic effects; as a particle's kinetic energy increases, its mass increases, requiring larger forces to achieve greater acceleration. The largest and most powerful machines in existence are the focus of experiments that can involve thousands of scientists from around the world. Fermilab's Tevatron in Illinois, which operated from 1983 to 2011, used a ring 3.9 miles (6.3 km) long to accelerate protons and antiprotons to energies up to 1 TeV ($10^{12} \times 1$ eV), where 1 eV is the energy gained by an electron accelerated across one volt—1 TeV is roughly the energy of motion of a flying mosquito.

In the late 1980s, at CERN in Switzerland, scientists armed with the Super Proton Synchrotron competed with Fermilab scientists using the Tevatron in the hunt for the top quark (the heaviest quark). Thanks to the sheer power of the Tevatron, Fermilab scientists were able to produce and detect the top quark in 1995, at a mass of approximately 176 GeV/c^2 (almost as heavy as a gold atom).

The Tevatron was knocked off its pedestal as the most powerful particle accelerator in 2009 with the first fully operational run of CERN's Large Hadron Collider (LHC). The LHC is a synchrotron,

The ATLAS calorimetry system at CERN, seen here during installation, measures the energy of particles after a collision by forcing most to stop and deposit their energy in the detector.

16.8 miles (27 km) long, spanning the border of Switzerland and France 328 ft (100 m) underground, and capable of accelerating two beams of protons to 99.9999991 percent the speed of light.

The LHC is a triumph of engineering; its groundbreaking technologies—deployed on an enormous scale—include 10,000 superconducting magnets, chilled to temperatures lower than those found in the wilderness of space. During operation, the site draws about one third of the total energy consumption of the nearby city of Geneva. A chain of booster accelerators (including the Super Proton Synchrotron) accelerate beams of charged particles to ever higher energies until they finally enter the LHC. Here particles collide head-on at four collision points, with combined energies of 13 TeV. A group of detectors records their disintegrations.

Recreating a primeval state
Among the experiments conducted at the LHC are attempts to recreate the conditions that existed at the very genesis of the universe. From what is known about its expansion, physicists can work back and predict that the early universe was unimaginably tiny, hot, and dense.

Under these conditions, elementary particles such as quarks and gluons may have existed in a sort of "soup" (quark–gluon plasma). As space expanded and cooled, they became tightly bound together, forming composite particles such as protons and neutrons. Smashing particles at near light speed can for an instant reinvent the universe as it was trillionths of a second after the Big Bang. ∎

John Cockcroft

Born in Yorkshire, England, in 1897, John Cockcroft served in World War I, then studied electrical engineering. He won a scholarship to Cambridge University in 1921, where Ernest Rutherford later supervised his doctorate. The Cockcroft-Walton accelerator he built at Cambridge with Ernest Walton won them the 1951 Nobel Prize in Physics.

In 1947, as director of the British Atomic Energy Research Establishment, Cockcroft oversaw the opening of the first nuclear reactor in western Europe—GLEEP at Harwell. In 1950,

Cockcroft's insistence that filters be installed in the chimneys at Windscale Piles in Cumbria limited the fallout in 1957 when one of the reactors caught fire.

Cockcroft served as president of the Institute of Physics and of the British Association for the Advancement of Science. He died in Cambridge in 1967.

Key works

1932 *Disintegration of Lithium by Swift Protons*
1932–36 "Experiments with High Velocity Positive Ions I–VI"

THE HUNT FOR THE QUARK

THE PARTICLE ZOO AND QUARKS

IN CONTEXT

KEY FIGURE
Murray Gell-Mann
(1929–2019)

BEFORE
1947 The subatomic kaon is discovered at the University of Manchester, exhibiting a much longer lifetime than predicted.

1953 Physicists propose the property of "strangeness" to explain the unusual behavior of kaons and other particles.

1961 Gell-Mann proposes the "Eightfold Way" to organize subatomic particles.

AFTER
1968 Scattering experiments reveal pointlike objects inside the proton, proving it is not a fundamental particle.

1974 Experiments produce J/ψ particles, which contain charm quarks.

1995 The discovery of the top quark completes the quark model.

By the end of World War II, physicists had discovered the proton, neutron, and electron, and a handful of other particles. In the following years, however, discoveries in cosmic rays (high-energy particles moving through space) and particle accelerators caused the number of particles to balloon into a chaotic "particle zoo."

Physicists were particularly confused by families of particles called kaons and lambda baryons, which decay far more slowly than expected. In 1953, American physicist Murray Gell-Mann and Japanese physicists Kazuhiko Nishijima and Tadao Nakano independently proposed that a fundamental quality called "strangeness" could explain these long-observed lifetimes. Strangeness is conserved in strong and electromagnetic interactions but not in weak interactions, so particles with strangeness can only decay via the weak interaction. Gell-Mann used strangeness and charge to classify subatomic particles into families of mesons (typically lighter) and baryons (typically heavier).

Quark theory

In 1964, Gell-Mann proposed the concept of the quark—a fundamental particle, which could explain the properties of the new mesons and baryons. According to quark theory, quarks come in six

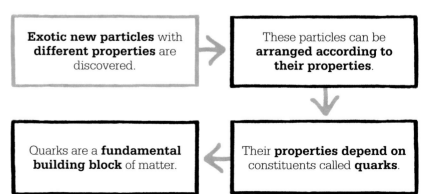

Exotic new particles with **different properties** are discovered. → These particles can be **arranged according to their properties**.

Quarks are a **fundamental building block** of matter. ← Their **properties depend on** constituents called **quarks**.

See also: Models of matter 68–71 ▪ Subatomic particles 242–243 ▪ Antimatter 246 ▪ The strong force 247 ▪ Force carriers 258–259 ▪ The Higgs boson 262–263

Nucleus

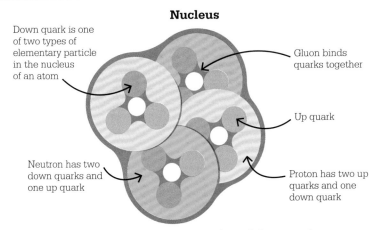

Down quark is one of two types of elementary particle in the nucleus of an atom

Gluon binds quarks together

Up quark

Neutron has two down quarks and one up quark

Proton has two up quarks and one down quark

Protons and neutrons each contain up quarks and down quarks. A proton has two up quarks and one down quark, and a neutron has one up quark and two down quarks. The quarks are bound together by gluons, the carrier particles of the strong force.

"flavors" with different intrinsic properties. Gell-Mann initially described up, down, and strange quarks, and charm, top, and bottom quarks were added later. Different quarks (and their antimatter equivalents, antiquarks) are bound by the strong force and are found in composite particles such as protons and neutrons. Evidence for this model of the nucleus was found at the Stanford Linear Accelerator Laboratory (SLAC) in California, when pointlike objects were discovered inside protons in 1968. Further SLAC experiments produced evidence for the other quarks.

The Standard Model

Quarks play an important role in the Standard Model of particle physics, which was developed in the 1970s to explain the electromagnetic, strong, and weak forces—as well as their carrier particles (which are always bosons) and the fundamental particles of matter (fermions). The quarks are one of the two groups of

fermions, the other being the leptons. Like the quarks, leptons come in six flavors, which are related in pairs, or "generations." These are the electron and electron neutrino, which are the lightest and most stable leptons; the muon and muon neutrino; and the tau and tau neutrino, which are the heaviest and most unstable leptons. Unlike quarks, leptons are unaffected by the strong force.

Quarks and leptons interact through three of the four fundamental forces: strong, electromagnetic, and weak. (The fourth fundamental force, gravity, behaves differently.) In the Standard Model, these forces are represented by their carrier particles—gluons, photons, and W and Z bosons.

The final element of the Standard Model is the Higgs boson, an elementary particle that gives all particles their mass. The Standard Model has brought order to the "particle zoo," although it remains a work in progress. ∎

Murray Gell-Mann

Born in New York in 1929 to a family of Jewish immigrants, Murray Gell-Mann began attending college at the age of just 15. In 1955, he moved to the California Institute of Technology (CIT), where he taught for almost 40 years. His interests expanded far beyond physics to include literature, history, and natural history, but in 1969 he was awarded the Nobel Prize in Physics for his work on the theory of elementary particles.

Later in his career, Gell-Mann became interested in the theory of complexity, cofounding the Santa Fe Institute for research into this area in 1984 and writing a popular book about the theory (*The Quark and the Jaguar*). At the time of his death in 2019, he held positions at CIT, the University of Southern California, and the University of New Mexico.

Key works

1994 *The Quark and the Jaguar: Adventures in the Simple and the Complex*
2012 *Mary McFadden: A Lifetime of Design, Collecting, and Adventure*

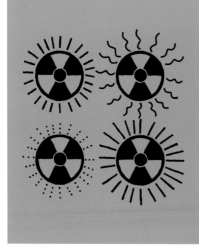

IDENTICAL NUCLEAR PARTICLES DO NOT ALWAYS ACT ALIKE

FORCE CARRIERS

IN CONTEXT

KEY FIGURE
Chien-Shiung Wu
(1912–1997)

BEFORE
1930 Austrian-born physicist Wolfgang Pauli proposes the existence of a neutrino to explain how energy and other quantities can be conserved in beta decay.

1933 Enrico Fermi lays out his theory for the weak interaction to explain beta decay.

AFTER
1956 American physicists Clyde Cowan and Frederick Reines confirm that both electrons and neutrinos are emitted during beta decay.

1968 The electromagnetic and weak forces are unified in "electroweak theory."

1983 W and Z bosons are discovered at the Super Proton Synchrotron, a particle accelerator machine at CERN.

From around 1930, scientists began to unpick the process of nuclear decay. Beta decay had puzzled earlier researchers, as energy seemed to disappear, in violation of the law of conservation of energy. Over the next five decades, leading physicists would discover the bearers of the missing energy, detail the changes that transformed elements within unstable atomic nuclei, and identify and observe the force carriers that mediate (transmit) the weak interaction in nuclear decay.

In 1933, Enrico Fermi proposed that beta radiation emerges from the nucleus as a neutron turns into a proton, emitting an electron and a further neutral particle that carries away some energy. (Fermi called this neutral particle a

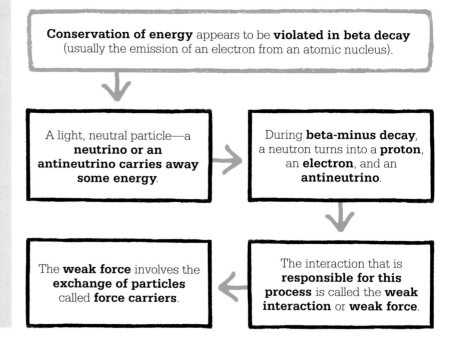

Conservation of energy appears to be **violated in beta decay** (usually the emission of an electron from an atomic nucleus).

A light, neutral particle—a **neutrino or an antineutrino carries away some energy**.

During **beta-minus decay**, a neutron turns into a **proton**, an **electron**, and an **antineutrino**.

The **weak force** involves the **exchange of particles** called **force carriers**.

The interaction that is **responsible for this process** is called the **weak interaction** or **weak force**.

See also: Quantum field theory 224–225 ▪ Nuclear rays 238–239 ▪ Antimatter 246 ▪ The particle zoo and quarks 256–257 ▪ Massive neutrinos 261 ▪ The Higgs boson 262–263

neutrino, but it was later identified as an antineutrino, the antiparticle of the neutrino.) In the other main type of beta decay—beta-plus decay—a proton becomes a neutron, emitting a positron and a neutrino. The force responsible for such decays was named the weak interaction—now recognized as one of four fundamental forces of nature.

The rule-breaking force

The strong and electromagnetic interactions conserve parity; their effect on a system is symmetric, producing a kind of mirror image. Suspecting this was not true of the weak interaction, fellow physicists Chen Ning Yang and Tsung-Dao Lee asked Chien-Shiung Wu to investigate. In 1956—the same year that Clyde Cowan and Frederick Reines confirmed the existence of neutrinos—Wu got to work at the US National Bureau of Standards low-temperature laboratory in Washington, D.C. She aligned the spins (internal angular momentum) of the nuclei in a sample of cobalt-60, and watched as it

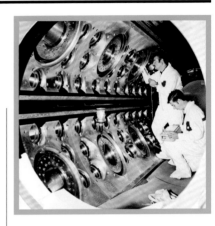

Gargamelle, seen here in 1970 was designed to detect neutrinos and antineutrinos. Through its portholes, cameras could track the trail of charged particles.

decayed. She saw that electrons emitted during beta radiation had a preference for a specific direction of decay (rather than being randomly emitted in all directions equally), showing that the weak interaction violates conservation of parity.

The weak interaction involves the exchange of force-carrier particles: W^+, W^-, and Z^0 bosons. All force-carrying particles are bosons (integer spin), while the key building blocks of matter, such as quarks and leptons, are all fermions (half-integer spin) and obey different principles. Uniquely, the weak interaction can also change the "flavor," or properties, of quarks. During beta-minus decay, a "down" quark changes into an "up" quark, transforming a neutron into a proton, and emits a virtual W^- boson, which decays into an electron and antineutrino. The W

and Z bosons are very heavy, so processes occurring through them tend to be very slow.

In 1973, the interactions of the bosons were observed at CERN's Gargamelle bubble chamber, which captured tracks providing the first confirmation of the weak neutral-current interaction. This was interpreted as a neutrino taking momentum away after being produced in the exchange of a Z boson. The W and Z bosons themselves were first observed in 1983 using the high-energy Super Proton Synchrotron (SPS) at CERN. ▪

Chien-Shiung Wu

Born in 1912, in Liuhe, near Shanghai in China, Chien-Shiung Wu developed a passion for physics after reading a biography of Polish physicist and chemist Marie Curie. She studied physics at National Central University in Nanjing, then moved to the US in 1936, gaining her PhD at the University of California-Berkeley.

Wu joined the Manhattan Project in 1944 to work on uranium enrichment. After World War II, she became a professor at Columbia University and focused on beta decay. Wu's collaborators Tsung-Dao Lee and Chen Ning

Yang received the 1957 Nobel Prize in Physics for discovering parity violation; her contribution was not acknowledged. She was finally honored in 1978, when she received the Wolf Prize for scientific achievement. Wu died in New York in 1997.

Key works

1950 "Recent investigation of the shapes of beta-ray spectra"
1957 "Experimental test of parity conservation in beta decay"
1960 *Beta decay*

NATURE IS ABSURD

QUANTUM ELECTRODYNAMICS

IN CONTEXT

KEY FIGURES
Shin'ichirō Tomonaga
(1906–1979), **Julian
Schwinger** (1918–1994)

BEFORE
1865 James Clerk Maxwell
lays out the electromagnetic
theory of light.

1905 Albert Einstein
publishes a paper describing
special relativity.

1927 Paul Dirac formulates a
quantum mechanical theory
of charged objects and the
electromagnetic field.

AFTER
1965 Tomonaga, Schwinger,
and Richard Feynman share
the Nobel Prize in Physics for
their work on quantum
electrodynamics.

1973 The theory of quantum
chromodynamics is developed,
with color charge as the source
of the strong interaction
(strong nuclear force).

The emergence of quantum
mechanics, which describes
the behavior of objects
at atomic and subatomic scales,
forced a transformation of many
branches of physics.

Paul Dirac proposed a quantum
theory of electromagnetism in 1927,
but models describing encounters
between electromagnetic fields and
high-speed particles—which obey
the laws of special relativity—broke
down. This led to an assumption
that quantum mechanics and
special relativity were not
compatible. In the 1940s, Shin'ichirō
Tomonaga, Richard Feynman, and
Julian Schwinger proved that
quantum electrodynamics (QED)
could be made consistent with
special relativity. In fact, QED was
the first theory to combine quantum
mechanics and special relativity.

In classical electrodynamics,
electrically charged particles exert
forces through the fields they
produce. In QED, however, the
forces between charged particles
arise from the exchange of virtual,
or messenger, photons—particles
that come into existence
momentarily and affect the motion
of "real" particles as they are
released or absorbed.

QED has been used to model
phenomena that previously defied
explanation. One example is the
Lamb shift—the difference in
energy between two energy levels
of a hydrogen atom. ∎

Following his contributions to
QED, Shin'ichirō Tomonaga received
the Nobel Prize in Physics, the Japan
Academy Prize, and many other awards.

See also: Force fields and Maxwell's equations 142–147 ▪ Particles and waves
212–215 ▪ Quantum field theory 224–225 ▪ The particle zoo and quarks 256–257

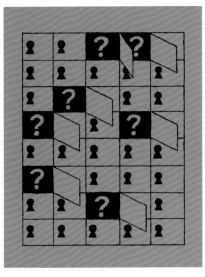

THE MYSTERY OF THE MISSING NEUTRINOS

MASSIVE NEUTRINOS

IN CONTEXT

KEY FIGURE
Masatoshi Koshiba (1926–)

BEFORE
1956 American physicists Clyde Cowan and Frederick Reines publish the results of their experiment confirming the existence of neutrinos.

1970 "Homestake" experiments begin in the US, detecting only one-third of the predicted number of solar neutrinos.

1987 Researchers at locations in Japan, the US, and Russia detect a record batch of 25 neutrinos, originating from a supernova in the Large Magellanic Cloud.

AFTER
2001 Scientists at Sudbury Neutrino Observatory, Canada, find further evidence for neutrino oscillation.

2013 Results from the T2K experiment confirm the neutrino oscillation theory.

Since the 1920s, physicists have known that nuclear fusion causes the sun and other stars to shine. They went on to predict that this process releases particles called neutrinos, which shower down on Earth from space.

Neutrinos have been compared to ghosts due to the challenge of detecting them. They are chargeless, almost without mass, and do not interact through the strong nuclear or electromagnetic forces, allowing them to pass through Earth unnoticed. Neutrinos come in three types, or flavors: electron, muon, and tau neutrinos.

In 1985, Japanese physicist Masatoshi Koshiba built a neutrino detector in a zinc mine. Detectors surrounded a vast water tank, picking up flashes of light as neutrinos interacted with the nuclei of water molecules. Koshiba confirmed that there seemed to be far fewer solar neutrinos reaching Earth than predicted. In 1996, he led the construction of an even larger detector (Super-Kamiokande), which allowed his team to solve

> Now that neutrino astrophysics is born, what should we do next?
> **Masatoshi Koshiba**

this problem. Observations of atmospheric neutrinos with the detector proved that neutrinos can switch between flavors in flight, a process called neutrino oscillation. This means that an electron neutrino created in the sun may change to a muon or tau neutrino and so elude detectors that are sensitive only to electron neutrinos. The discovery implied that neutrinos have mass, challenging the Standard Model, the theory of fundamental forces and particles. ■

See also: Heisenberg's uncertainty principle 220–221 ■ Particle accelerators 252–255 ■ The particle zoo and quarks 256–257 ■ Force carriers 258–259

I THINK WE HAVE IT

THE HIGGS BOSON

IN CONTEXT

KEY FIGURE
Peter Higgs (1929–)

BEFORE
1959 Sheldon Glashow in the US and Abdus Salam in Pakistan propose that the electromagnetic and weak forces merge in intense heat.

1960 Japanese–American physicist Yoichiro Nambu conceives his theory of symmetry breaking, which can be applied to the problem of W and Z boson mass.

AFTER
1983 W and Z bosons are both confirmed at the CERN Super Proton Synchroton.

1995 Fermilab discovers the top quark with a mass of 176 GeV/c^2, which matches Higgs field theory predictions.

2012 CERN confirms the discovery of the Higgs boson in its ATLAS and CMS detectors.

The Standard Model, completed in the early 1970s after decades of research, explained much about particle physics with a handful of fundamental forces and particles. Questions remained, however. While two of the forces—the electromagnetic force and the weak interaction—could be modeled as merging into a single electroweak force at intense temperatures, one aspect of this theory did not match reality. It implied that all electroweak force carriers are massless. While this is true for the photon, W and Z bosons are conspicuously massive.

In 1964, three groups of physicists—Peter Higgs in the UK; Robert Brout and François Englert in Belgium; and Gerald Guralnik, C. Richard Hagen, and Tom Kibble in the US—had proposed that the weak bosons might interact with a field which gives them mass; this

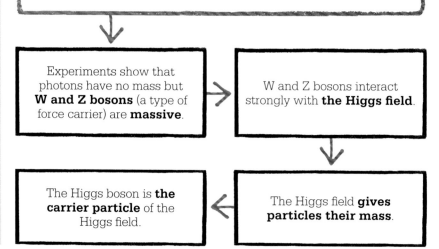

Electroweak theory predicts that all **force carriers** (particles that excite forces between other particles) **have no mass**.

Experiments show that photons have no mass but **W and Z bosons** (a type of force carrier) are **massive**.

W and Z bosons interact strongly with **the Higgs field**.

The Higgs field **gives particles their mass**.

The Higgs boson is **the carrier particle** of the Higgs field.

The Higgs field, imagined in this illustration, is a field of energy thought to exist throughout the universe; within it the Higgs boson interacts continuously with other particles.

it, and the search continued, influencing the design of the world's most powerful particle accelerator, CERN's Large Hadron Collider (LHC), which started up in 2008.

As physicists calculated that only one in 10 billion proton-proton collisions in the LHC would produce a Higgs boson, researchers used CERN's detectors to sift through the remnants of hundreds of trillions of collisions for hints of the particle. In 2012, CERN announced the discovery of a boson of mass around $126\,\text{GeV}/c^2$ (gigaelectron volts divided by the speed of light squared)—almost certainly the Higgs boson. CERN physicists were elated, while Higgs was moved to tears at the validation of his 50-year-old theory. The Standard Model is now complete, but physicists continue to explore the possibility that other types of Higgs bosons exist beyond it. ▪

became known as the Higgs field. According to this theory, the Higgs field began permeating throughout the universe soon after the Big Bang; the more a particle interacts with it, the more mass it acquires. Photons do not interact with the field, allowing them to travel at light speed. Meanwhile, W and Z bosons interact strongly with the field. At ordinary temperatures, this makes them heavy and sluggish with short ranges. As particles that interact with the Higgs field accelerate, they gain mass and therefore require more energy to push them further, preventing them from reaching the speed of light.

Near-mythical importance

Scientists realized that the only way to prove the Higgs field theory was to find an excitation of the field in the form of a heavy particle, known as the Higgs boson. The quest took on a near-mythical importance, gaining the boson the nickname of "God particle" in the 1980s, which Higgs and his colleagues disliked. No particle detector at the time could detect

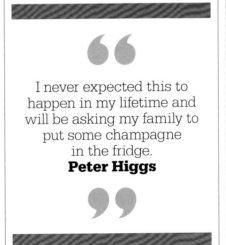

" I never expected this to happen in my lifetime and will be asking my family to put some champagne in the fridge.
Peter Higgs

Peter Higgs

Born in 1929 in Newcastle upon Tyne, Higgs was the son of a BBC sound engineer. Frequent house moves disrupted the boy's early education, but at Cotham Grammar School in Bristol he was inspired by the work of a former pupil, the theoretical physicist Paul Dirac. He then went to King's College London, and was awarded his PhD in 1954. After various academic posts, Higgs chose to stay at the University of Edinburgh. On one of many walks in the Highlands, he began to formulate his theory for the origin of mass, which gained him renown, although he readily acknowledged its many other contributors.

In 2013, Higgs shared the Nobel Prize in Physics with François Englert. He remains Emeritus professor at the University of Edinburgh.

Key works

1964 "Broken symmetries, and the masses of gauge bosons"
1966 "Spontaneous symmetry breakdown without massless bosons"

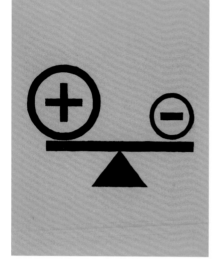

WHERE HAS ALL THE ANTIMATTER GONE?
MATTER–ANTIMATTER ASYMMETRY

Physicists have long puzzled over why the universe is made up almost entirely of matter. P (parity) symmetry, the idea that nature cannot distinguish between left and right, suggests that the Big Bang should have produced matter and antimatter in equal amounts. The first clue came in 1956, when an experiment showed that electrons emanating through beta decay in a weak interaction had a preferred direction. The following year, to maintain symmetry, Soviet physicist Lev Landau proposed CP symmetry— combining P symmetry with C (conservation of charge) symmetry, so that a particle and its oppositely-charged antiparticle would behave like mirror images.

Mirror universe or not?
In 1964, when experiments showed that neutral K-mesons violate CP conservation when they decay, physicists made a last-ditch bid to save symmetry. They incorporated time-reversal symmetry, giving CPT symmetry, which can be preserved if spacetime extends backward beyond the Big Bang into a mirror universe of antimatter.

In 1967, Soviet physicist Andrei Sakharov proposed instead that an imbalance of matter and antimatter could have evolved if CP violation occurred in the early universe. It has to be sufficient to prompt asymmetry, or else physics beyond the Standard Model will be required. Significant CP violation has now been shown, supporting Sakharov's ideas. ∎

Physicists began to think that they may have been looking at the wrong symmetry all along.
Ulrich Nierste
German theoretical physicist

STARS GET BORN AND DIE
NUCLEAR FUSION IN STARS

IN CONTEXT

KEY FIGURE
Hans Bethe (1906–2005)

BEFORE
1920 British physicist Arthur Eddington suggests that stars primarily get their energy from hydrogen–helium fusion.

1931 Deuterium, a stable isotope of hydrogen, is detected by American chemist Harold C. Urey and associates.

1934 Australian physicist Mark Oliphant demonstrates deuterium fusion, discovering tritium, a radioactive isotope of hydrogen, in the process.

AFTER
1958 The first tokamak (T-1) begins reactions in the Soviet Union, but loses energy through radiation.

2006 The ITER agreement is signed in Paris by seven ITER members, funding a long-term international project to develop nuclear fusion energy.

The idea that stars are endlessly powered from the fusion of hydrogen into helium excited eminent physicists in the early 20th century. By the mid-1930s, they had demonstrated nuclear fusion in the laboratory. Among the leading theorists was German-born Hans Bethe, who realized that the stars (including the sun) release energy through proton–proton chain reactions.

Fusion occurs only in extreme environments. Positively charged nuclei strongly repel each other but, with sufficient energy, they can be pushed close enough together to overcome this repulsion and fuse, forming heavier nuclei. As they fuse, binding energy is released.

Toward controlled fusion

In 1951, Bethe's work in the US led to the successful testing of the first hydrogen bomb, in which fission was used to induce nuclear fusion and release a lethal blast of energy.

Harnessing such power with controlled fusion reactions that gradually release energy for

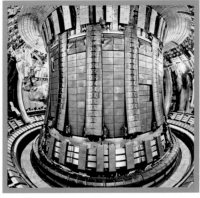

The JET tokamak reactor in the UK is the world's largest, most successful fusion facility. It is central to European research and the international ITER project advancing fusion science.

electricity has proved more difficult, due to the immense temperatures required (around 40 million Kelvin) and the challenge of containing such hot materials. The leading candidate for a fusion reactor— the tokamak—confines the hot, charged gas using magnetic fields.

The quest continues as fusion is thought safer than fission, with less radioactivity and nuclear waste. ∎

See also: Generating electricity 148–151 ▪ The strong force 247 ▪ Nuclear bombs and power 248–251 ▪ Particle accelerators 252–255

RELATIVI

THE UNIV

OUR PLACE IN
THE COSMOS

The Greek philosopher Aristotle describes a **static and eternal universe** where a **spherical Earth** is surrounded by concentric rings of planets and stars.

Persian astronomer Abd al-Rahman al-Sufi makes the first recorded observation of **Andromeda**, describing the galaxy as a "**small cloud**."

Galileo explains his principle of relativity: the **laws of physics are the same** whether a person is **stationary** or **moving at a constant velocity**.

Albert Einstein's **theory of special relativity** shows how **space** and **time change**, depending on the speed of one object relative to another.

4TH CENTURY BCE   **964**   **1632**   **1905**

c. 150 CE   **1543**   **1887**

Ptolemy creates a mathematical model of the universe, showing **Earth as a stationary sphere at the center**, with other known bodies orbiting around it.

Nicolaus Copernicus provides a model for a **heliocentric universe**, in which Earth revolves around the sun.

In the US, Albert Michelson and Edward Morley prove that **light moves at a constant speed** regardless of the motion of the observer.

Ancient civilizations questioned what the movement of stars across the night sky meant for humanity's existence and place within the universe. The seeming vastness of their planet led most to think that Earth must be the largest and most important object in the cosmos, and that everything else revolved around it. Among these geocentric views, Ptolemy of Alexandria's model from the 2nd century CE was so convincing that it dominated astronomy for centuries.

The advent of the telescope in 1608 showed that Polish astronomer Nicolaus Copernicus had been right to question Ptolemy's view. In Italy, Galileo Galilei observed four moons orbiting Jupiter in 1610, and so provided proof that other bodies orbited other worlds.

During the Scientific Revolution, physicists and astronomers began to study the motion of objects and light through space. Space was imagined to be a rigid grid that spanned the universe, and it was assumed that a length measured on Earth would be the same on any planet, star (including the sun), or neutron star (a small star created by the collapse of a giant star). Time, too, was thought to be absolute, with a second on Earth equivalent to one anywhere else in the universe. But it soon became clear that this was not the case.

A ball thrown from a moving vehicle will appear to a person standing on the roadside to have been given a boost of speed from the vehicle's movement. Yet if light is shone from a moving vehicle, the light does not receive a boost in speed—it still travels at the speed of light. Nature behaves strangely at close to the speed of light in order to impose a universal speed limit. Establishing how this works required a drastic rethinking of both space and time.

Special relativity

Early in the 20th century, space and time were discovered to be flexible: a meter or a second have different lengths in different places in the universe. German-born physicist Albert Einstein introduced the world to these ideas in his 1905 theory of special relativity. He explained how, from the perspective of an observer on Earth, objects moving through space at close to the speed of light seem to contract, and how time appears to run more slowly for these objects. Two years later, in 1907,

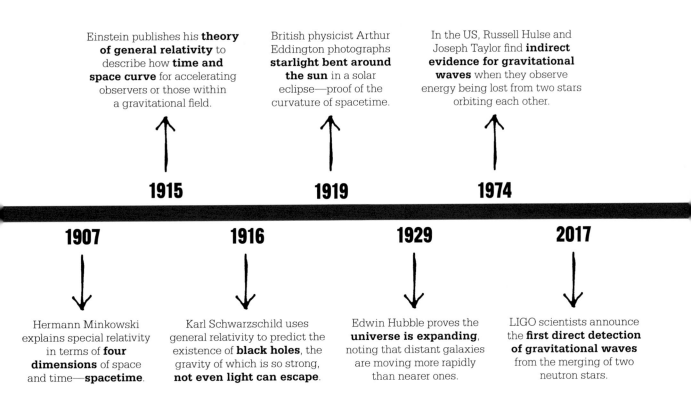

Einstein publishes his **theory of general relativity** to describe how **time and space curve** for accelerating observers or those within a gravitational field.

British physicist Arthur Eddington photographs **starlight bent around the sun** in a solar eclipse—proof of the curvature of spacetime.

In the US, Russell Hulse and Joseph Taylor find **indirect evidence for gravitational waves** when they observe energy being lost from two stars orbiting each other.

1915   **1919**   **1974**

1907   **1916**   **1929**   **2017**

Hermann Minkowski explains special relativity in terms of **four dimensions** of space and time—**spacetime**.

Karl Schwarzschild uses general relativity to predict the existence of **black holes**, the gravity of which is so strong, **not even light can escape**.

Edwin Hubble proves the **universe is expanding**, noting that distant galaxies are moving more rapidly than nearer ones.

LIGO scientists announce the **first direct detection of gravitational waves** from the merging of two neutron stars.

German mathematician Hermann Minkowski suggested that special relativity makes more sense if time and space are sewn together into a single fabric of "spacetime."

The theory of special relativity had far-reaching consequences. The notion that energy and mass were simply two forms of the same thing, as described by Einstein's formula $E = mc^2$, led to the discovery of the nuclear fusion that powers stars, and ultimately to the development of the atomic bomb.

In 1915, Einstein extended his ideas to include objects moving at changing speed and through gravitational fields. The theory of general relativity describes how spacetime can bend, just as a cloth might stretch if a heavy weight is placed upon it. German physicist Karl Schwarzschild went a step further, predicting the existence of extremely massive objects that could curve spacetime so much at a single point that nothing, not even light, could move fast enough to escape. Advances in astronomy have since provided evidence for these "black holes," and it is now thought that the largest stars become black holes when they die.

Beyond the Milky Way

By the early 1920s, astronomers were able to measure accurately the distance to stars in the night sky and the speed at which they are moving relative to Earth. This revolutionized how people perceived the universe and our place within it. At the start of the 20th century, astronomers believed that everything existed within 100,000 light years of space, that is, within the Milky Way. But American astronomer Edwin Hubble's discovery of another galaxy—the Andromeda Galaxy—in 1924 led to the realization that the Milky Way is just one of billions of galaxies that lie well beyond 100,000 light years away. What is more, these galaxies are moving apart, leading astronomers to believe that the universe began at a single point 13.8 billion years ago and exploded forth in a so-called Big Bang.

Astrophysicists today have far to go in their quest to understand the universe. Strange, invisible dark matter and dark energy make up 95 percent of the known universe, seeming to exert control without showing their presence. Likewise, the workings of a black hole and the Big Bang remain a mystery. But astrophysicists are getting closer. ∎

THE WINDINGS OF THE HEAVENLY BODIES

THE HEAVENS

IN CONTEXT

KEY FIGURE
Ptolemy (c. 100–c. 170 CE)

BEFORE
2137 BCE Chinese astronomers produce the first known record of a solar eclipse.

4th century BCE Aristotle describes Earth as a sphere at the center of the universe.

c. 130 BCE Hipparchus compiles his catalog of stars.

AFTER
1543 Nicolaus Copernicus suggests that the sun, not Earth, is at the center of the universe.

1784 French astronomer Charles Messier creates a database of other star clusters and nebulae in the Milky Way.

1924 American astronomer Edwin Hubble shows that the Milky Way is one of many galaxies in the universe.

Since time immemorial, humans have gazed at the night sky in awe, captivated by the motions of the sun, moon, and stars. One of the earliest examples of primitive astronomy is Britain's Stonehenge, a stone circle that dates back to about 3000 BCE. While the true purpose of these huge stones is not clear, it is believed that at least some were designed to align with the motion of the sun through the sky, and possibly also with the motion of the moon. Many other such monuments exist around the world.

Early humans linked objects in the night sky with gods and spirits on Earth. They believed heavenly bodies had an effect on aspects of their lives, such as the moon being linked to the fertility cycle. Some civilizations, including the Incas in the 15th century, assigned patterns, known as constellations, to stars that appeared regularly in the sky.

Stonehenge, a prehistoric monument in Wiltshire, southwest England, may have been built so that ancient people could track the motions of the sun through the sky.

See also: The scientific method 20–23 ▪ The language of physics 24–31 ▪ Models of the universe 272–273 ▪ Discovering other galaxies 290–293

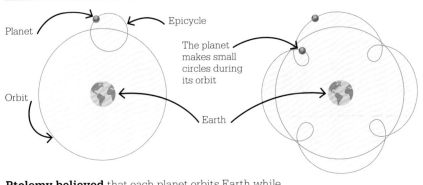

Planet

Epicycle

The planet makes small circles during its orbit

Orbit

Earth

Ptolemy believed that each planet orbits Earth while at the same time moving around a sub-orbit, or epicycle. He thought that this explained the unpredictable "retrograde" movements of the stars and planets.

Watching the sky

It was not just the motions of stars and planets that so intrigued early astronomers, but also short-lived events. Chinese astronomers kept records of Halley's Comet appearing in the night sky as far back as 1000 BCE, denoting it a "guest star." They also recorded supernovae, or exploding stars, most notably one that led to the formation of the Crab Nebula in 1054 CE. Reports from the time suggest the supernova was bright enough to be seen in the daytime for about a month.

In the 4th century BCE, the Greek philosopher Aristotle postulated that Earth was the center of the universe, with all other bodies— such as the moon and the planets— revolving around it. In about 150 CE, Ptolemy, an astronomer from Alexandria, furthered Aristotle's geocentric (Earth-centered) theory with his attempt to explain in mathematical terms the seemingly irregular motions of stars and planets in the night sky. Observing the back-and-forth movement of some planets, and the fact that others barely seemed to move at all, he concluded that heavenly bodies moved in a range of complex circular orbits and epicycles ("sub-orbits"), with Earth stationary at the center.

Earth-centered universe

Ptolemy based many of his calculations on the observations of Hipparchus, a Greek astronomer from the 2nd century BCE. In Ptolemy's best-known work, the *Almagest*, which sets out his geocentric theory, he used Hipparchus's notes on the motions of the moon and sun to calculate the positions of the moon, sun, planets, and stars at different times, and to predict eclipses.

Ptolemy's complex model of the universe, known as the Ptolemaic system, dominated astronomical thinking for centuries. It would take until the 16th century for Polish astronomer Nicolaus Copernicus to suggest that the sun, not Earth, was at the center of the universe. Although his model was initially derided—and Italian astronomer Galileo Galilei was put on trial in 1633 for supporting it—Copernicus was eventually proved right. ▪

Ptolemy

Claudius Ptolemaeus, known as Ptolemy, lived from about 100 CE to 170 CE. Little is known about his life, except that he lived in the city of Alexandria, in the Roman province of Egypt, and wrote about a variety of subjects, including astronomy, astrology, music, geography, and mathematics. In *Geography*, he listed the latitudes and longitudes of many of the places in the known world, producing a map that could be duplicated. His first major work on astronomy was the *Almagest*, in which he cataloged 1,022 stars and 48 constellations, and attempted to explain the movement of stars and planets in the night sky. His geocentric model of the universe persisted for centuries. Despite its inaccuracies, Ptolemy's work was hugely influential in understanding how things move in space.

Key works

c. 150 CE *Almagest*
c. 150 CE *Geography*
c. 150–170 CE *Handy Tables*
c. 150–170 CE *Planetary Hypotheses*

EARTH IS NOT THE CENTER OF THE UNIVERSE

MODELS OF THE UNIVERSE

IN CONTEXT

KEY FIGURE
Nicolaus Copernicus
(1473–1543)

BEFORE
6th century BCE The Greek philosopher Pythagoras says that Earth is round, based on the moon also being round.

3rd century BCE Eratosthenes measures the circumference of Earth with great accuracy.

2nd century CE Ptolemy of Alexandria asserts that Earth is the center of the universe.

AFTER
1609 Johannes Kepler describes the motions of the planets around the sun.

1610 Galileo Galilei observes moons orbiting Jupiter.

1616 Nicolaus Copernicus's work *De Revolutionibus Orbium Coelestium* (*On the Revolutions of the Heavenly Spheres*) is banned by the Catholic Church.

The **irregular movements of stars and planets** cannot be explained simply with Earth at the center of the universe.

↓

Earth may **seem to be stationary**, but it is in fact **rotating**, which explains why stars appear to cross the sky.

↓

Viewed from Earth, the **other planets** sometimes appear to be **moving backward**, but this is an illusion caused by the fact that **Earth itself is moving**.

↓

Copernicus believed **the sun** is stationary near the **center of the universe**, and Earth and the other planets revolve around it.

Today, the idea that Earth is flat seems comical. Yet early depictions of the world, including in ancient Egypt, show that this is what people believed. After all, Earth seemed to stretch into the distance, and from its surface it did not appear to curve. Earth was thought to be a circular disk floating on an ocean, with a dome above—the heavens—and the underworld below.

It was not until the 6th century BCE that Greek philosophers realized that Earth was round. Pythagoras first worked out that the moon was spherical, observing that the line between day and night on the moon is curved. This led him to suppose that Earth was also spherical. Aristotle later added to this by noting Earth's curved shadow during a lunar eclipse, and the changing positions of constellations. Eratosthenes went even further. He saw that the sun cast different shadows in the cities of Syene and Alexandria, and used this knowledge to work out the circumference of the planet. He calculated it to be between 24,000 and 29,000 miles (38,000 and 47,000 km)—not far off its true value of 24,901 miles (40,075 km).

> The passage of time has revealed to everyone the truths that I previously set forth.
> **Galileo Galilei**

Much study also went into working out Earth's place within the known universe. In the 2nd century CE, Ptolemy of Alexandria described Earth as a stationary sphere at the center of the universe, with all other known bodies orbiting around it. Five centuries earlier, the Greek astronomer Aristarchus of Samos had suggested that the sun was the center of the universe, but Ptolemy's geocentric (Earth-centered) view was the more widely accepted of the two.

Jupiter is now known to have 79 moons, but when Galileo Galilei first observed four moons orbiting the planet in 1610, he proved Copernicus's theory that not everything orbits Earth.

Sun-centered universe
The idea of a universe with the sun at its center remained dormant until the early 16th century, when Nicolaus Copernicus wrote a short manuscript, the *Commentariolus*, and circulated it among friends. In it, he proposed a heliocentric (sun-centered) model, with the sun near the center of the known universe, and Earth and other planets orbiting in a circular motion around it. He also suggested that the reason why the sun rose and set was due to the rotation of Earth.

The Catholic Church initially accepted Copernicus's work, but later banned it following Protestant accusations of heresy. Nonetheless, the heliocentric approach caught on. German astronomer Johannes Kepler published his laws of motion in 1609 and 1619, which showed that the planets moved around the

sun in not quite perfect circles, moving more quickly when they were closer to the sun and more slowly when they were further away. When Galileo Galilei spotted four moons orbiting Jupiter, this was proof that other bodies orbited other worlds—and that Earth was not the center around which everything revolved. ■

Nicolaus Copernicus

Nicolaus Copernicus was born on February 19, 1473, in Thorn (now Torun), Poland. His father, a wealthy merchant, died when Nicolaus was 10 years old, but the boy received a good education thanks to his uncle. He studied in Poland and Italy, where he developed interests in geography and astronomy.

Copernicus went on to work for his uncle, who was the bishop of Ermland in northern Poland, but after the bishop's death in 1512, he devoted more time to astronomy. In 1514, he distributed a handwritten pamphlet (the

Commentariolus), in which he proposed that the sun, not Earth, was near the center of the known universe. Although he finished a longer work titled *De Revolutionibus Orbium Coelestium* (*On the Revolutions of the Heavenly Spheres*) in 1532, he did not publish it until 1543, two months before he died.

Key works

1514 *Commentariolus*
1543 *On the Revolutions of the Heavenly Spheres*

NO TRUE TIMES OR TRUE LENGTHS
FROM CLASSICAL TO SPECIAL RELATIVITY

IN CONTEXT

KEY FIGURE
Hendrik Lorentz (1853–1928)

BEFORE
1632 Galileo Galilei posits that a person in a windowless room cannot tell whether the room is moving at a constant speed or not moving at all.

1687 Isaac Newton devises his laws of motion, using key parts of Galileo's theory.

AFTER
1905 Albert Einstein publishes his theory of special relativity, showing that the speed of light is always constant.

1915 Einstein publishes his theory of general relativity, explaining how the gravity of objects bends spacetime.

2015 Astronomers in the US and Europe find gravitational waves—ripples in spacetime predicted by Einstein a century earlier.

The ideas of relativity—peculiarities in space and time—are widely attributed to Albert Einstein in the early 20th century. Yet earlier scientists had also wondered whether everything they saw was as it seemed.

Galilean relativity

Back in 1632, in Renaissance Italy, Galileo Galilei had suggested that it was impossible to know whether a room was at rest or moving at a constant speed if there were objects moving within it. This idea is known as Galilean relativity, and others attempted to expand on it in subsequent years. One approach was to note that the laws of physics are the same in all inertial frames of reference (those that are moving at a constant velocity).

Dutch physicist Hendrik Lorentz produced a proof for this in 1892. His set of equations, known as the Lorentz transformations, showed how mass, length, and time change as a spatial object approaches the speed of light, and that the speed of light is constant in a vacuum.

Galileo used the example of a ship traveling at a constant speed on a flat sea. A passenger dropping a ball below deck would not be able to tell whether the ship was moving or stationary.

Lorentz's work paved the way for Einstein's theory of special relativity, in which he showed that not only are the laws of physics the same whether a person is moving at a constant speed or not moving, but that the speed of light is the same if it is measured in either scenario. This idea brought a whole new understanding of the universe. ∎

See also: Laws of motion 40–45 ▪ Laws of gravity 46–51 ▪ The speed of light 275 ▪ Special relativity 276–279 ▪ The equivalence principle 281

THE SUN AS IT WAS ABOUT EIGHT MINUTES AGO
THE SPEED OF LIGHT

The speed of light has long vexed humanity. In ancient Greece, the philosopher Empedocles thought that it must take a while for light from the sun to reach Earth, whereas Aristotle wondered if it had a speed at all.

Measuring light speed
Isaac Beeckman and Galileo Galilei made the first serious attempts to measure the speed of light, in the 17th century. Both relied on human sight, and were inconclusive. In 1850, Hippolyte Fizeau and Léon Foucault independently produced the first true measurements, using a rotating cog and a rotating mirror respectively to chop, or interrupt, a light beam. In Foucault's case, he calculated the speed of light from the angle between the light going to and from the rotating mirror and the speed of the mirror's rotation.

In the early 1880s, American physicist Albert Michelson improved Foucault's technique by reflecting a beam of light off two mirrors over a greater distance. He built an interferometer, a device that could split a beam of light in two, direct the two parts along different paths, and then recombine them. Noting the pattern of the returned light, he calculated the speed of light to be 299,853 km/s.

In 1887, Michelson and fellow American Edward Morley set up an experiment to measure Earth's motion through the "ether" that light had long been thought to travel through. They found no evidence of such an ether, but recorded increasingly precise values for a constant speed of light. ∎

> Light thinks it travels
> faster than anything
> but it is wrong.
> **Terry Pratchett**
> **British novelist**

See also: Focusing light 170–175 ▪ Lumpy and wavelike light 176–179
▪ Diffraction and interference 180–183 ▪ The Doppler effect and redshift 188–191

DOES OXFORD STOP AT THIS TRAIN?

SPECIAL RELATIVITY

IN CONTEXT

KEY FIGURE
Albert Einstein (1879–1955)

BEFORE
1632 Galileo Galilei puts forward his relativity hypothesis.

1687 Isaac Newton sets out his laws of motion.

1861 Scottish physicist James Clerk Maxwell formulates his equations describing electromagnetic waves.

AFTER
1907 Hermann Minkowski presents the idea of time as the fourth dimension in space.

1915 Albert Einstein includes gravity and acceleration in his theory of general relativity.

1971 To demonstrate time dilation due to general and special relativity, atomic clocks are flown around the world.

R elativity has deep roots. In 1632, Galileo Galilei imagined a traveler inside a windowless cabin on a ship sailing at a constant speed on a perfectly smooth sea. Was there any way for the traveler to determine whether the ship was moving without going on deck? Was there any experiment that, if carried out on a moving ship, would give a different result from the same experiment carried out on land? Galileo concluded that there was not. Provided the ship moved with a constant speed and direction, the results would be the same. No unit of measurement is absolute—all units are defined

See also: From classical to special relativity 274 ▪ The speed of light 275 ▪ Curving spacetime 280 ▪ The equivalence principle 281 ▪ Paradoxes of special relativity 282–283 ▪ Mass and energy 284–285

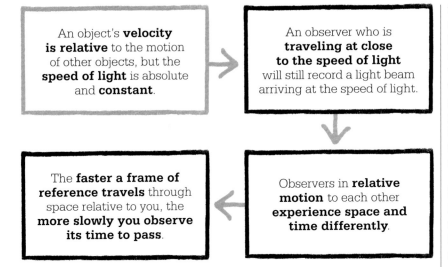

> An object's **velocity is relative** to the motion of other objects, but the **speed of light** is absolute and **constant**.

> An observer who is **traveling at close to the speed of light** will still record a light beam arriving at the speed of light.

> Observers in **relative motion** to each other **experience space and time differently**.

> The **faster a frame of reference travels** through space relative to you, the **more slowly you observe its time to pass**.

relative to something else. In order to measure anything, be it time, distance, or mass, there has to be something to measure it against.

How fast people perceive an object to be moving depends on their speed relative to that object. For example, if a person on a fast-moving train tosses an apple to another passenger, it travels from one to the other at a few miles per hour, but to an observer standing by the tracks, the apple and the passengers all flash by at a hundred miles per hour.

Frames of reference

The idea that motion has no meaning without a frame of reference is fundamental to Albert Einstein's theory of special relativity. The theory is special in that it concerns itself with the special case of objects moving at a constant velocity relative to one another. Physicists call this an inertial frame of reference. As pointed out by Isaac Newton, an inertial state is the default for any object not being acted on by a force. Inertial motion is motion in a straight line at a uniform speed. Before Einstein, Newton's idea of absolute motion—that an object could be said to be moving (or to be at rest) without reference to anything else—held sway. Special relativity would put an end to this.

Principle of Relativity

As was well known to scientists of the 19th century, an electric current is generated if a magnet is moved inside a coil of wire, and also if the coil is moved and the magnet stays still. Following British physicist Michael Faraday's discoveries in the 1820s and 1830s, scientists had assumed that there were two different explanations for this phenomenon—one for the moving coil and one for the moving magnet. Einstein was having none of it. In his 1905 paper "On the Electrodynamics of Moving Bodies," he said it did not matter which was moving—it was their movement relative to each other

that generated the current. Setting out his "Principle of Relativity," he declared: "The same laws of electrodynamics and optics will be valid for all frames of reference for which the laws of mechanics hold good." In other words, the laws of physics are the same in all inertial frames of reference. It was similar to the conclusion that Galileo had reached in 1632.

Light is a constant

In James Clerk Maxwell's equations of 1865 for calculating the variables in an electromagnetic field, the speed of an electromagnetic wave, approximately 186,000 miles/s (300,000 km/s), is a constant defined by the properties of the vacuum of space through which it moves. Maxwell's equations hold true in any inertial frame. Einstein declared that, while some things may be relative, the speed of light is absolute and constant: it travels at a constant velocity, independent of everything else, including the motion of the light source, and is not measured relative to anything else. This is what makes light fundamentally different from matter—all speeds below that »

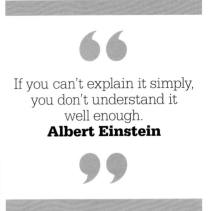

> *If you can't explain it simply, you don't understand it well enough.*
> **Albert Einstein**

of light are relative to the frame of reference of the observer, and nothing can travel faster than light. The consequence of this is that two observers moving relative to each other will always measure the speed of a beam of light as being the same—even if one is moving in the same direction as the light beam and one is moving away from it. In Galilean terms, this made little sense.

A matter of time

The idea that the speed of light is constant for all observers is central to special relativity. Everything else derives from this one deceptively simple fact. Einstein saw that there was a fundamental connection between time and the velocity of light. It was this insight that led him to the completion of his theory. Even if an observer is traveling at close to the speed of light, she will still record a light beam arriving at the speed of light. For this to occur, said Einstein, time for the observer has to be running more slowly.

Newton thought that time was absolute, flowing without reference to anything else at the same steady pace wherever in the universe it was measured. Ten seconds for one observer is the same as ten seconds for another, even if one is standing still and the other is rushing by in the fastest available spacecraft.

Newtonian physics states that velocity equals distance traveled divided by the time taken to cover that distance—as an equation: $v = d/t$. So, if the speed of light, v, always remains the same, whatever the other two values, then it follows that d and t, distance, or space, and time, must change.

Time, Einstein argued, passes differently in all moving frames of reference, which means that observers in relative motion (in different moving reference frames) will have clocks that run at different rates. Time is relative. As Werner Heisenberg said of Einstein's discovery, "This was a change in the very foundation of physics."

According to special relativity, the more quickly a person travels through space, the more slowly he travels through time. This phenomenon is called time dilation. Scientists at CERN's Large Hadron Collider, where particles are smashed together at near light speed velocities, have to take the effects of time dilation into account when interpreting the results of their experiments.

Contracting space

Einstein asked himself a question: if I held a mirror while traveling at the speed of light, would I see my reflection? How would light reach the mirror if the mirror was moving at light speed? If the speed of light is a constant, then, no matter how fast he was moving, the light going from Einstein to the mirror and back again would always be traveling at 300,000 km/s because the speed of light does not change.

In order for light to reach the mirror, not only does time have to be slowed down, but also the

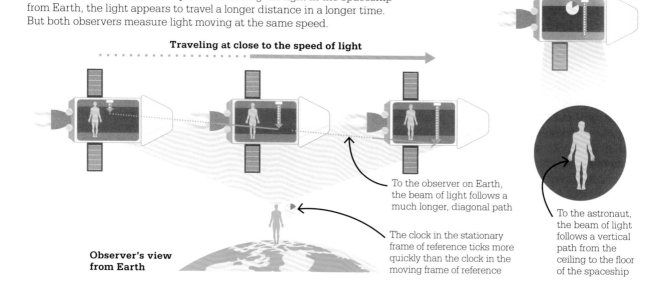

Inside a spaceship moving at near light speed, an astronaut using a clock to measure the speed of light finds that it travels a relatively short distance in a short time. For a person watching the light in the spaceship from Earth, the light appears to travel a longer distance in a longer time. But both observers measure light moving at the same speed.

Astronaut's view

Traveling at close to the speed of light

To the observer on Earth, the beam of light follows a much longer, diagonal path

The clock in the stationary frame of reference ticks more quickly than the clock in the moving frame of reference

To the astronaut, the beam of light follows a vertical path from the ceiling to the floor of the spaceship

Observer's view from Earth

> Einstein, in the special theory of relativity, proved that different observers, in different states of motion, see different realities.
> **Leonard Susskind**
> **American physicist**

distance traveled by the beam of light has to decrease. At approximately 99.5 percent of light speed the distance is reduced by a factor of 10. This shrinkage only takes place in the direction of motion and will only be apparent to an observer who is at rest with respect to the moving object. The crew of a spaceship traveling at close to the speed of light would not perceive any change in their spaceship's length; rather, they would see the observer appear to contract as they streaked by.

One consequence of the contraction of space is to shorten the time it would take a spacecraft to travel to the stars. Imagine there is a cosmic railroad network with tracks stretching from star to star. The faster a spacecraft travels, the shorter the track appears to become, and therefore the shorter the distance to be covered to reach its destination. At 99.5 percent of light speed, the journey to the

Light travels at a constant speed, so the speed of light from a car's headlights will not increase when the vehicle accelerates, nor will it decrease when the car slows down.

nearest star would take around five months in spaceship time. However, for an observer back on Earth, the trip would appear to take more than four years.

Einstein's equation
One of the most famous equations in all of physics is derived from special relativity. Einstein published it as a sort of short postscript to his special theory. Known sometimes as the law of mass–energy equivalence, $E = mc^2$ effectively says that energy (E) and mass (m) are two aspects of the same thing. If an object gains or loses energy, it loses or gains an equivalent amount of mass in accordance with the formula.

For example, the faster an object travels, the greater is its kinetic energy, and the greater also is its mass. The speed of light (c) is a big number—squared, it is a very big number indeed. This means that when even a tiny amount of matter is converted into the equivalent amount of energy the yield is colossal, but it also means that there has to be an immense input of energy to see an appreciable increase in mass. ∎

Albert Einstein

Albert Einstein was born in Ulm, Germany, on March 14, 1879. Allegedly slow to learn to talk, Einstein declared later that "I very rarely think in words." At age four or five, Albert was fascinated by the invisible forces that made a compass needle move; he said that it awakened his curiosity about the world. He began violin lessons at the age of six, sparking a love of music that lasted throughout his life. The story that Einstein was bad at math is not true—he was an able student.

In 1901, Einstein acquired Swiss citizenship and became a technical assistant in the Swiss Patent Office where, in his free time, he produced much of his best work. In 1933, he emigrated to the US to take up the position of professor of theoretical physics at Princeton, and was made a US citizen in 1940. He died on April 18, 1955.

Key works

1905 "On a Heuristic Viewpoint Concerning the Production and Transformation of Light"
1905 "On the Electrodynamics of Moving Bodies"

A UNION OF SPACE AND TIME
CURVING SPACETIME

IN CONTEXT

KEY FIGURE
Hermann Minkowski
(1864–1909)

BEFORE
4th century BCE Euclid's work in geometry shows how the ancient Greeks tried to apply mathematics to physical space.

1637 French philosopher René Descartes develops Cartesian coordinates—using algebra to calculate positions.

1813 German mathematician Carl Friedrich Gauss suggests the idea of non-Euclidean spaces, which do not conform to Euclidean geometry.

AFTER
1915 Albert Einstein develops his theory of general relativity with the help of Hermann Minkowski's work.

1919 Arthur Eddington sees curved spacetime in action by noting the changed positions of stars during an eclipse.

The world seems to conform to certain geometric rules. It is possible to calculate the coordinates of a point, for example, or map out a particular shape. This basic understanding of the world, using straight lines and angles, is known as Euclidean space, named after Euclid of Alexandria.

With the dramatic evolution of physics in the early 20th century, the need for a new way to understand the universe grew. German mathematician Hermann Minkowski realized that much of the work of physicists was more easily understood when considered in four dimensions. In Minkowski "spacetime," three coordinates describe where a point is in space, and the fourth gives the time at which an event happened there.

Minkowski noted in 1908 that Earth, and the universe, are curved and so do not follow straight lines. Similar to how an aircraft follows a curved path over Earth rather than a straight one, light itself curves around the universe. This means that coordinates in spacetime cannot be accurately measured using straight lines and angles, and rely on detailed calculations called non-Euclidean forms of geometry.

These can be useful when, for example, calculating distances on the surface of Earth. In Euclidean space, the distance between two points would be calculated as though Earth's surface were flat. But in non-Euclidean spaces, the curvature of the planet has to be accounted for, using what are known as geodesics ("great arcs") to get a more accurate value. ∎

From this moment, space by itself and time by itself shall sink into the background.
Hermann Minkowski

See also: Measuring distance 18–19 ▪ The language of physics 24–31 ▪ Measuring time 38–39 ▪ Special relativity 276–279

GRAVITY IS EQUIVALENT TO ACCELERATION
THE EQUIVALENCE PRINCIPLE

IN CONTEXT

KEY FIGURE
Albert Einstein (1879–1955)

BEFORE
c. 1590 Galileo Galilei shows that two falling objects accelerate at the same speed regardless of their mass.

1609 German astronomer Johannes Kepler describes what would happen if the moon's orbit were stopped and it dropped toward Earth.

1687 Isaac Newton's theory of gravitation includes the idea of the equivalence principle.

AFTER
1964 Scientists drop test masses of aluminum and gold to prove the equivalence principle on Earth.

1971 American astronaut David Scott drops a hammer and feather on the moon to show that they fall at the same rate, as predicted by Galileo centuries earlier.

lbert Einstein's theory of special relativity describes how objects experience space and time differently, depending on their motion. An important implication of special relativity is that space and time are always linked in a four-dimensional continuum called spacetime. His later theory of general relativity describes how spacetime is warped by massive objects. Mass and energy are equivalent, and the warping they cause in spacetime creates the effects of gravity.

Einstein based his 1915 general relativity theory on the equivalence principle—the idea that inertial mass and gravitational mass have the same value. This was first noted by Galileo and Isaac Newton in the 17th century, and then developed by Einstein in 1907. When a force is applied to an object, the inertial mass of that object can be worked out by measuring its acceleration. The gravitational mass of an object can be calculated by measuring the force of gravity. Both calculations will produce the same number.

In an accelerating spaceship, a dropped ball would behave in exactly the same way as a ball dropped in Earth's gravitational field.

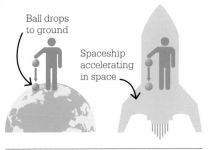

Ball drops to ground

Spaceship accelerating in space

If a person in a stationary spaceship on Earth drops an object, the measured mass of the object will be the same as if that person is in an accelerating spaceship in space. Einstein said it is impossible to tell whether a person is in a uniform gravitational field or accelerating through space using this approach.

Einstein went on to imagine what a beam of light would look like for the person inside the spaceship. He concluded that powerful gravity and extreme acceleration would have the same effect—the beam of light would curve downward. ∎

See also: Free falling 32–35 ▪ Laws of gravity 46–51 ▪ Special relativity 276–279 ▪ Curving spacetime 280 ▪ Mass and energy 284–285

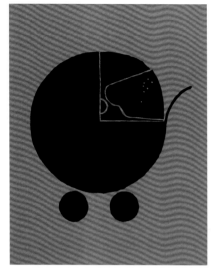

WHY IS THE TRAVELING TWIN YOUNGER?

PARADOXES OF SPECIAL RELATIVITY

IN CONTEXT

KEY FIGURE
Paul Langevin (1872–1946)

BEFORE
1905 Albert Einstein posits
that a moving clock will
experience less time than
a stationary clock.

1911 Einstein goes on
to suggest that a moving
person will be younger
than a stationary person.

AFTER
1971 American physicists
Joseph Hafele and Richard
Keating prove that Einstein's
theory of time dilation is
correct by taking atomic
clocks on board airplanes and
comparing them with atomic
clocks on the ground.

1978 The first GPS satellites
are also found to experience
time dilation.

2019 NASA launches an
atomic clock designed to
be used in deep space.

Albert Einstein's theories of relativity have prompted scientists to explore how light and other objects behave as they reach the extremities of time, space, and motion—for example, when a moving object approaches the speed of light. But they have also resulted in some interesting thought experiments that appear, at first, to be unsolvable.

One of the best-known thought experiments is the "twin paradox." This was first put forward by French physicist Paul Langevin in 1911. Building on Einstein's

Returning to Earth,
having aged by two years,
[the traveler] will climb out
of his vehicle and find
our globe aged by at least
two hundred years …
Paul Langevin

work, Langevin suggested the paradox based on a known consequence of relativity—time dilation. This means that any object moving faster than an observer will be seen by the observer to experience time more slowly. The faster the object is moving, the more slowly it will be seen to experience time.

Langevin wondered what might happen if there were identical twin brothers on Earth, and one of them was sent into space on a very fast trip to another star and back. The twin on Earth would say that the traveler was moving, so the traveler would be the younger of the two when he returned. But the traveler would argue that the twin on Earth was moving, and that he was stationary. He would therefore say that the Earth twin would be younger.

There are two solutions to this apparent paradox. The first is that the traveler must change his velocity in order to return home, while the twin on Earth stays at a constant velocity—so the traveler will be younger. The other solution is that the traveler leaves Earth's reference frame, whereas the twin on Earth

See also: Measuring time 38–39 ▪ From classical to special relativity 274 ▪ The speed of light 275 ▪ Special relativity 276–279 ▪ Curving spacetime 280

remains there, meaning that the traveler dictates the events that will happen.

The barn-pole paradox

In the "barn-pole paradox," a pole vaulter runs through a barn at nearly the speed of light, with a pole twice the length of the barn. While the pole vaulter is inside the barn, both barn doors are closed simultaneously, and then opened again. Is it possible for the pole to fit inside the barn while the doors are closed?

From the reference frame of someone watching, as something approaches the speed of light, it appears to shorten, in a process known as length contraction. However, from the pole vaulter's frame of reference, the barn is moving toward her at the speed of light; she is not moving. So the barn would be contracted, and the pole would not fit.

The solution is that there are some inconsistencies that enable the pole to fit. From the observer's point of reference, the pole is small enough to fit inside before the barn doors open again to allow the pole out. From the pole vaulter's perspective, the doors do not close and open simultaneously, but rather one after the other, allowing the pole to enter at one end and escape from the other. ▪

Paul Langevin

Paul Langevin was born in Paris on January 23, 1872. After studying science in the French capital, he moved to England to study under J.J. Thomson at Cambridge University's Cavendish Laboratory. He then returned to the Sorbonne in Paris to get a doctorate in physics in 1902. He became professor of physics at the Collège de France in 1904, and then director of the School of Physics and Chemistry in 1926. He was elected to the Academy of Sciences in 1934.

Langevin was best known for his work on magnetism, but also built on the work of Pierre Curie at the end of World War I to devise a way to find submarines using echolocation. He also helped to spread Albert Einstein's theories in France. In the 1930s, he was arrested by France's Vichy government for opposing fascism, and spent most of World War II under house arrest. Langevin died on December 19, 1946, aged 74.

Key works

1908 "On the Theory of Brownian Motion"
1911 "The Evolution of Space and Time"

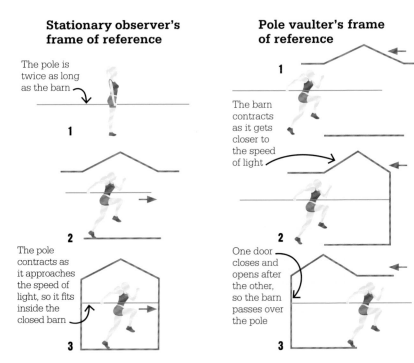

Stationary observer's frame of reference

The pole is twice as long as the barn

1

The pole contracts as it approaches the speed of light, so it fits inside the closed barn

2

3

Pole vaulter's frame of reference

1

The barn contracts as it gets closer to the speed of light

2

One door closes and opens after the other, so the barn passes over the pole

3

The space around a moving object contracts as the object nears the speed of light. A stationary observer watching the pole reaching the barn would see it contract, allowing it to fit inside. For the pole vaulter, the barn contracts, but as the doors do not open and close simultaneously, it can pass over her.

EVOLUTION OF THE STARS AND LIFE

MASS AND ENERGY

The famous equation $E = mc^2$ has taken various forms over the years, and its impact on physics is hard to overstate. It was devised by Albert Einstein in 1905 and concerned the mass–energy equivalence—that energy (E) is equal to mass (m) multiplied by the square of the speed of light (c^2). According to Einstein's theory of relativity, the equation can be used to calculate the energy from a given mass and work out any changes that have occurred in a nuclear reaction.

Chain reaction
Einstein's equation showed that mass and energy were linked in ways that had not been thought possible, and that a small loss of mass could be accompanied by a huge release of energy. One of the major implications of this was in

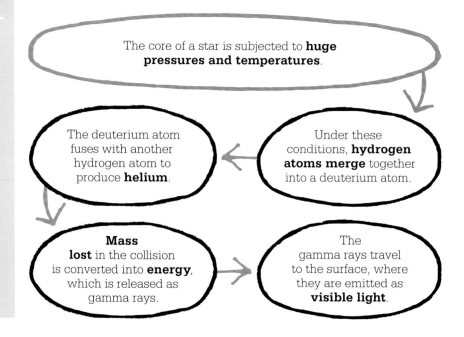

The core of a star is subjected to **huge pressures and temperatures**.

Under these conditions, **hydrogen atoms merge** together into a deuterium atom.

The deuterium atom fuses with another hydrogen atom to produce **helium**.

Mass lost in the collision is converted into **energy**, which is released as gamma rays.

The gamma rays travel to the surface, where they are emitted as **visible light**.

See also: Energy and motion 56–57 ▪ Light from the atom 196–199 ▪ Nuclear bombs and power 248–251 ▪ The speed of light 275 ▪ Special relativity 276–279

understanding how stars produce energy. Until the pioneering work of British physicist Arthur Eddington in the early 20th century, scientists were at a loss to explain how stars like the sun radiated so brightly.

When an unstable (radioactive) atom is hit by another particle, such as a neutron, it can break apart into new particles. These can hit other atoms and cause a chain reaction, in which some of the mass lost in each collision is converted into new particles, and the rest is released as energy. In 1920, Eddington suggested that certain elements within stars could be undergoing a similar process to produce energy.

Nuclear reactions

Eddington proposed that hydrogen was being turned into helium and other heavier elements by stars— a process that is now called nuclear fusion—and that this could account for their energy output (the visible light they produce). Using Einstein's equation, he described how the intense heat and pressure at the core of a star could enable nuclear reactions to occur, destroying mass and releasing energy.

Today, knowledge of how mass can be turned into energy has allowed physicists to replicate it in nuclear reactors, which use the process of nuclear fission—the splitting of a heavy atom into two lighter ones. It also led to the creation of the atomic bomb, in which an unstoppable chain reaction occurs, releasing a powerful and deadly amount of energy. Particle accelerators such as the Large Hadron Collider rely on Einstein's equation to smash particles together and create new ones. The more energy there is in a collision, the greater the mass of the particles that are created.

Physicists are now trying to recreate the process of nuclear fusion, which would produce more energy than nuclear fission but requires conditions so extreme that they are difficult to replicate. It is hoped that in the coming decades nuclear fusion could be a viable source of power on Earth. ▪

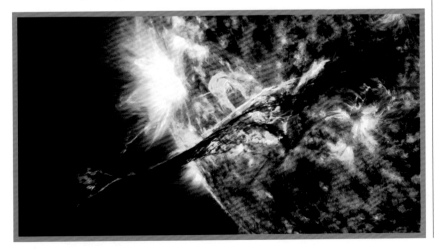

The sun releases vast amounts of energy by turning hydrogen into helium in the process of nuclear fusion. It is said to have enough hydrogen to keep it shining for the next 5 billion years.

Arthur Eddington

Arthur Eddington was born in 1882 in Westmorland (now Cumbria), UK. He studied in Manchester and then at Trinity College, Cambridge, winning honors for his work. From 1906 to 1913, he was chief assistant at the Royal Observatory of Greenwich.

Eddington's parents were Quakers, so he was a pacifist during World War I. He made important contributions to science from 1914 onward, and was the first to expound on Einstein's relativity theory in the English language. In 1919, he traveled to Príncipe, off the west coast of Africa, to observe a solar eclipse and prove the idea of gravitational lensing (that large masses bend light, as predicted in Einstein's theory of general relativity) in the process. He published articles on gravity, spacetime, and relativity, and produced his masterwork on nuclear fusion in stars in 1926. He died in 1944, aged 61.

Key works

1923 *The Mathematical Theory of Relativity*
1926 *The Internal Constitution of the Stars*

WHERE SPACETIME SIMPLY ENDS

BLACK HOLES AND WORMHOLES

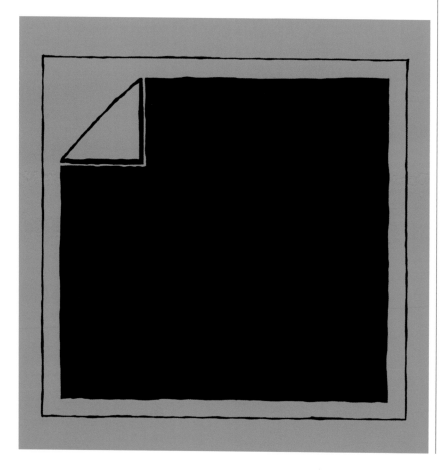

IN CONTEXT

KEY FIGURE
Karl Schwarzschild
(1873–1916)

BEFORE
1784 John Michell proposes
the idea of "dark stars,"
which trap light with their
intense gravity.

1796 French scholar Pierre-
Simon Laplace predicts the
existence of large invisible
objects in the universe.

AFTER
1931 Indian–American
astrophysicist Subrahmanyan
Chandrasekhar proposes a
mass limit for the formation
of a black hole.

1971 The first black hole,
Cygnus X-1, is indirectly
discovered in the Milky Way.

2019 A global team of
astronomers reveals the
first image of a black hole.

T he idea of black holes and
wormholes has been talked
about for centuries, but
only recently have scientists begun
to understand them—and with
them, the wider universe. Today,
scientists are starting to appreciate
just how impressive black holes are,
and have even managed to take a
picture of one—indisputable proof
that they exist.

British clergyman John Michell
was the first to think about black
holes, suggesting in the 1780s that
there might be stars with gravity so
intense that nothing, not even light,
could escape. He called these "dark
stars," and said that while they
would be essentially invisible, it

See also: Free falling 32–35 ▪ Laws of gravity 46–51 ▪ Particles and waves 212–215 ▪ From classical to special relativity 274 ▪ Special relativity 276–279 ▪ Curving spacetime 280 ▪ The equivalence principle 281 ▪ Gravitational waves 312–315

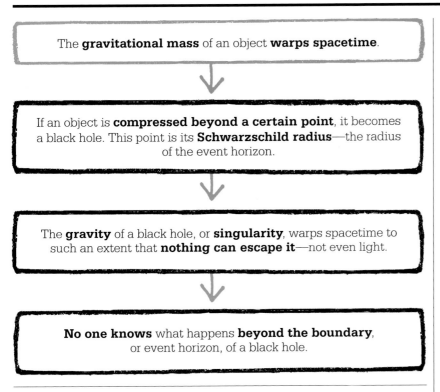

The **gravitational mass** of an object **warps spacetime**.

If an object is **compressed beyond a certain point**, it becomes a black hole. This point is its **Schwarzschild radius**—the radius of the event horizon.

The **gravity** of a black hole, or **singularity**, warps spacetime to such an extent that **nothing can escape it**—not even light.

No one knows what happens **beyond the boundary**, or event horizon, of a black hole.

should, in theory, be possible to see their gravitational effects on other objects.

In 1803, however, British physician Thomas Young demonstrated that light is a wave, rather than a particle, prompting scientists to question whether it would be affected by gravity at all. Michell's idea was swept aside and largely forgotten about until 1915, when Albert Einstein came up with his general theory of relativity—and suggested that light itself could be bent by the mass of an object.

No escape

The following year, German physicist Karl Schwarzschild took Einstein's idea to its extreme. He suggested that if a mass were condensed enough, it would create a gravitational field from which not

even light could escape, with its edge being what is called the event horizon. Similar to Michell's "dark star" idea, this later became known as a "black hole."

Schwarzschild produced an equation that allowed him to calculate the Schwarzschild radius (the radius of the event horizon) for any given mass. If an object is compressed into a volume so dense that it fits within that radius, it will warp spacetime to such an extent that nothing can escape its gravitational pull—it will collapse into a singularity, or the center of a black hole. Schwarzschild, however, did not think that a singularity could really exist. He saw it as a somewhat theoretical point where some property is infinite, which, in the case of a black hole, is the density of the matter.

Even so, Schwarzschild's work showed that black holes could mathematically exist, something that many scientists had not thought possible. Astronomers today use his equations to estimate the masses of black holes, albeit not very precisely, since their rotation and electric charge cannot be accounted for. Back in the early 20th century, Schwarzschild's findings allowed scientists to start contemplating what might happen near or even inside a black hole.

Mass limit

One important factor was working out how black holes formed, something set forth in 1931 by Subrahmanyan Chandrasekhar. He used Einstein's theory of special relativity to show that there is a mass limit below which a star at the end of its life will collapse into a stable, smaller dense star known as a white dwarf. If the remains of the star's mass are above this limit, which Chandrasekhar calculated as 1.4 times the mass of the sun, it will collapse further into either a neutron star or a black hole. »

The final fate of massive stars is to collapse behind an event horizon to form a 'black hole' which will contain a singularity.
Stephen Hawking

Black holes

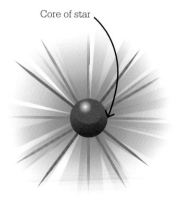

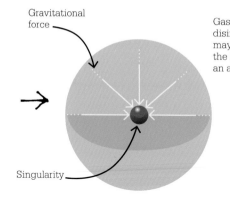

When a massive star dies, it collapses, unable to resist the crushing force of its own gravity. This causes a supernova explosion, and the star's outer parts are blasted into space.

If the core that remains after the supernova is still massive (more than 1.4 times the mass of the sun), it keeps shrinking and collapses under its own weight into a point of infinite density—a singularity.

The singularity is now so dense that it distorts the spacetime around it so that not even light can escape. This black hole is pictured in two dimensions as an infinitely deep hole called a gravity well.

In 1939, American physicists Robert Oppenheimer and Hartland Snyder put forward a more modern-looking idea of a black hole. They described how the sort of body Schwarzschild had envisaged would be detectable only by its gravitational influence. Others had already considered the peculiar physics that would take place inside a black hole, including Belgian astronomer Georges Lemaître in 1933. His "Alice and Bob" thought experiment supposed that if Bob were to witness Alice falling into a black hole, she would appear to freeze at its edge—the invisible boundary called the event horizon—but she would experience something entirely different as she fell within its bounds.

It would take until 1964 for the next major development to come, when British physicist Roger Penrose suggested that if a star imploded with enough force, it would always produce a singularity like the one Schwarzschild had proposed. Three years later, the term "black hole" was born during a talk by American physicist John Wheeler. He had already come up with a term to describe theoretical tunnels in spacetime: "wormholes."

Many physicists were, by now, considering the idea of wormholes. The notion of a white hole, in which the event horizon would not stop light from escaping, but rather stop it from entering, progressed into an idea that black holes and white holes could be linked. Using exotic matter, which involves negative energy densities and negative pressure, it was suggested that information could pass through a wormhole from one end to the other, perhaps between a black and a white hole, or even two black holes, across vast distances of space and time.

In reality, however, the existence of wormholes remains questionable. Although scientists think they may occur on a microscopic level, attempts to draw up larger versions have so far been unsuccessful. Nonetheless, the idea of them persists today—and with it, the very vague possibility that humans could traverse large distances in the universe with ease.

Looking for proof
While theories abounded, no one had yet been able to detect a black hole, let alone a wormhole. But in 1971, astronomers observed an odd source of X-rays in the Cygnus constellation. They suggested these X-rays, dubbed Cygnus X-1, were the result of a bright blue star being torn apart by a large and dark object. At last, astronomers had witnessed their first black hole, not by seeing it directly, but by seeing its effect on a nearby object.

From there, black hole theory soared in popularity. Physicists produced novel ideas about how black holes might behave—among them Stephen Hawking, who in 1974 came up with the idea that black holes might emit particles (now known as Hawking radiation). He suggested that the strong gravity of a black hole could produce pairs of both particles and antiparticles, something that is thought to happen in the quantum world. While one of the particle–antiparticle pair would be pulled into the black hole, the other would escape, carrying with it information about the event horizon, the strange boundary beyond which no mass or light can escape.

Supermassive black holes

Black holes of different shapes and sizes were imagined, too. Stellar black holes, such as Cygnus X-1, were thought to contain between about 10 and 100 times the mass of the sun squashed into an area dozens of miles across. But it was also thought that black holes could merge into a supermassive black hole, containing millions or billions of times the mass of the sun packed into an area millions of miles across.

Nearly every galaxy is thought to have a supermassive black hole at its core, possibly surrounded by a swirling accretion disk of superheated dust and gas, which can be seen even at great distances. Around the supermassive black hole believed to be at the center of the Milky Way, astronomers have seen what they think are stars orbiting it—the very idea suggested by Michell back in the 18th century.

The greatest moment in black hole theory's history so far arrived in April 2019, when astronomers from an international collaboration called the Event Horizon Telescope (EHT) revealed the first-ever image of a black hole. Using multiple telescopes around the globe to create a virtual super telescope, they were able to photograph the supermassive black hole in a galaxy called Messier 87, 53 million light-years away. The image lived up to all predictions, with a ring of light surrounding a dark center.

When the universe ends in trillions of years, it is thought that black holes will be the only thing to remain as entropy—the unavailable energy in the universe—increases. Finally, they, too, will evaporate. ∎

In this illustration, the black hole Cygnus X-1 is pulling material from a blue companion star toward it. The material forms a red and orange disk that rotates around the black hole.

Karl Schwarzschild

Born in Frankfurt, Germany, in 1873, Karl Schwarzschild showed an early talent for science, publishing a paper on the orbits of celestial objects when he was 16. He became a professor at the University of Göttingen in 1901, and director of the Astrophysical Observatory at Potsdam in 1909.

One of Schwarzschild's main contributions to science came in 1916, when he produced the first solution of Einstein's equations of gravity from the theory of general relativity. He showed that objects of a high enough mass would have an escape velocity (the velocity needed to escape its gravitational pull) higher than the speed of light. This was the first step in understanding black holes, and ultimately led to the idea of an event horizon. Schwarzschild died from an autoimmune disease in 1916, while serving on the Russian front in World War I.

Key work

1916 *On the Gravitational Field of a Mass Point after Einstein's Theory*

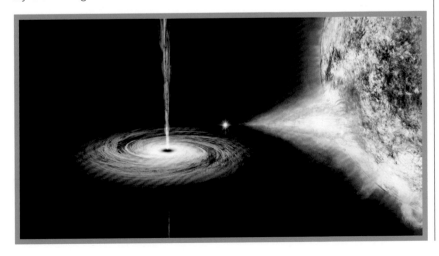

THE FRONTIER OF THE KNOWN UNIVERSE

DISCOVERING OTHER GALAXIES

IN CONTEXT

KEY FIGURES
Henrietta Swan Leavitt
(1868–1921),
Edwin Hubble (1889–1953)

BEFORE
964 Persian astronomer Abd
al-Rahman al-Sufi is the first
person to observe Andromeda,
although he does not realize it
is another galaxy.

1610 Galileo Galilei proposes
that the Milky Way is made of
many stars after observing it
through his telescope.

AFTER
1953 Galaxies close to Earth
are found to be part of a
supercluster, called the Virgo
Cluster, by French astronomer
Gérard de Vaucouleurs.

2016 A study led by American
astrophysicist Christopher
Conselice reveals that the
known universe contains
two trillion galaxies.

A fter the publication of
Copernicus's heliocentric
model of the universe in
1543, which placed the sun rather
than Earth at its center, further
attempts to understand the size
and structure of the universe
made little progress. It was nearly
400 years before astrophysicists
realized that not only was the sun
not the center of the universe, as
Copernicus had thought, but it was
not even the center of our galaxy,
the Milky Way.

In the 1920s, the discovery by
Edwin Hubble that the Milky Way
is just one of many galaxies in the
universe marked a significant leap
forward in astronomical knowledge.

See also: The Doppler effect and redshift 188–191 ▪ Models of the universe 272–273 ▪ The static or expanding universe 294–295

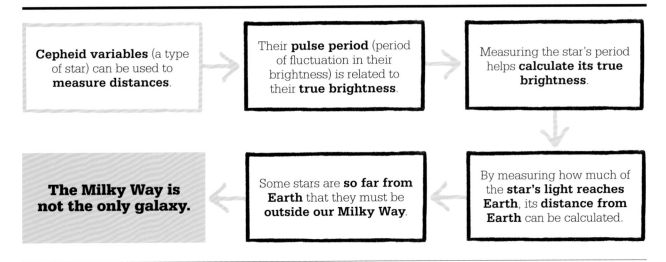

Cepheid variables (a type of star) can be used to **measure distances**.

Their **pulse period** (period of fluctuation in their brightness) is related to their **true brightness**.

Measuring the star's period helps **calculate its true brightness**.

By measuring how much of the **star's light reaches Earth**, its **distance from Earth** can be calculated.

Some stars are **so far from Earth** that they must be **outside our Milky Way**.

The Milky Way is not the only galaxy.

Key to this breakthrough was a new way of measuring distances in space made possible by the work of American astronomer Henrietta Swan Leavitt in 1908.

Determining distance

In the early 1900s, establishing a star's distance from Earth was no easy task. Within the Milky Way, distances between objects could be measured using parallax, a process that employs basic trigonometry. By measuring the angles of the sight lines to an object from both sides of Earth's orbit around the sun, for example, astronomers can calculate the angle where the two sight lines converge and thus the distance to that object. However, for distances greater than 100 light years, and outside the Milky Way, parallax is too imprecise. Early 20th-century astronomers needed another method.

At the time, Leavitt was working as a "computer," the name given to data processors at Harvard College Observatory in Cambridge, Massachusetts. Edward Charles

Pickering, who ran the observatory, had asked Leavitt to measure the brightness of the stars on a collection of photographic plates showing the Magellanic Clouds, now known to be small galaxies outside the Milky Way.

In the course of this work, Leavitt discovered a class of very bright stars called Cepheid variables whose luminosity fluctuates as they pulsate. Each of these stars

pulsates in a regularly recurring cycle, or period, dependent upon physical changes within the star. Compressed by gravity, the star becomes smaller, more opaque, and gradually hotter as the light energy trapped inside the star starts to build. Eventually, the extreme heat causes the gas in the outer layers to expand and the star becomes more transparent, allowing the light »

Cepheid variable stars pulsate—expand and contract over a regular cycle—resulting in varying temperature and brightness. Astronomers can plot their changing brightness over time on a light curve.

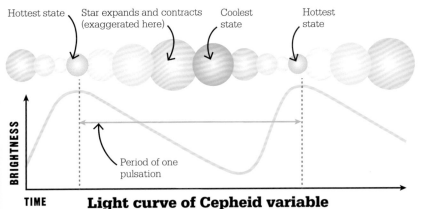

Hottest state
Star expands and contracts (exaggerated here)
Coolest state
Hottest state

BRIGHTNESS

Period of one pulsation

TIME **Light curve of Cepheid variable**

Since the [Cepheid] variables are probably at nearly the same distance from the Earth, their periods are apparently associated with their actual emission of light.
Henrietta Swan Leavitt

energy to pass through. However, as the star expands, it also becomes cooler, and gravitational forces eventually outweigh the outward pressure of expansion. At this point, the star becomes smaller and the process repeats.

Leavitt noticed that the cycles of the stars varied between two and sixty days, and the brightest stars spent longer at their maximum luminosity. Because all the stars were in the Magellanic Clouds, Leavitt knew that any differences between the stars'

luminosities had to be due to their intrinsic brightness and not to how far they were from Earth.

A new tool

Leavitt published her findings in 1912, creating a chart showing the periods of 25 Cepheid variables, indicating their "true luminosity," and how bright they appeared from Earth, their "apparent luminosity." The significance of Leavitt's discovery was that once the distance of one Cepheid variable had been calculated using parallax, the distance of Cepheid variables of comparable pewriod in stars beyond the limits of parallax could be calculated by comparing the "true luminosity" established by its period with the "apparent luminosity"—the amount it had dimmed by the time its light reached Earth.

Within a year of Leavitt's discovery, Danish astronomer Ejnar Hertzsprung created a luminosity scale of Cepheid variables. These stars were designated as "standard candles"—benchmarks for working out the cosmic distances of objects. Over the next decade, other astronomers began using Leavitt's and Hertzsprung's work to calculate the distance to stars.

The Great Debate

Using Cepheid variables was not a perfect method for ascertaining distance—it could not account for light-absorbing cosmic dust, which distorts apparent luminosity—but it led to a fundamental change in our understanding of the universe. In 1920, the National Academy of

Edwin Hubble uses the 100-inch (2.5-meter) Hooker telescope at Mount Wilson Observatory, California. In 1924, Hubble announced that he had used it to see beyond the Milky Way.

An image taken by the Hubble Space Telescope shows RS Puppis, a Cepheid variable with a phenomenon known as a light echo—when light from the star is reflected from a cosmic dust cloud.

Sciences in the US held a debate on the nature and extent of the universe. American astronomers Heber Curtis and Harlow Shapley argued for and against the existence of other galaxies in what was known as "the Great Debate" or the "Shapley–Curtis Debate." Shapley asserted that the Milky Way was a single galaxy and that the "spiral nebulae" (spiral-shaped concentrations of stars) such as Andromeda were gas clouds within our own galaxy, which he believed was much larger than most astronomers thought. He argued that if Andromeda was a separate galaxy, its distance from Earth would be too great to be acceptable. He also contended that the sun was not at the center of the Milky Way, but on the outer edges of the Milky Way.

Curtis argued that spiral nebulae were separate galaxies. One of his arguments was that exploding stars (known as novae) in these nebulae looked similar to those in our own galaxy, and there were a much larger number in some locations—such as

Andromeda—than in others. This suggested that more stars existed and hinted at the presence of separate galaxies, which he called "island universes." At the same time, Curtis placed the sun at the center of our own galaxy.

The Great Debate is a testament to how incomplete our understanding of the universe was just 100 years ago, as is Curtis's incorrect theory about the sun's position in the Milky Way. (Shapley's theory that the sun is on the outer edges of the Milky Way is close to being correct.)

Hubble breakthrough

In 1924, American astronomer Edwin Hubble settled the Shapley-Curtis Debate by taking some measurements using Cepheid variables. During the debate, Shapley had surmised that the Milky Way was about 300,000 light years across, ten times larger than Curtis's estimated 30,000 light years. Hubble, using Cepheid variables that he had located in Andromeda, calculated that Andromeda was 900,000 light years away (now revised to 2.5 million light years)—far beyond Curtis

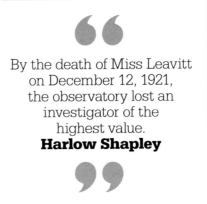

> By the death of Miss Leavitt on December 12, 1921, the observatory lost an investigator of the highest value.
> **Harlow Shapley**

and Shapley's estimates for the perceived size of the Milky Way. Curtis was therefore correct. Hubble's calculations proved that not only was the Milky Way much larger than thought, but Andromeda was not a nebula at all, but another galaxy—the Andromeda Galaxy. It was the first discovery of a known galaxy outside our own.

Expansion and collision

Hubble's discovery, made possible by the work of Leavitt, ultimately led to the knowledge that the Milky Way is just one of many

galaxies, each containing millions to hundreds of billions of stars. It is currently estimated that there are about two trillion galaxies in the universe, and that the first was created just a few hundred million years after the Big Bang, which occurred 13.8 billion years ago. The age of our own Milky Way is thought to be almost as ancient, at around 13.6 billion years old.

In 1929, Hubble also discovered that the universe appeared to be expanding, and that almost all galaxies are moving away from each other, with increasing velocity as the distances become greater. One notable exception is Andromeda, which is now known to be moving on a collision course toward our galaxy—at a rate of 68 miles (110 km) per second. That is relatively slowly compared to its distance, but it means that in about 4.5 billion years, the Milky Way and Andromeda will collide and form a single galaxy, which astrophysicists have nicknamed Milkomeda. This event will not be too dramatic, however: even though the two galaxies will merge, it is unlikely that any two stars will collide. ■

Henrietta Swan Leavitt

Born in Lancaster, Massachusetts, in 1868, Leavitt studied at Oberlin College, Ohio, before attending the Society for the Collegiate Instruction of Women (now called Radcliffe College) in Cambridge, Massachusetts. She developed an interest in astronomy after taking a course in the subject. Around the same time, an illness led to hearing loss, which grew worse over the course of her life.

After graduating in 1892, Leavitt worked at the Harvard Observatory, then led by American astronomer Edward Charles Pickering. She joined a

group of women known as the "Harvard computers" whose role was to study photographic plates of stars. Initially, she was unpaid, but she later earned 30 cents an hour. As part of her work she found 2,400 variable stars, and in so doing discovered Cepheid variables. Leavitt died of cancer in 1921.

Key works

1908 *1777 variables in the Magellanic Clouds*
1912 *Periods of 25 Variable Stars in the Small Magellanic Cloud*

THE FUTURE OF THE UNIVERSE
THE STATIC OR EXPANDING UNIVERSE

IN CONTEXT

KEY FIGURE
Alexander Friedmann
(1888–1925)

BEFORE
1917 Albert Einstein introduces the cosmological constant to suggest the universe is static.

AFTER
1929 American astronomer Edwin Hubble proves the universe is expanding—he finds that distant galaxies are moving more rapidly than those closer to us.

1931 Einstein accepts the theory of an expanding universe.

1998 Scientists from two independent projects discover the universe's expansion is accelerating.

2013 Dark energy is calculated to make up 68 percent of the universe, and is closely linked to the cosmological constant.

When Albert Einstein came up with his theory of general relativity in 1915, there was one problem. He was adamant that the universe was static and everlasting, but according to his calculations, this should mean that the universe would eventually collapse on itself due to gravity. To get around this problem, Einstein proposed the idea of a cosmological constant, represented by the Greek letter lambda (Λ). This was a measure of the "vacuum energy" in space.

Einstein's general theory of relativity laid out a set of "field equations" that showed how the curvature of spacetime was related to the mass and energy moving through it. But when he applied these equations to the universe, he found it should be either expanding or contracting—neither of which he believed could be true.

Instead, Einstein believed the universe was everlasting, and added in his "cosmological member" (now known as the cosmological constant). This allowed for a universe that could overcome the effects of gravity, and, instead of collapsing on itself, would remain static.

In 1922, Russian mathematician Alexander Friedmann came to a different conclusion. He showed that the universe was homogeneous:

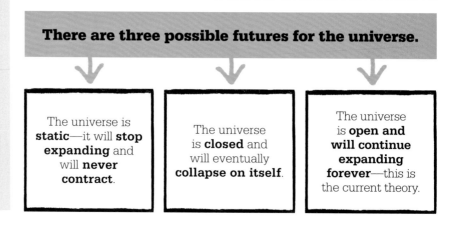

There are three possible futures for the universe.

The universe is **static**—it will **stop expanding** and will **never contract**.

The universe is **closed** and will eventually **collapse on itself**.

The universe is **open and will continue expanding forever**—this is the current theory.

See also: From classical to special relativity 274 ▪ Curving spacetime 280 ▪ Mass and energy 284–285 ▪ The Big Bang 296–301 ▪ Dark energy 306–307

The shape of the universe

If the density of the universe is exactly the same as a critical value, it is "flat." In a flat universe, parallel lines never meet. The 2-D analogy of this model is a plane (a flat surface).

If the universe is denser than a critical value, it is positively curved or "closed" and is finite in mass and extent. The 2-D analogy is a spherical surface where parallel lines converge.

If the universe is less dense than a critical value, it is negatively curved or "open" and therefore infinite. The 2-D analogy is a saddle-shaped surface where parallel lines diverge.

it was identical wherever you were in the universe and whichever way you looked, so the universe could not be static. All galaxies were moving apart from each other, but depending on which galaxy you were in, all other galaxies would also be moving away from you. Therefore you might think you were at the center of the universe—but any other observer in another galaxy would believe the same to be true of their position.

Models for the universe
As a result of his work, Friedmann came up with three models for the universe, varying according to the value of the cosmological constant. One model was that gravity would cause the expansion to slow, ultimately reversing and ending in a "Big Crunch." The second was that the universe would eventually become static once expansion stopped. A third model was that the expansion would continue forever at a faster and faster rate.

Although Einstein initially derided Friedmann's ideas in 1923, by the following year he had accepted them. However, it would be 1931 before he truly accepted that the universe was expanding—two years after American astronomer Edwin Hubble had produced evidence that the universe was indeed expanding. Hubble had noted the stretched light—or redshift—of distant galaxies, which could be used to measure the distance of faraway objects. He found that galaxies that were more distant were moving faster than closer galaxies, proof that the universe itself was expanding.

Following Hubble's discovery, the idea of a cosmological constant was deemed to be an error on Einstein's part. However, in 1998, scientists discovered that the universe was expanding at an accelerated rate. The cosmological constant would become crucial to understanding dark energy and lead to the term's reintroduction. ■

Alexander Friedmann

Alexander Friedmann was born in St. Petersburg, Russia, in 1888. His father was a ballet dancer and his mother a pianist. In 1906, Friedmann was admitted to the St. Petersburg State University to study mathematics. Here he also applied himself to studying quantum theory and relativity and earned his master's degree in pure and applied mathematics in 1914.

In 1920, after serving as an aviator and instructor in World War I, Friedmann began research based on Einstein's general theory of relativity. He used Einstein's field equations to develop his own idea of a dynamic universe, countering Einstein's belief that the universe was static. In 1925, Friedmann was appointed director of Main Geophysical Observatory in Leningrad, but died later that year, at the age of just 37, from typhoid fever.

Key works

1922 "On the Curvature of Space"
1924 "On the Possibility of a World with Constant Negative Curvature of Space"

THE COSMIC EGG, EXPLODING AT THE MOMENT OF CREATION

THE BIG BANG

IN CONTEXT

KEY FIGURE
Georges Lemaître
(1894–1966)

BEFORE
1610 Johannes Kepler surmises that the universe is finite, because the night sky is dark and not lit by an infinite number of stars.

1687 Isaac Newton's laws of motion explain how things move through the universe.

1929 Edwin Hubble discovers that all galaxies are moving away from each other.

AFTER
1998 Leading astronomers announce that the expansion of the universe is accelerating.

2003 NASA's WMAP probe finds evidence for "inflation," or the burst of expansion just after the Big Bang, by mapping out tiny temperature fluctuations across the sky.

Lemaître uses mathematics to show that Einstein's theory of general relativity means **the universe must be expanding**.

→

Hubble provides experimental evidence that **galaxies are moving apart**, and that those most distant are moving the fastest.

↓

The universe must have **started from a single point**—Lemaître's primeval atom.

←

Cosmic microwave background radiation (CMBR) shows the **residual heat** left over after the Big Bang, suggesting it really did occur.

For more than two millennia, humanity's greatest minds have pondered the origins of the universe and our place within it. For centuries, many believed that some sort of deity created the universe, and Earth was at its center, with all the stars traveling around it. Few suspected that Earth was not even the center of its own solar system, which itself orbits one of hundreds of billions of stars in the universe. Today's theory about the origins of the cosmos dates from the early 1930s, when Belgian astronomer Georges Lemaître first suggested what is now known as the Big Bang.

Lemaître's cosmic egg

In 1927, Lemaître had proposed that the universe was expanding; four years later he developed the idea, explaining that the expansion had

---

Georges Lemaître

Born in 1894 in Charleroi, Belgium, Georges Lemaître served in World War I as an artillery officer, then entered a seminary and was ordained as a priest in 1923. He studied in the UK at the University of Cambridge's solar physics laboratory for a year, and in 1924 joined the Massachusetts Institute of Technology in the US. By 1928, he was a professor of astrophysics at the Catholic University of Leuven, in Belgium.

Lemaître was familiar with the works of Edwin Hubble and others who discussed an expanding universe, and published work developing this idea in 1927. The 1931 theory that Lemaître is best known for—the idea that the universe had expanded from a single point—was initially dismissed, but was finally proven to be correct shortly before his death in 1966.

Key works

1927 "Discussion on the evolution of the universe"
1931 "The primeval atom hypothesis"

See also: The Doppler effect and redshift 188–191 ▪ Seeing beyond light 202–203 ▪ Antimatter 246 ▪ Particle accelerators 252–255 ▪ Matter–antimatter asymmetry 264 ▪ The static or expanding universe 294–295 ▪ Dark energy 306–307

begun a finite amount of time ago at a single point from what he called the "primeval atom" or "cosmic egg." Lemaître, a priest, did not think this idea was at odds with his faith; he declared an equal interest in seeking truth from the point of view of religion and from that of scientific certainty. He had partly derived his theory of an expanding universe from Einstein's theory of general relativity, but Einstein dismissed the idea of expansion or contraction for lack of evidence of large-scale motion. He had earlier added a term called the "cosmological constant" to his field equations for general relativity to ensure that they allowed for a static universe.

In 1929, however, American astronomer Edwin Hubble made a discovery that supported the idea of an expanding universe. By observing the change in light as objects moved away from Earth—known as redshift—Hubble could calculate how fast a galaxy moves away and thus its distance. All galaxies appeared to move away from Earth, with those furthest away moving the fastest. Lemaître, it seemed, was on the right track.

The steady-state model
Despite such evidence, Hubble and Lemaître still faced stiff competition from the steady-state

The best proof of the Big Bang is cosmic microwave background radiation (CMBR), seen here in an all-sky image of the minute fluctuations in the temperature of CMBR, taken by NASA's WMAP satellite between 2003 and 2006. The variations from dark blue (cool) to red (hot) correspond to changes in density in the early universe.

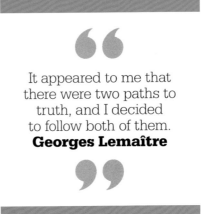

> It appeared to me that there were two paths to truth, and I decided to follow both of them.
> **Georges Lemaître**

theory. In this model, the universe had always existed. Matter formed continually in the space between other galaxies as they drifted apart, and the continuous creation of matter and energy—at a rate of one particle of hydrogen per cubic meter every 300,000 years—kept the universe in balance. The hydrogen would form into stars and thus give rise to heavier elements, and then planets, more stars, and galaxies. The idea was championed by British astronomer Fred Hoyle

who, on the radio in 1949, derided the competing theory of Hubble and Lemaître as being a "big bang." The catchy name stuck to describe the ideas widely accepted today.

Residual heat
A year earlier, Ukrainian physicist George Gamow and American cosmologist Ralph Alpher had published "The origin of chemical elements" to explain conditions immediately after an exploding primeval atom and the distribution of particles through the universe. The paper accurately predicted cosmic microwave background radiation (CMBR)—the residual heat left over from the Big Bang. In 1964, the ideas received a huge boost when Arno Penzias and Robert Wilson detected CMBR by chance while attempting to use a large antenna to conduct radio astronomy.

The presence of CMBR in the universe all but ruled out the steady-state theory. It pointed to a much hotter period in the universe's history, where matter was clumping together to form »

galaxies, suggesting that the universe had not always been the same. Its evident expansion and the fact that galaxies were once much closer together posed a problem for steady-state theorists, who believed that the density of matter in the universe is constant and unvarying over distance and time. There were later attempts to reconcile steady-state theory with CMBR and other discoveries, but to little avail. The Big Bang theory is now the preeminent explanation for how our universe began.

The universe in infancy

Finding CMBR proved crucial because it helped develop a picture of how the universe probably evolved, although Big Bang theory cannot describe the exact moment of its creation 13.8 billion years ago or what came before it (if anything).

American physicist Alan Guth was among the cosmologists who developed Big Bang theory in the latter parts of the 20th century. In 1980, he proposed that cosmic "inflation" occurred a tiny fraction of a second (10^{-35} seconds, roughly a trillionth of a trillionth of a trillionth of a second) into the life of the universe. From its initial, infinitely hot and dense point of "singularity," the universe began

The Big Bang was not an explosion *in* space; it was more like an explosion *of* space.
Tamara Davis
Australian astrophysicist

Big Bang theory holds that the universe evolved from an infinitely dense and hot primordial "singularity" (Lemaître's "primeval atom"), which rapidly expanded, giving off vast amounts of heat and radiation.

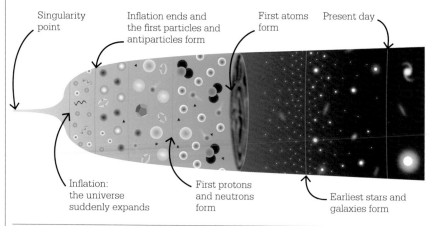

Singularity point

Inflation ends and the first particles and antiparticles form

First atoms form

Present day

Inflation: the universe suddenly expands

First protons and neutrons form

Earliest stars and galaxies form

to expand faster than the speed of light. Guth's theory, though still unproven, helps explain why the universe cooled and also why it appears to be uniform, with matter and energy evenly distributed.

Immediately after the Big Bang, physicists believe the universe was pure energy and the four fundamental forces (gravity, the electromagnetic force, the strong force, and the weak interaction) were unified. Gravity split away, and matter and energy were in an interchangeable "mass–energy" state. At the start of inflation, the strong nuclear force broke away, and a huge amount of mass–energy formed. Photons—particles of light primed with electromagnetic energy—dominated the universe. Toward the end of inflation, still trillionths of a second after the Big Bang, a hot quark–gluon plasma emerged—a sea of particles and antiparticles that continually swapped mass for energy in collisions of matter and antimatter. For reasons still unknown, this process created more matter than antimatter, and matter became the main component of the universe.

Still fractions of a second after the Big Bang, the electromagnetic force and the weak interaction separated and the universe cooled enough for quarks and gluons to bind together to form composite particles—protons, neutrons, antiprotons, and antineutrons. Within three minutes, proton–neutron collisions produced the first atomic nuclei, some fusing into helium and lithium. Neutrons were absorbed in these reactions, but many free protons remained.

Opaque to transparent

The early universe was opaque and remained so for a few hundred thousand years. Hydrogen nuclei made up almost three quarters of its mass; the rest was helium nuclei with trace amounts of lithium and deuterium nuclei. Around 380,000 years after the Big Bang, the universe had cooled and expanded enough for nuclei to capture free electrons and form the first hydrogen, helium, deuterium, and lithium atoms. Freed from their interaction with free electrons and nuclei, photons could move freely through space as radiation, and the universe left its dark ages and became transparent. CMBR is the residual light from this period.

Stars and galaxies evolve

Astronomers now believe that the first stars formed hundreds of millions of years after the Big Bang. As the universe became transparent, dense clumps of neutral hydrogen gas formed and grew under the force of gravity as more matter was pulled in from areas of lower density. When the clumps of gas reached a high enough temperature for nuclear fusion to occur, the earliest stars appeared.

Physicists modeling these stars believe they were so huge, hot, and luminous—30 to 300 times as big as the sun and millions of times brighter—that they prompted fundamental changes in the universe. Their ultraviolet light re-ionized hydrogen atoms back into electrons and protons, and when these relatively short-lived stars exploded into supernovae, after around a million years, they created new heavier elements, such as uranium and gold. About one billion years after the Big Bang,

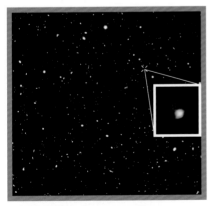

The furthest known galaxy, as seen by the Hubble space telescope, was probably formed about 400 million years after the Big Bang. It is shown here as it appeared 13.4 billion years ago.

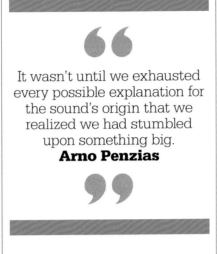

> It wasn't until we exhausted every possible explanation for the sound's origin that we realized we had stumbled upon something big.
> **Arno Penzias**

the next generation of stars, which contained heavier elements and were longer-lived, grouped together due to gravity, forming the first galaxies. These galaxies began to grow and evolve, some colliding with each other to create more and more stars within them of all shapes and sizes. The galaxies then drifted further apart as the universe expanded over the next few billion years. There were fewer collisions, and the universe became relatively more stable, as it is today. Filled with trillions of galaxies and spanning billions of light years, it is still expanding. Some scientists believe that it will expand forever, until everything spreads out into nothingness.

An observable timeline

Big Bang theory has enabled scientists to get a firm grasp on the origins of our universe, providing a timeline that stretches back 13.8 billion years to its very first moments. Crucially, most of its predictions are testable. Physicists can recreate the conditions after the Big Bang took place, while CMBR offers direct observation of an era that started when the universe was a mere 380,000 years old. ∎

Cosmic microwave background "noise"

In 1964, American astronomers Arno Penzias (pictured below, right) and Robert Wilson (left) were working at the Holmdel Horn Antenna at Bell Telephone Laboratories in New Jersey. The large horn-shaped telescope was designed to pick up incredibly sensitive detections of radio waves.

The two astronomers were looking for neutral hydrogen (HI)—atomic hydrogen with one proton and one electron, which is abundant in the universe but rare on Earth—but were hampered by a strange background noise. Wherever they pointed the antenna, the universe sent back the equivalent of TV static. Having first thought that birds or faulty wiring could be the cause, they then consulted other astronomers. They realized they were picking up cosmic microwave background radiation (CMBR), residual heat from the Big Bang, first predicted in 1948. The momentous discovery earned the pair the Nobel Prize in Physics in 1978.

VISIBLE MATTER ALONE IS NOT ENOUGH

DARK MATTER

IN CONTEXT

KEY FIGURES
Fritz Zwicky (1898–1974),
Vera Rubin (1928–2016)

BEFORE
17th century Isaac Newton's
theory of gravity leads some
to wonder if there are dark
objects in the universe.

1919 British astronomer
Arthur Eddington proves that
massive objects can warp
spacetime and bend light.

1919 Fritz Zwicky, a maverick
Swiss astronomer, puts
forward the existence of dark
matter for the first time.

AFTER
1980s Astronomers identify
more galaxies believed to be
full of dark matter.

2019 The search for dark
matter continues, with no
definitive results so far.

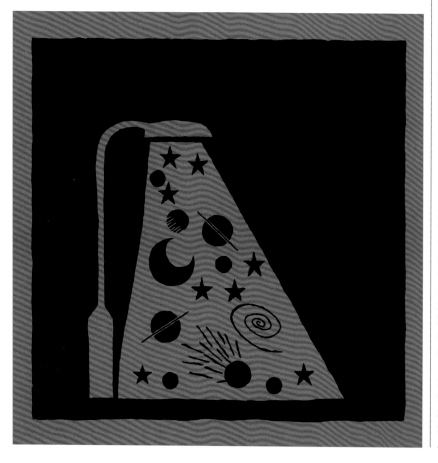

The universe as it seems to be does not make sense. Looking at all the visible matter, galaxies should not exist— there simply is not enough gravity to hold them all together. But there are trillions of galaxies in the universe, so how can this be? This question has plagued astronomers for decades, and the solution is no less vexing—matter that cannot be seen or detected, better known as dark matter.

The idea of invisible matter in the universe stretches back to the 17th century, when English scientist Isaac Newton first put forward his theory of gravity. Astronomers around this time

> There has to be a lot of mass to make the stars orbit so rapidly, but we can't see it. We call this invisible mass dark matter.
> **Vera Rubin**

began to wonder if there could be dark objects in the universe that reflected no light, but could be detected by their gravitational effects. This gave rise to the idea of black holes, and the idea of a dark nebula that absorbed rather than reflected light was proposed in the 19th century.

It would take until the following century, however, for a greater shift in understanding how things worked. As astronomers began to study more and more galaxies, they started to have more and more questions about how the galaxies were able to exist in the first place. Ever since, the hunt has been on for mysterious dark matter and dark energy (invisible energy driving the expansion of the universe)—and astronomers today are getting close.

Invisible universe

Albert Einstein's general theory of relativity was key to understanding gravity and ultimately dark matter. It suggested that light itself could be bent by the gravitational mass of large objects, with such objects actually warping spacetime.

In 1919, Arthur Eddington set out to prove Einstein's theory. He had measured the positions of stars earlier that year, before traveling to the island of Príncipe, off the west coast of Africa, to view the same stars during an eclipse. He was able to work out that the positions of the stars had changed slightly as their light bent around the large mass of the sun—an effect now known as gravitational lensing.

More than a decade later, in 1933, Fritz Zwicky made a startling discovery. While studying a cluster of galaxies called the Coma Cluster, he calculated that the mass of the galaxies in the cluster must be far greater than the observable matter in the form of stars in order to hold the cluster together. This led him to reason there was unseen matter, or *dunkle materie* ("dark matter"), holding the cluster together.

Other astronomers began to apply the same methods to other galaxies and clusters, and reached the same conclusion. There simply was not enough matter to hold everything together, so either the laws of gravity were wrong, or there was something going on that they just could not see. It could not be something like a dark nebula, which can be seen by the light it absorbs. Rather, it had to be something else entirely.

Galactic spin

American astronomer Vera Rubin was able to shed light on the problem. In the late 1970s, while working at the Carnegie Institution of Washington, she and her fellow American colleague Kent Ford were bemused to discover that the Andromeda Galaxy was not rotating as it should be. From »

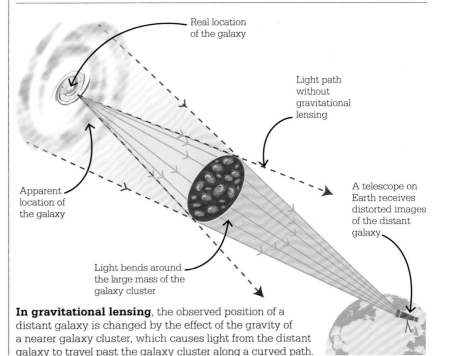

In gravitational lensing, the observed position of a distant galaxy is changed by the effect of the gravity of a nearer galaxy cluster, which causes light from the distant galaxy to travel past the galaxy cluster along a curved path.

Real location of the galaxy

Light path without gravitational lensing

A telescope on Earth receives distorted images of the distant galaxy

Apparent location of the galaxy

Light bends around the large mass of the galaxy cluster

> In a spiral galaxy, the ratio of dark-to-light matter is about a factor of ten. That's probably a good number for the ratio of our ignorance-to-knowledge.
> **Vera Rubin**

their observations, they found that the edges of the galaxy were moving at the same speed as the center of the galaxy. Imagine an ice skater spinning with her arms outstretched—her hands are moving faster than her body. If the ice skater then pulls in her arms, she spins faster as her mass moves toward her center. But this was not the case with the Andromeda Galaxy.

At first the two astronomers did not realize the implications of what they were seeing. Gradually, however, Rubin began to work out that the mass of the galaxy

was not concentrated at its center, but was spread across the galaxy. This would explain why the orbital speed was similar throughout the galaxy—and the best way to explain this observation was if there were a halo of dark matter surrounding the galaxy and holding it together. Rubin and Ford had, indirectly, found the first evidence for an invisible universe.

Hide and seek

Following Rubin and Ford's discovery, astronomers started to appreciate the scale of what they were witnessing. In the 1980s, building on Eddington's work showing that large masses could bend spacetime, many instances of gravitational lensing caused by dark matter were spotted. From these, astronomers calculated that a vast 85 percent of the mass in the Universe was dark matter.

In the 1990s, astronomers noted something unusual about the expansion of the universe: it was accelerating. Gravity, as they understood it, should have meant that the expansion would at some point slow down. In order to explain this observation, astronomers came up with something called "dark

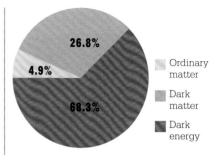

Visible matter—the atoms that make up stars and planets—represents a tiny proportion of the universe. Most of the universe's energy density is composed of invisible dark matter and dark energy.

energy," which makes up about 68 percent of the mass-energy content of the universe.

Together, dark matter and dark energy make up 95 percent of the known universe, with visible matter—the things that can be seen—accounting for just 5 percent. With so much dark matter and dark energy present, it ought to be easy to find. They are called "dark" for a reason, however—no direct evidence has yet been found to confirm that either dark matter or dark energy exists.

Astronomers are fairly certain that dark matter is some sort of particle. They know it interacts with gravity because they can see its effect on galaxies on a huge scale. But strangely, it appears to have no interactions with regular matter; if it had, it would be possible to see these interactions taking place everywhere. Instead, astronomers think that dark matter passes straight through ordinary matter, making it incredibly difficult to detect.

That hasn't stopped scientists from trying to detect dark matter, however, and they believe they are getting closer. One of the candidate particles for dark matter is called a weakly interacting massive particle

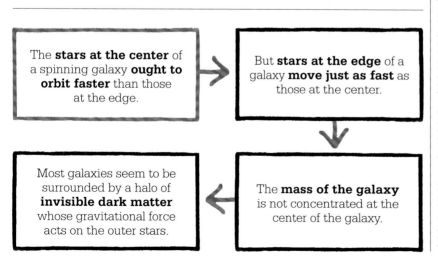

The **stars at the center** of a spinning galaxy **ought to orbit faster** than those at the edge.

But **stars at the edge** of a galaxy **move just as fast** as those at the center.

The **mass of the galaxy** is not concentrated at the center of the galaxy.

Most galaxies seem to be surrounded by a halo of **invisible dark matter** whose gravitational force acts on the outer stars.

(WIMP), and while such particles are incredibly difficult to find, it is theoretically not impossible that they exist.

The great unknown

In their search for dark matter, scientists have tried building vast detectors, filling them with liquid, and burying them underground. The idea is that if a dark matter particle passed through one of these detectors, as seems likely to be happening all the time, it would leave a noticeable trace. Examples of such efforts include the Gran Sasso National Laboratory in Italy, and the LZ Dark Matter Experiment in the US.

Physicists at CERN's Large Hadron Collider (LHC) near Geneva, Switzerland, have also tried to find dark matter, looking for any dark matter particles that might be produced when other particles are smashed together at very high speeds. So far, however, no such evidence has been found, although it is hoped that future upgrades to the LHC may result in success.

Today the hunt for dark matter continues in earnest, and while astronomers are fairly certain it is there, much remains unknown.

It is not clear if dark matter is a single particle or many, nor if it experiences forces like regular matter does. It may also be composed of a different particle called an axion, which is much lighter than a WIMP.

Ultimately, there is still much to learn about dark matter, but it has clearly had a major impact on astronomy. Its indirect discovery has led astronomers to reason that there is a vast, unseen universe that simply cannot be measured at the moment, and while that might seem daunting, it is fascinating, too. The hope in the coming years is that dark matter particles will finally be detected. Then, in a similar way to how the detection of gravitational waves has changed astronomy, astronomers will be able to probe dark matter and find out what it really is, and understand with certainty exactly what effect it has on the universe. Until then, everyone is in the dark. ■

While mapping the Andromeda Galaxy's mass, Vera Rubin and Kent Ford realized that the mass was spread across the galaxy, and so must be held together by a halo of invisible matter.

Vera Rubin

Vera Rubin was born in Philadelphia, Pennsylvania, in 1928. She developed an interest in astronomy at an early age and pursued science at college, despite being told by her high school teacher to choose a different career. She was rejected by the Princeton University astrophysics graduate program as it did not admit women, and instead enrolled at Cornell University.

In 1965, Rubin joined the Carnegie Institution of Washington, where she opted for the uncontroversial field of mapping the mass of galaxies. She won numerous prizes for the discovery that mass was not concentrated at a galaxy's center, hinting at the existence of dark matter. The first woman to be allowed to observe at the Palomar Observatory, Rubin was an ardent champion of women in science. She died in 2016.

Key works

1997 *Bright Galaxies, Dark Matters*
2006 "Seeing Dark Matter in the Andromeda Galaxy"

AN UNKNOWN INGREDIENT DOMINATES THE UNIVERSE
DARK ENERGY

IN CONTEXT

KEY FIGURE
Saul Perlmutter (1959–)

BEFORE
1917 Albert Einstein proposes a cosmological constant to counteract gravity and keep the universe static.

1929 Edwin Hubble finds proof that the universe is expanding.

1931 Einstein calls the cosmological constant his "greatest blunder."

AFTER
2001 Results show that dark energy probably makes up a large portion of the energy–mass content of the universe.

2011 Astronomers find indirect evidence for dark energy in the cosmic microwave background (CMB).

2013 Models of dark energy are refined, showing it to be very similar to Einstein's predicted cosmological constant.

Until the early 1990s, no one was really sure what the fate of the universe would be. Some thought it would expand forever, others that it would become static, and others still that it would collapse into itself. But in 1998, two teams of American astrophysicists—one led by Saul Perlmutter, and the other by Brian Schmidt and Adam Riess—set out to measure the rate of expansion of the universe. Using powerful telescopes, they observed very distant Type 1a supernovae (see flowchart, below). To their surprise, they saw that the supernovae were fainter than they had expected, with a more reddish hue, and so must be further away. The two teams reached the same conclusion:

If a white dwarf (the remnant core of a star) orbits a giant star, accreting stellar material, it can cause a **Type 1a supernova**.

A Type 1a supernova has a **known brightness**, so its apparent magnitude (brightness as seen from Earth) shows how **far away** it is.

By measuring the brightness and redshift of each supernova, its **distance** and **speed relative to Earth** can be calculated.

The light from distant supernovae has **taken longer** than expected to reach Earth, so **cosmic expansion** must be **accelerating**.

There is an **invisible force** at work, pushing the universe apart.

See also: The Doppler effect and redshift 188–191 ▪ From classical to special relativity 274 ▪ Mass and energy 284–285 ▪ The static or expanding universe 294–295 ▪ The Big Bang 296–301 ▪ Dark matter 302–305 ▪ String theory 308–311

Saul Perlmutter

Saul Perlmutter was born on September 22, 1959, in Illinois, and grew up near Philadelphia, Pennsylvania. He earned a bachelor's degree in physics from Harvard University in 1981, and a doctorate in physics from the University of California, Berkeley, in 1986. He was made a professor of physics there in 2004.

In the early 1990s, Perlmutter became intrigued by the idea that supernovae could be used as standard candles (objects of known brightness that can be used to measure distances across space) to measure the expansion of the universe. He had to work hard to earn time on large telescopes, but his efforts paid off, and Perlmutter was awarded the Nobel Prize in Physics in 2011, alongside Brian Schmidt and Adam Riess.

Key works

1997 "Discovery of a Supernova Explosion at Half the Age of the Universe and its Cosmological Implications"
2002 "The Distant Type Ia Supernova Rate"

the supernovae were moving at a faster rate than would be expected if gravity were the only force acting on them, so cosmic expansion must be accelerating over time.

Mysterious force

This discovery went against the idea that gravity should eventually pull everything together again. It became apparent that the overall energy content of the universe must be dominated by something else entirely—a constant, invisible force that works in the opposite way to

> "If you're puzzled about what dark energy is, you're in good company."
> **Saul Perlmutter**

gravity and pushes matter away. This mysterious force was named "dark energy." If there were such an energy field pervading the universe, Perlmutter, Schmidt, and Riess thought that it could explain the expansion.

Albert Einstein had come up with a similar concept in 1917. His cosmological constant was a value that counteracted gravity and allowed the universe to remain static. But when the universe was shown to be expanding, Einstein declared that the constant was a mistake and dropped it from his theory of relativity.

Today, dark energy is still thought to be the most likely cause of cosmic expansion, although it has never been observed directly. In 2011, however, while studying the remnants of the Big Bang (known as the cosmic microwave background, or CMB), scientists suggested that a lack of large-scale structure to the universe hinted at the existence of dark energy, which would act against gravity to prevent large structures of matter forming.

Astronomers now believe that dark energy makes up a huge portion of the mass–energy content of the universe—about 68 percent—which could have big implications. It is possible that the universe will continue expanding at an ever-increasing rate, until galaxies are moving apart faster than the speed of light, and eventually disappearing from view. Stars in each galaxy might then do the same, followed by planets, and then matter, leaving the universe as a dark and endless void trillions of years from now. ▪

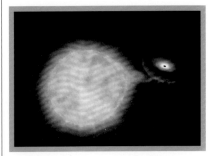

The gravity of this white dwarf is pulling material away from a nearby giant star. When its mass has reached about 1.4 times the current mass of the sun, a Type 1a supernova will occur.

THREADS IN A TAPESTRY

STRING THEORY

IN CONTEXT

KEY FIGURE
Leonard Susskind (1940–)

BEFORE
1914 The idea of a fifth dimension is touted to explain how gravity works alongside electromagnetism.

1926 Swedish physicist Oscar Klein develops ideas of extra unobservable dimensions.

1961 Scientists devise a theory to unify electromagnetism and the weak nuclear force.

AFTER
1975 Abraham Pais and Sam Treiman coin the term "Standard Model."

1995 American physicist Edward Witten develops M-theory, which includes 11 dimensions.

2012 The Large Hadron Collider detects the Higgs boson.

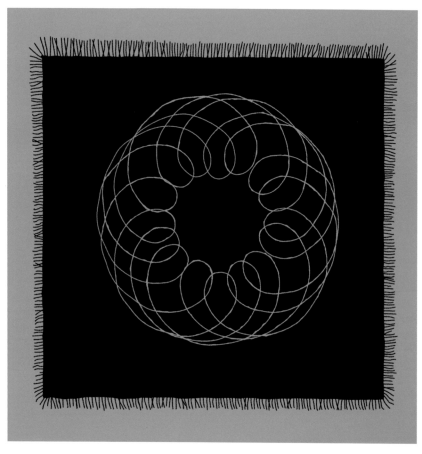

article physicists use a theory called the "Standard Model" to explain the universe. Developed in the 1960s and 1970s, this model describes the fundamental particles and forces of nature that make up everything and hold the universe together.

One problem with the Standard Model, however, is that it does not fit with Albert Einstein's theory of general relativity, which relates gravity (one of the four forces) to the structure of space and time, treating them as a four-dimensional entity called "spacetime." The Standard Model cannot be reconciled with the curvature of spacetime according to general relativity.

See also: Laws of gravity 46–51 ▪ Heisenberg's uncertainty principle 220–221 ▪ Quantum entanglement 222–223 ▪ The particle zoo and quarks 256–257 ▪ Force carriers 258–259 ▪ The Higgs boson 262–263 ▪ The equivalence principle 281

The **Standard Model of particle physics** can explain everything except gravity.

String theory presents elementary particles as **tiny strands of energy**, each with its own distinctive **vibration**.

Each of these particles, according to **supersymmetry**, has a corresponding **superpartner**.

The properties of one vibrating string correspond to those of the **graviton**, the predicted **force carrier of gravity**.

This could be the **missing link** between Einstein's **theory of relativity** and the **Standard Model**.

But **string theory cannot be tested** as the energy required to do so is higher than we can create.

Quantum mechanics, on the other hand, explains how particles interact at the smallest of levels—on an atomic scale—but cannot account for gravity. Physicists have tried to unite the two theories, but in vain—the issue remains that the Standard Model is only able to explain three of the four fundamental forces.

Particles and forces

In particle physics, atoms consist of a nucleus made up of protons and neutrons, surrounded by electrons. The electron—and the quarks that make up the protons and neutrons—are among the 12 fermions (matter particles): elementary, or fundamental, particles that are the smallest known building blocks of the universe. Fermions are subdivided into quarks and leptons. Alongside these fermions, there are bosons (force carrier particles) and four forces of nature: electromagnetism, gravity, the strong force, and the

weak force. Different bosons are responsible for carrying the different forces between fermions. The Standard Model allows physicists to describe what is known as the Higgs field—a field of energy thought to pervade the entire universe. The interaction of particles within the Higgs field gives them their mass, and a measurable boson called the Higgs boson is the force carrier for the Higgs field. But none of the

known bosons is the force carrier for gravity, leading scientists to come up with a hypothetical, yet-to-be-detected particle called the graviton.

In 1969, in an attempt to explain the nuclear force, which binds protons and neutrons within the nuclei of atoms, American physicist Leonard Susskind developed the idea of string theory. Coincidentally, American–Japanese physicist Yoichiro Nambu and Danish »

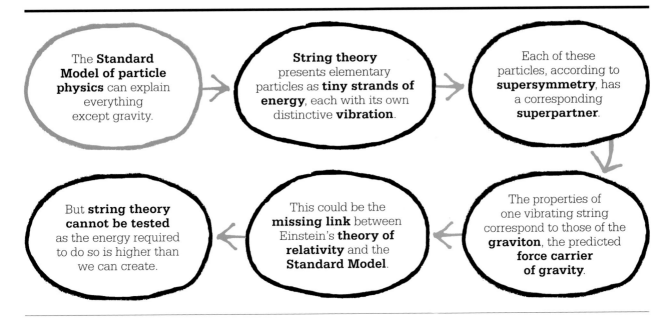

Quark

Proton

Nucleus

Atom

Neutron

Electron

Vibrating string

Each string has its own distinctive vibration

According to string theory, elementary particles—such as electrons and the quarks that make up protons and neutrons—are strings or filaments of energy. Each string vibrates at a different frequency, and the vibrations correspond to the speed, spin, and charge of the particles.

physicist Holger Nielsen independently conceived the same idea at the same time. According to string theory, particles—the building blocks of the universe—are not pointlike, but rather tiny, one-dimensional, vibrating strands of energy, or strings, which give rise to all forces and matter. When strings collide, they combine and vibrate together briefly before separating again.

Early models of string theory were problematic, however. They explained bosons but not fermions, and required certain hypothetical particles, known as tachyons, to travel faster than light. They also required many more dimensions than the familiar four of space and time.

Supersymmetry

To get around some of these early problems, scientists came up with the principle of supersymmetry. In essence, this suggests that the universe is symmetric, giving each of the known particles from the Standard Model an undetected partner, or "superpartner"—so each fermion, for example, is paired with a boson, and vice versa.

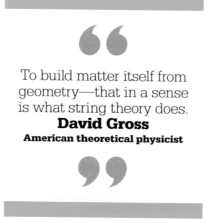

> To build matter itself from geometry—that in a sense is what string theory does.
> **David Gross**
> **American theoretical physicist**

When the Higgs boson, predicted by British physicist Peter Higgs in 1964, was eventually detected in 2012 by CERN's Large Hadron Collider, it was lighter than expected. Particle physicists had thought that its interactions with Standard Model particles in the Higgs field, giving them their mass, would make it heavy. But that was not the case. The idea of superpartners, particles that could potentially cancel out some of the effects of the Higgs field and produce a lighter Higgs boson, enabled scientists to address this problem. It also allowed them to

find that three of the four forces of nature, namely electromagnetism, the strong force, and the weak force, may have existed at the same energies at the Big Bang—a key step toward unifying these forces in a Grand Unified Theory.

Together, string theory and supersymmetry gave rise to superstring theory, in which all fermions and bosons and their superpartners are the result of vibrating strings of energy. In the 1980s, American physicist John Schwarz and British physicist Michael Green developed the idea that elementary particles such as electrons and quarks are outward manifestations of "strings" vibrating on the quantum gravity scale.

In the same way that different vibrations of a violin string produce different notes, properties such as mass result from different vibrations of the same kind of string. An electron is a piece of string that vibrates in a certain way, while a quark is an identical piece of string that vibrates in a different way.

In the course of their work, Schwarz and Green realized that string theory predicted a massless particle akin to the hypothetical

Leonard Susskind

Born in New York City in 1940, Leonard Susskind is currently the Felix Bloch Professor of Physics at Stanford University in California. He earned his PhD from Cornell University, New York, in 1965, before joining Stanford University in 1979.

In 1969, Susskind came up with the theory for which he is best known—string theory. His mathematical work showed that particle physics could be explained by vibrating strings at the smallest level. He developed his idea further in the 1970s, and in 2003, coined the phrase "string

theory landscape." This radical idea was intended to highlight the large number of universes that could possibly exist, forming a mind-boggling "megaverse"—including, perhaps, other universes with the necessary conditions for life to exist. Susskind remains highly regarded in the field today.

Key works

2005 *The Cosmic Landscape*
2008 *The Black Hole War*
2013 *The Theoretical Minimum*

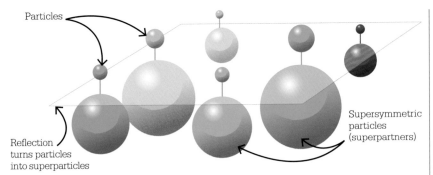

Particles

Reflection
turns particles
into superparticles

Supersymmetric
particles
(superpartners)

According to supersymmetry, every boson, or force carrier particle, has a massive "superpartner" fermion, or matter particle, and every fermion has a massive "superpartner" boson. Superstring theory describes superpartners as strings that vibrate in higher octaves, like the harmonics of a violin. Some string theorists predict that superpartners may have masses up to 1,000 times greater than that of their corresponding particles, but no supersymmetric particles have yet been found.

graviton. The existence of such a particle could explain why gravity is so weak compared with the other three forces, as gravitons would leak in and out of the 10 or so dimensions required by string theory. Here, at last, appeared to be something Einstein had long sought, a theory that could describe everything in the universe—a "Theory of Everything."

A unifying theory

Physicists hunting for an all-encompassing theory encounter problems when considering black holes, where general relativity theory meets quantum mechanics in trying to explain what happens when a vast amount of matter is packed into a very small area. Under general relativity, the core of a black hole, known as its singularity, could be said to have essentially zero size. But under quantum mechanics, that does not hold true because nothing can be infinitely small. According to the uncertainty principle, devised by German physicist Werner Heisenberg in 1927, it is simply not possible to reach infinitely small levels because a particle can always exist in multiple states.

Key quantum theories such as superposition and entanglement also dictate that particles can be in two states at once. They must produce a gravitational field, which would be consistent with general relativity, but under quantum theory that does not seem to be the case.

If superstring theory can resolve some of these problems, it might be the unifying theory physicists have been looking for. It might even be possible to test superstring theory by colliding particles together. At higher energies, scientists think they could potentially see gravitons dissipating into other dimensions, providing a key piece of evidence for the theory. But not everyone is convinced.

Unraveling the idea

Some scientists, such as American physicist Sheldon Glashow, believe the pursuit of string theory is futile because no one will ever be able to prove whether or not the strings it describes truly exist. They involve energies so high (beyond the measurement called Planck energy) that they are impossible for humans to detect, and may remain so for the foreseeable future. Being

unable to devise an experiment that can test string theory leads scientists like Glashow to question whether it belongs in science at all. Others disagree, and note that experiments are currently underway to try to look for some of these effects and provide an answer. The Super-Kamiokande experiment in Japan, for example, could test aspects of string theory by looking for proton decay—the hypothesized decay of a proton over extremely long timescales—which is predicted by supersymmetry.

Superstring theory can explain much of the unknown universe, such as why the Higgs boson is so light and why gravity is so weak, and it may help shed light on the nature of dark energy and dark matter. Some scientists even think that string theory could provide information about the fate of the universe, and whether or not it will continue to expand indefinitely. ∎

The walls of the Super-Kamiokande neutrino observatory in Japan are lined with photomultipliers that detect light emitted by neutrinos interacting with the water inside the tank.

RIPPLES IN SPACETIME
GRAVITATIONAL WAVES

IN CONTEXT

KEY FIGURES
Barry Barish (1936–),
Kip Thorne (1940–)

BEFORE
1915 Einstein's theory of
general relativity provides some
evidence for the existence of
gravitational waves.

1974 Scientists indirectly
observe gravitational waves
while studying pulsar stars.

1984 In the US, the Laser
Interferometer Gravitational-
Wave Observatory Project
(LIGO) is set up to detect
gravitational waves.

AFTER
2017 Scientists detect
gravitational waves from
two merging neutron stars.

2034 The LISA gravitational
wave mission is expected to
launch into space in 2034 to
study gravitational waves.

In 2016, scientists made an
announcement that promised
to revolutionize astronomy:
in September 2015, a team of
physicists had found the first direct
evidence for gravitational waves,
ripples in spacetime caused by the
merging or collision of two objects.
Up to this point, knowledge of the
universe and how it works was
derived mainly through what could
be seen, in the form of light waves.
Now, scientists had a new way to
probe black holes, stars, and other
wonders of the cosmos.

The idea of gravitational waves
had already existed for more than a
century. In 1905, French physicist
Henri Poincaré initially postulated

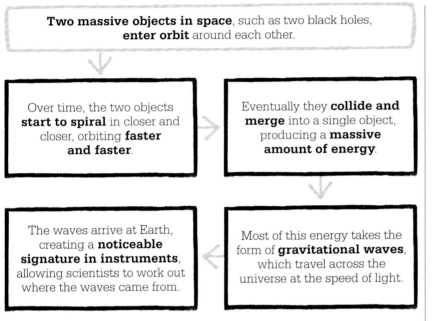

Two massive objects in space, such as two black holes, **enter orbit** around each other.

Over time, the two objects **start to spiral** in closer and closer, orbiting **faster and faster**.

Eventually they **collide and merge** into a single object, producing a **massive amount of energy**.

Most of this energy takes the form of **gravitational waves**, which travel across the universe at the speed of light.

The waves arrive at Earth, creating a **noticeable signature in instruments**, allowing scientists to work out where the waves came from.

---

the theory that gravity was transmitted on a wave, which he called *l'onde gravifique*—the gravitational wave. A decade later, Albert Einstein took this idea to another level in his theory of general relativity, where he proposed that gravity was not a force but rather a curvature of spacetime, caused by mass, energy, and momentum.

Einstein showed that any massive object causes spacetime to bend, which can in turn bend light itself. Whenever a mass moves and changes this distortion, it produces waves that travel out from the mass at the speed of light. Any moving mass produces these waves, even everyday things such as two people spinning in a circle, although such waves are too small to be detected.

Despite his theories, Einstein wrestled with his own belief in gravitational waves, noting in 1916 that "there are no gravitational waves analogous to light waves."

He returned to the idea in a new paper on gravitational waves in 1918, suggesting they might exist but that there was no way to ever measure them. By 1936 he had swung back again and stated the waves did not exist at all.

Searching for the unseeable
It was not until the 1950s that physicists started to realize that gravitational waves might well be real. A series of papers underlined that general relativity had indeed predicted the existence of the waves, as a method of transferring energy via gravitational radiation.

The detection of gravitational waves, however, posed huge challenges for scientists. They were fairly certain that the waves were present throughout the universe, but they needed to devise an experiment sensitive enough to detect them. In 1956, British physicist Felix Pirani showed that

gravitational waves would move particles as they passed by, which, in theory, could be detectable. Scientists then set about devising experiments that might be able to measure these disturbances.

The earliest efforts at detection were unsuccessful, but in 1974, American physicists Russell Hulse and Joseph Taylor found the first indirect evidence for gravitational waves. They observed a pair of orbiting neutron stars (small stars created by the collapse of giant stars), one of which was a pulsar (a rapidly spinning neutron star). As they spun around each other and drew closer, the stars were losing energy consistent with the radiation of gravitational waves. The discovery took scientists one step closer to proving the existence of gravitational waves.

The wave machines
Although such large and fast-moving masses as neutron stars orbiting each other produced potentially the largest gravitational waves, those waves would produce an incredibly small effect. But from the late 1970s, a number of physicists, including American–German Rainer Weiss »

The wave is assumed to propagate with the speed of light.
Henri Poincaré

and Ronald Drever, from Scotland, began to suggest that it might be possible to detect the waves by using laser beams and an instrument called an interferometer. In 1984, American physicist Kip Thorne joined with Weiss and Drever to establish the Laser Interferometer Gravitational-Wave Observatory (LIGO), with the aim of creating an experiment that would be able to detect gravitational waves.

In 1994, work began in the US on two interferometers, one at Hanford, Washington, and the other at Livingstone, Louisiana. Two machines were necessary to verify that any detected wave was gravitational rather than a random, local vibration. The research was led by American physicist Barry Barish, who became director of LIGO in 1997 and created the LIGO Scientific Collaboration, a team of 1,000 scientists around the world. This global collaboration gave the project renewed impetus, and by 2002 scientists had completed the two LIGO machines. Each consisted of two steel tubes, 4 ft (1.2 m) wide and 2.5 miles (4 km) long, protected inside a concrete shelter.

Because gravitational waves interact with space by compressing and stretching it in a perpendicular

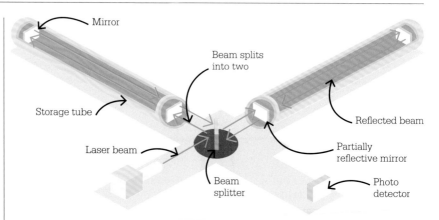

LIGO uses laser beams to detect gravitational waves. One beam is fired at a partially reflective mirror, which splits the beam down two storage tubes set at a right angle. Each beam passes through another partially reflective mirror and bounces between this and a mirror at the end of the tube, with some light from each tube meeting at the beam splitter. If there are no gravitational waves, the beams cancel each other out. If waves are present, they cause interference between the beams, creating flickering light that registers on a photo detector.

direction, the tubes were built at right angles to each other, in an L shape. In theory, a wave that distorts spacetime would change the length of each tube, stretching one and compressing the other repeatedly until the wave had passed. To measure any minute change, a laser beam was split in two, then shone down each tube. Any incoming gravitational wave would cause the beams of light to reflect back at different times, as spacetime itself is stretched and shortened. By measuring this change, the scientists hoped to work out where the gravitational waves were coming from and what had caused them.

Cataclysmic ripples

Over the next eight years, no waves were recorded, a situation complicated by the machines picking up interference, such as wind noise and even the sound of trains and logging machinery. In 2010, it was decided to completely overhaul both LIGO machines,

and in September 2015 they were turned back on, now able to scan a much greater area of space. Within days, the new, more sensitive instruments had picked up tiny split-second ripples in spacetime, reaching Earth from a cataclysmic event somewhere deep in the universe.

Scientists were able to work out that these gravitational waves had been produced by two black holes colliding about 1.3 billion light years from Earth and creating 50 times more power than all the stars in the universe at that moment. Known as stellar black holes, they each had an estimated mass of 36 and 29 times the mass of the sun, forming a new black hole that was 62 times the sun's mass. The remaining mass—three times that of the sun—was catapulted into space almost entirely as gravitational waves. By measuring the signals received from the two LIGO sites—backed up by the Virgo interferometer in Italy— scientists were now able to look

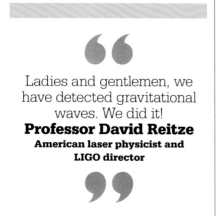

Ladies and gentlemen, we have detected gravitational waves. We did it!
Professor David Reitze
American laser physicist and LIGO director

back in time to the origin of gravitational waves and study new portions of the universe that were simply inaccessible before. They could do this by matching the signals they detected with the patterns they expected to see from different spacetime events.

Since the initial detection of gravitational waves in 2015 and the breakthrough announcement of their discovery in 2016, many more potential gravitational wave signals have been found. Most have been black hole mergers, but in 2017, LIGO scientists made their first confirmed detection of gravitational waves produced from the collision of two neutron stars, around 130 million years ago.

Signals from afar
LIGO and Virgo continue to detect gravitational waves on an almost weekly basis. The equipment is regularly upgraded as lasers become more powerful and mirrors more stable. Objects barely the size of a city can be detected across the universe, being squashed together in dramatic events that push the boundaries of physics. The discovery of gravitational waves is helping scientists probe the very nature of

the universe, revealing ever more about its origins and expansion, and even, potentially, its age.

Pushing space boundaries
Astronomers are also working on new experiments that could probe gravitational waves in even greater detail. One of these is a mission, organized by the European Space Agency (ESA), called LISA (Laser Interferometer Space Antenna). LISA, due to launch in 2034, will consist of three spacecraft flown in a triangle, each separated by 1.5 million miles (2.5 million km). Lasers will be fired between the craft and the signals produced studied for any evidence of minuscule movement that could be the ripples of gravitational waves.

LISA's observations over such a vast area will allow scientists to detect gravitational waves from a variety of other objects, such as supermassive black holes, or even from the dawn of the universe. The secrets of deepest space may eventually be unlocked. ∎

When two neutron stars collide, they release visible gamma rays and invisible gravitational waves, which reach Earth at virtually the same moment millions of years later.

Kip Thorne

Born in Utah in 1940, Kip Thorne graduated in physics from the California Institute of Technology (Caltech) in 1962 before completing a doctorate at Princeton University in 1965. He returned to Caltech in 1967, where he is now the Feynman Professor of Theoretical Physics, Emeritus.

Thorne's interest in gravitational waves led to the founding of the LIGO project, which he has supported by identifying distant sources of the waves and devising techniques for extracting information from them. For his work on gravitational waves and LIGO, Thorne was awarded the Nobel Prize in Physics in 2017, alongside collaborators Rainer Weiss and Barry Barish.

Thorne also lends his physics knowledge to the arts: the 2014 film *Interstellar* was based on his original concept.

Key works

1973 *Gravitation*
1994 *Black Holes and Time Warps*
2014 *The Science of* Interstellar

DIRECTORY

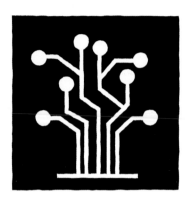

As Isaac Newton himself memorably put it, in a letter to Robert Hooke in 1675, "If I have seen further, it is by standing on the shoulders of Giants." Ever since its beginnings in Mesopotamia, in the fourth millennium BCE, the story of science has been one of collaboration and continuity. From natural philosophers to inventors, experimenters, and, more recently, professional scientists, far more people have made important contributions to this story than could possibly be explored in detail across the preceding chapters of this book. The following directory therefore attempts to offer at least an outline of some other key figures in our still-unfolding quest to understand the way our universe works—from the tiniest nucleus to the furthest galaxy.

ARCHIMEDES
c. 287 BCE–c. 212 BCE

Born in the Greek colony of Syracuse on the Mediterranean island of Sicily, Archimedes was one of the most important engineers, physicists, and mathematicians of the ancient world. Little is known of his early life, but he may have studied at Alexandria, Egypt, in his youth. Although later acclaimed for his groundbreaking mathematical proofs, particularly in the field of geometry, in his lifetime he was famous mainly for inventions such as the "Archimedes screw" water pump and the compound pulley, as well as his famous discovery—upon which he is said to have shouted "Eureka!"—of Archimedes' principle, which describes the displacement of water. When the Romans invaded Sicily, Archimedes devised a number of ingenious weapons to defend his city. He was killed by a Roman soldier despite orders that he be captured alive.
See also: The scientific method 20–23 ▪ Fluids 76–79

HASAN IBN AL-HAYTHAM
c. 965–1040

Arabic scholar al-Haytham (sometimes known in the West as Alhazen) was born in Basra, now in Iraq, but spent much of his career in Cairo, Egypt. Little is known of his life but he initially worked as a civil engineer, and made contributions to medicine, philosophy, and theology, as well as physics and astronomy. Al-Haytham was one of the first proponents of the scientific method, devising experiments to prove or disprove hypotheses. He is best known for his *Book of Optics* (1021), which successfully combined various classical theories of light with observations of anatomy, to explain vision in terms of light rays reflected from objects, gathered by the eye, and interpreted by the brain.
See also: The scientific method 20–23 ▪ Reflection and refraction 168–169 ▪ Focusing light 170–175

AVICENNA
c. 980–1037

Ibn Sina (known in the West as Avicenna) was a Persian polymath born near Bukhara, in present-day Uzbekistan, into the family of a well-placed civil servant of the Persian Samanid dynasty. He displayed a talent for learning from his youth, absorbing the works of many classical as well as earlier Islamic scholars. Although best known today for his hugely influential works on medicine, he wrote widely across a range of fields, including physics. He developed a theory of motion that recognized the concept later known as inertia and the influence of air resistance, and also set out an early argument that the speed of light must be finite.
See also: Laws of motion 40–45 ▪ The speed of light 275

JOHANNES KEPLER
1571–1630

German mathematician and astronomer Johannes Kepler found fame in the 1590s for a theory that attempted to link the orbits of the planets to the geometry of mathematical "Platonic solids." He became assistant to the great Danish astronomer Tycho Brahe, who was compiling the most accurate catalog of planetary motions so far attempted. After Tycho's death in 1601, Kepler continued and developed his work. The success of his 1609 laws of planetary motion, which placed the planets on elliptical (rather than circular) orbits around the sun, completed the "Copernican revolution" and paved the way for Newton's more general laws of motion and universal gravitation.
See also: Laws of gravity 46–51 ▪ The heavens 270–271 ▪ Models of the universe 272–273

EVANGELISTA TORRICELLI
1608–1647

Born in the province of Ravenna, Italy, Torricelli showed his talent for mathematics at an early age, and was sent to be educated by monks at a local abbey. He later traveled to Rome, where he became secretary and unofficial student to Benedetto Castelli, a professor of mathematics and friend of Galileo Galilei. Torricelli became a disciple of Galilean science and shared many discussions with Galileo in the months before his death in 1642. The following year, Torricelli published a description of the first mercury barometer—the invention for which he is best known for today.
See also: The scientific method 20–23 ▪ Pressure 36 ▪ Fluids 76–79

GUILLAUME AMONTONS
1663–1705

Paris-born Guillaume Amontons was the son of a lawyer. He devoted himself to science after losing his hearing in childhood. Largely self-taught, he was a skilled engineer and inventor, devising improvements to various scientific instruments. While investigating the properties of gases, he discovered the relationships between temperature, pressure, and volume, though he was unable to quantify the precise equations of the later "gas laws." He is best remembered for his laws of friction, which describe the forces of static and sliding friction affecting bodies whose surfaces are in contact with each other.
See also: Energy and motion 56–57 ▪ The gas laws 82–85 ▪ Entropy and the second law of thermodynamics 94–99

DANIEL GABRIEL FAHRENHEIT
1686–1736

Born in Danzig (now Gdansk in Poland) to a German merchant family, Fahrenheit spent most of his working life in the Netherlands. Orphaned in 1701, he trained as a merchant before pursuing his scientific interests and mixing with many leading thinkers of the day, including Ole Rømer and Gottfried Leibniz. He lectured on chemistry and learned how to make delicate glass parts for scientific instruments such as thermometers. This led in 1724 to his concept of a standardized temperature scale (though Fahrenheit's own choice of "fixed" temperature points was later revised to make the scale that still bears his name more accurate).
See also: The scientific method 20–23 ▪ Heat and transfers 80–81

LAURA BASSI
1711–1778

The daughter of a successful lawyer in Bologna, Italy, Bassi benefited from a privately tutored education, in which she showed an early interest in physics. In her teens, she became fascinated by the still-controversial theories of Isaac Newton. After receiving a doctorate in philosophy at the age of 20, she became the first woman to hold a university chair in the study of science. Working at the University of Bologna, she met and married a fellow lecturer, Giuseppe Veratti, and the pair worked closely throughout the rest of their careers. Bassi devised many advanced experiments to demonstrate the accuracy of Newtonian physics and wrote widely on mechanics and hydraulics. In 1776, she was appointed as the university's professor of experimental physics.
See also: Laws of motion 40–45 ▪ Laws of gravity 46–51

WILLIAM HERSCHEL
1738–1822

William Herschel was a German-born astronomer who moved to England at the age of 19. He read widely on acoustics and optics, and a permanent musical appointment in Bath, from 1766, allowed him to pursue these interests in earnest. He built the finest reflecting telescopes of the time, and began to systematically study the stars, assisted from 1772 by his sister Caroline. His discovery of the planet Uranus in 1781 won him an appointment to the role of "King's Astronomer" for Britain's George III. In 1800, while measuring the properties of different colors of visible light, he discovered the existence of infrared radiation.
See also: Electromagnetic waves 192–195 ▪ The heavens 270–271 ▪ Models of the universe 272–273

PIERRE-SIMON LAPLACE
1749–1827

Growing up in Normandy, France, Laplace showed an early talent for mathematics. He entered the university at Caen at the age of 16, and went on to become a professor at the Paris Military School. Throughout a long career he not only produced important work in pure mathematics, but also applied it to areas ranging from the prediction of tides and the shape of Earth to the stability of planetary orbits and the history of the solar system. He was the first to suggest that our solar system formed from a collapsing cloud of gas and dust, and was also the first to put the objects we now call black holes on a mathematical footing.
See also: Laws of gravity 46–51 ▪ Models of the universe 272–273 ▪ Black holes and wormholes 286–289

SOPHIE GERMAIN
1776–1831

Born to a wealthy Parisian silk merchant, Germain had to fight the prejudices of her parents in order to follow her interest in mathematics. Initially self-taught, from 1794 she obtained lecture notes from the École Polytechnique, and had private tuition from Joseph-Louis Lagrange. Later, she also corresponded with Europe's leading mathematicians. Although best known for her work in mathematics, she also made

important contributions to the physics of elasticity when she won the Paris Academy's competition (inspired by the acoustic experiments of Ernst Chladni) to mathematically describe the vibration of elastic surfaces.
See also: Stretching and squeezing 72–75 ▪ Music 164–167

JOSEPH VON FRAUNHOFER
1787–1826

Apprenticed to a glassmaker in Germany after being orphaned at the age of 11, the young Fraunhofer saw his life transformed in 1801 after his master's workshop collapsed and he was pulled from the rubble by a rescue party of local dignitaries. The Prince Elector of Bavaria and other benefactors encouraged his academic leanings and ultimately enrolled him in a glassmaking institute, where he was able to pursue his studies. Fraunhofer's discoveries allowed Bavaria to become the leading center of glass manufacturing for scientific instruments. His inventions include diffraction grating, for dispersing light of different colors, and the spectroscope, for measuring the precise positions of different features in a spectrum.
See also: Diffraction and interference 180–183 ▪ The Doppler effect and redshift 188–191 ▪ Light from the atom 196–199

WILLIAM THOMSON, LORD KELVIN
1824–1907

Belfast-born William Thomson was one of the most important figures in 19th-century physics. After studying at the universities of Glasgow and Cambridge, UK, the talented Thomson returned to Glasgow as a professor at the age of just 22. His interests were wide-ranging—he helped establish the science of thermodynamics, calculated the great age of Earth, and investigated the possible fate of the universe itself. He found wider fame, however, as an electrical engineer

and made key contributions to the design of the first transatlantic telegraph cable, which was being planned in the 1850s. The project's eventual success in 1866 led to public acclaim, a knighthood, and eventually elevation to the peerage.
See also: Heat and transfers 80–81 ▪ Internal energy and the first law of thermodynamics 86–89 ▪ Heat engines 90–93

ERNST MACH
1838–1916

Born and raised in Moravia (now part of the Czech Republic), Austrian philosopher and physicist Mach studied physics and medicine at the University of Vienna. Initially interested in the Doppler effect in both optics and acoustics, he was encouraged by the invention of Schlieren photography (a method of imaging otherwise-invisible shock waves) to investigate the dynamics of fluids and the shock waves formed around supersonic objects. Although best known for this work (and the "Mach number" measurement of speeds relative to the speed of sound), he also made important contributions to physiology and psychology.
See also: Energy and motion 56–57 ▪ Fluids 76–79 ▪ The development of statistical mechanics 104–111

HENRI BECQUEREL
1852–1908

The third in a family of wealthy Parisian physicists, which would go on to include his son Jean, Becquerel pursued joint careers in engineering and physics. His work involved studies of subjects such as the plane polarization of light, geomagnetism, and phosphorescence. In 1896, news of Wilhelm Röntgen's discovery of X-rays inspired Becquerel to investigate whether phosphorescent materials, such as certain uranium salts, produced similar rays. He soon detected some kind of emission from the salts, but further tests showed

that uranium compounds emitted rays even if they were not phosphorescent. Becquerel was the first person to discover the existence of "radioactive" materials, for which he shared the 1903 Nobel Prize with Marie and Pierre Curie.
See also: Polarization 184–187 ▪ Nuclear rays 238–239 ▪ The nucleus 240–241

NIKOLA TESLA
1856–1943

A Serbian–American physicist and inventor, Nikola Tesla was a hugely important figure in the early establishment of widespread electrical power. After proving his engineering talent in Hungary, he was employed by Thomas Edison's companies in Paris and later in New York, before quitting to market his own inventions independently. One of these, an induction motor that could be powered by an alternating current (AC) system, proved hugely important to the widespread adoption of AC. Tesla's many other inventions (some of them ahead of their time) included wireless lighting and power, radio-controlled vehicles, bladeless turbines, and improvements to the AC system itself.
See also: Electric current and resistance 130–133 ▪ The motor effect 136–137 ▪ Induction and the generator effect 138–141

J.J. THOMSON
1856–1940

After demonstrating an unusual talent for science at an early age, Joseph John Thomson, who was born in Manchester, UK, was admitted to study at Owens College (now the University of Manchester) at the age of just 14. From there he moved on to the University of Cambridge, where he distinguished himself in mathematics before being appointed as Cavendish Professor of Physics in 1884. He is chiefly known today for his discovery of the electron in 1897, which he

identified by careful analysis of the properties of recently discovered "cathode rays." A few months after the discovery, he was able to show that the particles within the rays could be deflected in electric fields, and to calculate the ratio between their mass and electric charge.
See also: Atomic theory 236–237
- Subatomic particles 242–243

ANNIE JUMP CANNON
1863–1941

The eldest daughter of a US senator for Delaware, Jump Cannon learned about the stars at an early age from her mother, who subsequently encouraged her interest in science. She flourished in her studies, despite a bout of scarlet fever that left her almost entirely deaf. She joined the staff of Harvard College Observatory in 1896, to work on an ambitious photographic catalog of stellar spectra. Here, she manually classified some 350,000 stars, developing the classification system that is still widely used today and publishing the catalogs that would eventually reveal stellar composition.
See also: Diffraction and interference 180–183 - Light from the atom 196–199
- Energy quanta 208–211

ROBERT MILLIKAN
1868–1953

Robert Millikan was born in Illinois and studied classics at Oberlin College in Ohio, before being diverted into physics by a suggestion from his professor of Greek. He went on to earn a doctorate from Columbia University and began work at the University of Chicago. It was here in 1909 that he and graduate student Harvey Fletcher devised an ingenious experiment to measure the charge on the electron for the first time. This fundamental constant of nature paved the way for the accurate calculation of many other important physical constants.
See also: Atomic theory 236–237
- Subatomic particles 242–243

EMMY NOETHER
1882–1935

German mathematician Noether showed a gift for mathematics and logic at an early age and went on to earn a doctorate at the University of Erlangen, Bavaria, Germany, despite policies discriminating against female students. In 1915, the mathematicians David Hilbert and Felix Klein invited her to their prestigious department at the University of Göttingen to work on interpreting Einstein's theory of relativity. Here, Noether went on to make huge contributions to the foundations of modern mathematics. However, in physics she is best known for a proof, published in 1918, that the conservation of certain properties (such as momentum and energy) is linked to the underlying symmetry of systems and the physical laws that govern them. Noether's theorem and its ideas about symmetry underpin much of modern theoretical physics.
See also: The conservation of energy 55
- Force carriers 258–259 - String theory 308–311

HANS GEIGER
1882–1945

Geiger was a German physicist who studied physics and mathematics at the University of Erlangen in Bavaria, Germany, earning his doctorate and receiving a fellowship at the University of Manchester, UK, in 1906. From 1907, he worked under Ernest Rutherford at the university. Geiger had previously studied electrical discharge through gases, and the two men formulated a method of harnessing this process to detect otherwise-invisible radioactive particles. In 1908, under Rutherford's direction, Geiger and his colleague Ernest Marsden carried out the famous "Geiger–Marsden experiment"—showing how a few radioactive alpha particles fired at thin gold foil bounce back toward the source, and thereby demonstrating the existence of the atomic nucleus.
See also: Atomic theory 236–237
- The nucleus 240–241

LAWRENCE BRAGG
1890–1971

As the son of William Henry Bragg (1862–1942), professor of physics at the University of Adelaide in Australia, Lawrence Bragg took an early interest in the subject. After the family moved to the UK for William to take a chair at the University of Leeds, Lawrence enrolled at the University of Cambridge. It was here, as a postgraduate student in 1912, that he came up with an idea for settling the long-running debate about the nature of X-rays. He reasoned that if X-rays were electromagnetic waves rather than particles, they should produce interference patterns due to diffraction as they passed through crystals. Father and son developed an experiment to test the hypothesis, not only proving that X-rays are indeed waves, but also pioneering a new technique for studying the structure of matter.
See also: Diffraction and interference 180–183 - Electromagnetic waves 192–195

ARTHUR HOLLY COMPTON
1892–1962

Born into an academic family in Wooster, Ohio, Compton was the youngest of three brothers, all of whom obtained PhDs from Princeton University. After becoming interested in how X-rays could reveal the internal structure of atoms, he became head of the physics department of Washington University in St. Louis, in 1920. It was here in 1923 that his experiments led to the discovery of "Compton scattering"—the transfer of energy from X-rays to electrons that could only be explained if the X-rays had particle-like, as well as wavelike, properties. The idea of a particle-like aspect to electromagnetic radiation had been proposed by both Planck and Einstein, but Compton's discovery was the first incontrovertible proof.
See also: Electromagnetic waves 192–195 - Energy quanta 208–211
- Particles and waves 212–215

IRÈNE JOLIOT-CURIE
1897–1956

The daughter of Marie and Pierre Curie, Irène showed a talent for mathematics from an early age. After working as a radiographer during World War I, she completed her degree and continued her studies at the Radium Institute founded by her parents, where she met her husband-to-be, chemist Frédéric Joliot. Working together, they were the first to measure the mass of the neutron in 1933. They also studied what happened when lightweight elements were bombarded with radioactive alpha particles (helium nuclei), and discovered that the process created materials that were themselves radioactive. The Joliot-Curies had successfully created artificial radioactive isotopes, an achievement for which they were awarded the 1935 Nobel Prize in Chemistry.
See also: Nuclear rays 238–239
■ The nucleus 240–241 ■ Particle accelerators 252–255

LEO SZILARD
1898–1964

Born into a Jewish family in Budapest, Hungary, Szilard showed early talent by winning a national prize in mathematics at the age of 16. He finished his education and settled in Germany, before fleeing to the UK with the rise of the Nazi party. Here, he cofounded an organization to help refugee scholars, and formulated the idea of the nuclear chain reaction— a process that harnesses a cascade of neutron particles to release energy from atoms. After emigrating to the United States in 1938, he worked with others to make the chain reaction a reality as part of the Manhattan Project.
See also: Nuclear rays 238–239
■ The nucleus 240–241 ■ Nuclear bombs and power 248–251

GEORGE GAMOW
1904–1968

Gamow studied physics in his home city Odessa (then part of the USSR, but now in Ukraine), and later in Leningrad, where he became fascinated by quantum physics. In the 1920s, collaboration with Western colleagues led to successes such as his description of the mechanism behind alpha decay and radioactive "half-life." In 1933 he defected from the increasingly oppressive Soviet Union, eventually settling in Washington, D.C. From the late 1930s, Gamow renewed an early interest in cosmology, and in 1948, he and Ralph Alpher outlined what is now called "Big Bang nucleosynthesis"—the mechanism by which a vast burst of energy gave rise to the raw elements of the early universe in the correct proportions.
See also: Nuclear rays 238–239
■ The particle zoo and quarks 256–257
■ The static or expanding universe 294–295 ■ The Big Bang 296–301

J. ROBERT OPPENHEIMER
1904–1967

New York–born Oppenheimer's genius began to flourish during a postgraduate period at the University of Göttingen, Germany, in 1926, where he worked with Max Born and met many other leading figures in quantum physics. In 1942, Oppenheimer was recruited to work on calculations for the Manhattan Project, to develop an atomic bomb in the US. A few months later, he was chosen to head the laboratory where the bomb would be built. After leading the project to its conclusion at the end of World War II, when atomic bombs were dropped on Hiroshima and Nagasaki in Japan with devastating results, Oppenheimer became an outspoken critic of nuclear proliferation.
See also: The nucleus 240–241
■ Nuclear bombs and power 248–251

MARIA GOEPPERT MAYER
1906–1972

Born into an academic family in Katowice (now in Poland), Goeppert studied mathematics and then physics at the University of Göttingen in Germany. Her 1930 doctoral thesis predicted the phenomenon of atoms absorbing pairs of photons (later demonstrated in 1961). In 1930, she married and moved to the US with her American chemist husband, but found it difficult to obtain an academic position. From 1939, she worked at Columbia University, where she became involved in the separation of uranium isotopes needed for the atomic bomb during World War II. In the late 1940s, at the University of Chicago, she developed the "nuclear shell model," explaining why atomic nuclei with certain numbers of nucleons (protons and neutrons) are particularly stable.
See also: Nuclear rays 238–239
■ The nucleus 240–241 ■ Nuclear bombs and power 248–251

DOROTHY CROWFOOT HODGKIN
1910–1994

After learning Latin specifically to pass the entrance exam, Hodgkin studied at Somerville College, Oxford, before moving to Cambridge to carry out work on X-ray crystallography. During her doctorate, she pioneered methods of using X-rays to analyze the structure of biological protein molecules. Returning to Somerville, she continued her research and refined her techniques to work on increasingly complex molecules, including steroids and penicillin (both in 1945), vitamin B12 (in 1956, for which she was awarded the Nobel Prize in Chemistry in 1964), and insulin (finally completed in 1969).
See also: Diffraction and interference 180–183 ■ Electromagnetic waves 192–195

SUBRAHMANYAN CHANDRASEKHAR
1910–1995

Born in Lahore (which was then part of India but is now in Pakistan), Chandrasekhar obtained his first degree in Madras (now Chennai) and continued his studies at the University of Cambridge from 1930. His most famous work was on the physics of superdense stars such as white dwarfs. He showed that in stars with more than 1.44 times the mass of the sun, internal pressure would be unable to resist the inward pull of gravity—leading to a catastrophic collapse (the origin of neutron stars and black holes). In 1936, he moved to the University of Chicago, becoming a naturalized US citizen in 1953.

See also: Quantum numbers 216–217
- Subatomic particles 242–243
- Black holes and wormholes 286–289

RUBY PAYNE-SCOTT
1912–1981

Australian radio astronomer Ruby Payne-Scott was born in Grafton, New South Wales, and studied sciences at the University of Sydney, graduating in 1933. After early research into the effect of magnetic fields on living organisms, she became interested in radio waves, leading to her work at the Australian government's Radiophysics Laboratory on radar technology during World War II. In 1945, she coauthored the first scientific report linking sunspot numbers to solar radio emissions. She and her coauthors later established an observatory that pioneered ways of pinpointing radio sources on the sun, conclusively linking the radio bursts to sunspot activity.

See also: Electromagnetic waves 192–195
- Seeing beyond light 202–203

FRED HOYLE
1915–2001

Hoyle was born in Yorkshire, UK, and read mathematics at the University of Cambridge before working on the development of radar during World War II. Discussions with other scientists on the project fired an interest in cosmology, while trips to the US offered insight into the latest research in astronomy and nuclear physics. Years of work led in 1954 to his theory of supernova nucleosynthesis, explaining how heavy elements are produced inside heavyweight stars and scattered across the universe when they explode. Hoyle became famous as both an author and a popular scientist, although he espoused

some more controversial views, including his nonacceptance of the Big Bang theory.

See also: Nuclear fusion in stars 265
- The static or expanding universe 294–295
- The Big Bang 296–301

SUMIO IIJIMA
1939–

Born in Saitama Prefecture, Japan, Iijima studied electrical engineering and solid-state physics. In the 1970s, he used electron microscopy to study crystalline materials at Arizona State University, and continued to research the structures of very fine solid particles, such as the newly discovered "fullerenes" (balls of 60 carbon atoms), on his return to Japan in the 1980s. From 1987, he worked in the research division of electronics giant NEC, and it was here, in 1991, that he discovered and identified another form of carbon, cylindrical structures of immense strength, now known as nanotubes. The potential applications of this new material have helped to spur a wave of research in nanotechnology.

See also: Models of matter 68–71
- Nanoelectronics 158
- Atomic theory 236–237

STEPHEN HAWKING
1942–2018

Perhaps the most famous scientist of modern times, Hawking was diagnosed with motor neuron disease in 1963, while studying for his doctorate in cosmology at the University of Cambridge. His doctoral thesis in 1966 demonstrated that the Big Bang theory, in which the universe developed from an infinitely hot, dense point called a singularity, was consistent with general relativity (undermining one of the last major objections to the theory). Hawking spent his early career investigating black holes (another type of singularity), and in 1974 showed that they should emit a form of radiation. His later work addressed questions about the evolution of the universe, the nature of time, and the unification of quantum theory with

gravity. Hawking also became renowned as a science communicator after the 1988 publication of his book *A Brief History of Time*.

See also: Black holes and wormholes 286–289
- The Big Bang 296–301
- String theory 308–311

ALAN GUTH
1947–

Initially studying particle physics, Guth changed the focus of his work after attending lectures on cosmology at Cornell University in 1978–1979. He proposed a solution to some of the big unanswered questions about the universe, by introducing the idea of a brief moment of violent "cosmic inflation," which blew up a small part of the infant universe to dominate over all the rest, just a fraction of a second after the Big Bang itself. Inflation offers explanations for questions such as why the universe around us seems so uniform. Others have taken the idea as a springboard for suggesting our particular "bubble" is one of many in an "inflationary multiverse."

See also: Matter–antimatter asymmetry 264
- The static or expanding universe 294–295
- The Big Bang 296–301

FABIOLA GIANOTTI
1960–

After earning a PhD in experimental particle physics at the University of Milan, Gianotti joined CERN (the European Organization for Nuclear Research) where she was involved in the research and design of several experiments using the organization's various particle accelerators. Most significantly, she was project leader on the huge ATLAS experiment at the Large Hadron Collider, leading the data analysis that was able to confirm the existence of the Higgs boson, the final missing element in the Standard Model of particle physics, in 2012.

See also: Particle accelerators 252–255
- The particle zoo and quarks 256–257
- The Higgs boson 262–263

GLOSSARY

In this glossary, terms defined within another entry are identified with *italic* type.

Absolute zero The lowest possible temperature: 0K or −459.67°F (−273.15°C).

Acceleration The rate of change of *velocity*. Acceleration is caused by a *force* that results in a change in an object's direction and/or speed.

Air resistance The *force* that resists the movement of an object through the air.

Alpha particle A *particle* made of two *neutrons* and two *protons*, which is emitted during a form of *radioactive decay* called alpha decay.

Alternating current (AC) An *electric current* whose direction reverses at regular intervals. See also *Direct current (DC)*.

Angular momentum A measure of the rotation of an object, which takes into account its *mass*, shape, and *spin* speed.

Antimatter *Particles* and *atoms* that are made of *antiparticles*.

Antiparticle A *particle* that is the same as a normal particle except that it has an opposite *electric charge*. Every particle has an equivalent antiparticle.

Atom The smallest part of an *element* that has the chemical properties of that element. An atom was thought to be the smallest part of *matter*, but many subatomic *particles* are now known.

Beta decay A form of *radioactive decay* in which an atomic *nucleus* gives off beta *particles* (*electrons* or *positrons*).

Big Bang The event with which the universe is thought to have begun, around 13.8 billion years ago, exploding outward from a *singularity*.

Blackbody A theoretical object that absorbs all *radiation* that falls on it, radiates *energy* according to its temperature, and is the most efficient emitter of radiation.

Black hole An object in space that is so dense that light cannot escape its gravitational *field*.

Bosons Subatomic *particles* that exchange *forces* between other particles and provide particles with *mass*.

Circuit The path around which an *electric current* can flow.

Classical mechanics A set of laws describing the motion of bodies under the action of *forces*.

Coefficient A number or *expression*, usually a *constant*, that is placed before another number and multiplies it.

Conductor A substance through which heat or *electric current* flows easily.

Constant A quantity in a mathematical *expression* that does not vary—often symbolized by a letter such as *a*, *b*, or *c*.

Cosmic microwave background (CMB) Faint microwave *radiation* that is detectable from all directions. The CMB is the oldest radiation in the universe, emitted when the universe was 380,000 years old. Its existence was predicted by the *Big Bang* theory, and it was first detected in 1964.

Cosmic rays Highly energetic *particles*, such as *electrons* and *protons*, that travel through space at close to the speed of light.

Cosmological constant A term that Albert Einstein added to his *general relativity* equations, which is used to model the *dark energy* that is accelerating the expansion of the universe.

Dark energy A poorly understood force that acts in the opposite direction to *gravity*, causing the universe to expand. About three-quarters of the *mass–energy* of the universe is dark energy.

Dark matter Invisible *matter* that can only be detected by its gravitational effect on visible matter. Dark matter holds galaxies together.

Diffraction The bending of *waves* around obstacles and spreading out of waves past small openings.

Direct current (DC) An *electric current* that flows in one direction only. See also *Alternating current (AC)*.

Doppler effect The change in *frequency* of a *wave* (such as a light or sound wave) experienced by an observer in relative motion to the wave's source.

Elastic collision A collision in which no kinetic *energy* is lost.

Electric charge A property of subatomic *particles* that causes them to attract or repel one another.

Electric current A flow of objects with *electric charge*.

Electromagnetic force One of the four *fundamental forces* of nature. It involves the transfer of *photons* between *particles*.

Electromagnetic radiation A form of *energy* that moves through space. It has both an electrical and a magnetic field, which oscillate at right-angles to each other. Light is a form of electromagnetic radiation.

Electromagnetic spectrum The complete range of *electromagnetic radiation*. See also *Spectrum*.

Electroweak theory A theory that explains the *electromagnetic* and *weak nuclear force* as one "electroweak" force.

Electron A subatomic *particle* with a negative *electric charge*.

Electrolysis A chemical change in a substance caused by passing an *electric current* through it.

Element A substance that cannot be broken down into other substances by chemical reactions.

Energy The capacity of an object or system to do *work*. Energy can exist in many forms, such as potential energy (for example, the energy stored in a spring) and kinetic energy (movement). It can change from one form to another, but can never be created or destroyed.

Entanglement In quantum physics, the linking between *particles* such that a change in one affects the other no matter how far apart in space they may be.

Entropy A measure of the disorder of a system, based on the number of specific ways a particular system may be arranged.

Event horizon A boundary surrounding a *black hole* within which the gravitational pull of the black hole is so strong that light cannot escape. No information about the black hole can cross its event horizon.

Exoplanet A planet that orbits a star that is not our sun.

Expression Any meaningful combination of mathematical symbols.

Fermion A subatomic *particle*, such as an *electron* or a *quark*, that is associated with *mass*.

Field The distribution of a *force* across *spacetime*, in which each point can be given a value for that force. A gravitational field is an example of a field in which the force felt at a particular point

is inversely proportional to the square of the distance from the source of *gravity*.

Force A push or a pull, which moves or changes the shape of an object.

Fraunhofer lines Dark absorption lines found in the *spectrum* of the sun, first identified by German physicist Joseph von Fraunhofer.

Frequency The number of *waves* that pass a point every second.

Friction A *force* that resists or stops the movement of objects that are in contact with one another.

Fundamental forces The four *forces* that determine how matter behaves. The fundamental forces are the *electromagnetic force*, *gravity*, the *strong nuclear force*, and the *weak nuclear force*.

Galaxy A large collection of stars and clouds of gas and dust that is held together by *gravity*.

Gamma decay A form of radioactive decay in which an atomic nucleus gives off high-*energy*, short-*wavelength* gamma *radiation*.

General relativity A theoretical description of *spacetime* in which Einstein considered *accelerating* frames of reference. General relativity provides a description of *gravity* as the warping of spacetime by *energy*.

Geocentrism A historic model of the universe with Earth at its center. See also *Heliocentrism*.

Gluons *Particles* within *protons* and *neutrons* that hold *quarks* together.

Gravitational wave A distortion of *spacetime* that travels at the speed of light, generated by the *acceleration* of *mass*.

Gravity A *force* of attraction between objects with *mass*. Massless *photons* are also affected by gravity, which *general relativity* describes as a warping of *spacetime*.

Heat death A possible end state for the universe in which there are no temperature differences across space, and no *work* can be done.

Heliocentrism A model of the universe with the sun at its center.

Higgs boson A subatomic particle associated with the Higgs *field*, whose interaction with other *particles* gives them their *mass*.

Ideal gas A gas in which there are zero inter-*particle forces*. The only interactions between particles in an ideal gas are *elastic collisions*.

Inertia The tendency of an object to keep moving or remain at rest until a *force* acts on it.

Insulator A material that reduces or stops the flow of heat, electricity, or sound.

Interference The process whereby two or more *waves* combine, either reinforcing each other or canceling each other out.

Ion An *atom*, or group of atoms, that has lost or gained one or more of its *electrons* to become electrically charged.

Isotopes *Atoms* of the same *element* which have the same number of

protons, but a different number of *neutrons*.

Light year A unit of distance that is the distance traveled by light in one year, equal to 5,878 trillion miles (9,460 trillion km).

Leptons *Fermions* that are only affected by two of the *fundamental forces*—the *electromagnetic force* and the *weak nuclear force*.

Magnetism A *force* of attraction or repulsion exerted by magnets. Magnetism arises from the motion of *electric charges* or from the magnetic moment of *particles*.

Mass A property of an object that is a measure of the *force* required to accelerate it.

Matter Any physical substance. Our entire visible world is made of matter.

Molecule A substance made of two or more *atoms* that are chemically bonded to one another.

Momentum An object's *mass* multiplied by its *velocity*.

Neutrino An electrically neutral *fermion* that has a very small, as yet unmeasured, *mass*. Neutrinos can pass right through *matter* undetected.

Neutron An electrically neutral subatomic *particle* that forms part of an *atom*'s *nucleus*. A neutron is made of one up-*quark* and two down-quarks.

Nuclear fission A process whereby the *nucleus* of an *atom* splits into two smaller nuclei, releasing *energy*.

Nuclear fusion A process whereby atomic *nuclei* join together to form heavier nuclei, releasing *energy*. Inside stars such as the sun, this process involves the fusion of hydrogen nuclei to make helium.

Nucleus The central part of an *atom*. The nucleus consists of *protons* and *neutrons* and contains almost all of an atom's *mass*.

Optics The study of vision and the behavior of light.

Orbit The path of a body around another, more massive, body.

Particle A tiny speck of *matter* that can have *velocity*, position, *mass*, and *electric charge*.

Photoelectric effect The emission of *electrons* from the surfaces of certain substances when light hits them.

Photon The *particle* of light that transfers the *electromagnetic force* from one place to another.

Piezoelectricity Electricity that is produced by applying stress to certain crystals.

Plane A flat surface on which any two given points can be joined by a straight line.

Plasma A hot, electrically charged fluid in which the *electrons* are free from their *atoms*.

Polarized light Light in which the *waves* all oscillate in just one *plane*.

Positron The *antiparticle* counterpart of an *electron*, with the same *mass* but a positive *electric charge*.

Potential difference The difference in *energy* per unit of charge between two places in an electric *field* or circuit.

Pressure A continual *force* per unit area pushing against an object. The pressure of gases is caused by the movement of their *molecules*.

Proton A *particle* in the *nucleus* of an *atom* that has positive charge. A proton contains two up-*quarks* and one down-quark.

Quanta Packets of *energy* that exist in discrete units. In some systems, a quantum is the smallest possible amount of energy.

Quantum electrodynamics (QED) A theory that explains the interaction of subatomic *particles* in terms of an exchange of *photons*.

Quantum mechanics The branch of physics that deals with subatomic *particles* behaving as *quanta*.

Quark A subatomic *particle* that *protons* and *neutrons* are made from.

Radiation An electromagnetic *wave* or a stream of *particles* emitted by a radioactive source.

Radioactive decay The process in which unstable atomic *nuclei* emit *particles* or *electromagnetic radiation*.

Redshift The stretching of light emitted by *galaxies* moving away from Earth, due to the *Doppler effect*. This causes visible light to move toward the red end of the *spectrum*.

Refraction The bending of electromagnetic *waves* as they move from one medium to another.

Resistance A measure of how much a material opposes the flow of *electric current*.

Semiconductor A substance that has a *resistance* somewhere between that of a *conductor* and an *insulator*.

Singularity A point in *spacetime* with zero length.

Spacetime The three dimensions of space combined with a fourth dimension—time—to form a single continuum.

Special relativity Einstein's theory that an absolute time or absolute space are impossible. Special relativity is the result of considering that both the speed of light and the laws of physics are the same for all observers.

Spectrum The range of the *wavelengths* of *electromagnetic radiation*. The full spectrum ranges from gamma rays, with wavelengths shorter than an *atom*, to radio waves, whose wavelength may be many kilometers long.

Spin A quality of subatomic *particles* that is similar to *angular momentum*.

Standard Model The framework of *particle* physics in which there are 12 fundamental *fermions*—six *quarks* and six *leptons*.

String theory A theoretical framework of physics in which point-like *particles* are replaced by one-dimensional strings.

Strong nuclear force One of the four *fundamental forces*, which binds *quarks* together to form *neutrons* and *protons*.

Supernova The result of the collapse of a massive star, causing an explosion that may be billions of times brighter than the sun.

Superposition In quantum physics, the principle that, until it is measured, a *particle* such as an *electron* exists in all its possible states at the same time.

Thermodynamics The branch of physics that deals with heat and its relation to *energy* and *work*.

Time dilation The phenomenon whereby two objects moving relative to each other, or in different gravitational *fields*, experience a different rate of flow of time.

Uncertainty principle A property of quantum mechanics whereby the more accurately certain qualities, such as *momentum*, are measured, the less is known of other qualities such as position, and vice-versa.

Velocity A measure of an object's speed and direction.

Voltage A common term for electrical *potential difference*.

Wave An oscillation that travels through space, transferring *energy* from one place to another.

Wavelength The distance between two successive peaks or two successive troughs in a *wave*.

Weak nuclear force One of the four *fundamental forces*, which acts inside an atomic *nucleus* and is responsible for *beta decay*.

Work The *energy* transferred when a *force* moves an object in a particular direction.

INDEX

Page numbers in **bold** refer to main entries.

QUOTATIONS

The following primary quotations are attributed to people who are not the key figure for the relevant topic.

ACKNOWLEDGMENTS

Dorling Kindersley would like to thank Rose Blackett-Ord, Rishi Bryan, Daniel Byrne, Helen Fewster, Dharini Ganesh, Anita Kakar, and Maisie Peppitt for editorial assistance; Mridushmita Bose, Mik Gates, Duncan Turner, and Anjali Sachar for design assistance; Alexandra Beeden for proofreading; Helen Peters for indexing; and Harish Aggarwal (Senior DTP Designer), Priyanka Sharma (Jackets Editorial Coordinator), and Saloni Singh (Managing Jackets Editor).

PICTURE CREDITS